The aim of the book is to introduce the reader to the kinetic analysis of a wide range of biological processes at the molecular level. It is intended to show that the same approach can be used to resolve the number of steps in enzyme reactions, muscle contraction, visual perception and ligand binding to receptors which trigger other physiological processes. Attention is also given to methods for characterizing these steps in chemical terms. Although the treatment is mainly theoretical, a wide range of examples and experimental techniques is also introduced and a historical approach is used to demonstrate the development of the theory and experimental techniques of kinetic analysis in biology.

D0710832

KINETICS FOR THE LIFE SCIENCES

KINETICS FOR THE LIFE SCIENCES

Receptors, transmitters and catalysts

H. GUTFREUND
University of Bristol

CAMBRIDGE
UNIVERSITY PRESS

Published by the Press Syndicate of the University of Cambridge
The Pitt Building, Trumpington Street, Cambridge CB2 1RP
40 West 20th Street, New York, NY10011–4211, USA
10 Stamford Road, Oakleigh, Melbourne 3166, Australia

First published 1995

Printed in Great Britain by Biddles Ltd, Guildford & Kings Lynn

A catalog record for this book is available from the British Library

Library of Congress cataloguing in publication data

Gutfreund, H.
Kinetics for the life sciences : receptors, transmitters and
catalysts / H. Gutfreund.
        p.   cm.
Includes bibliographical references and index.
ISBN 0 521 48027 2. – ISBN 0 521 48586 X (pbk.)
1. Enzyme kinetics.   2. Chemical kinetics.   3. Biophysics.
I. Title.
QP601.3.G88   1995
574.19′2 – dc20   95-28 CIP

ISBN 0 521 48027 2 hardback
ISBN 0 521 48586 X paperback

SE

Contents

Preface

Charles Darwin wrote in a letter to a friend 'I have long since discovered that geologists never read each other's works, and that the only object in writing a book is proof of earnestness, and that you do not form your opinions without undergoing labour of some kind' (Darwin, 1887).

An author has to justify the addition of yet another volume to the overflowing shelves of academic libraries. There are, after all, a number of excellent books on chemical kinetics (for instance Moore & Pearson, 1981) and more than enough books on enzyme kinetics, even if mainly on steady states. However the kinetic treatment of other physiological systems is quite dispersed. The present volume is intended to cross fertilize ideas between different kinetic approaches. It is to be hoped that enzymologists and those working on muscle contraction or visual response at the molecular level will be inspired by the potential of methods not previously applied in their own field, even if some of them receive only scant attention. It is, of course, inevitable that in attempting to cover a wide range of systems and approaches, I shall write with more authority and in greater detail on some compared to others.

Communication between those interested in the use of kinetics for the study of different biological problems is too limited. On the one hand, the terminology of enzymology, such as the use and interpretation of Michaelis constants, has been adopted, warts and all, without modern input, by authors on membrane transport, neuroreceptors and signal transmission. On the other hand, quite sophisticated mathematical analyses of the kinetics of consecutive reactions in response to physiological stimuli by drugs, nerve impulses, etc., were available in the literature long before enzymologists became involved in the interpretation of sequential processes. Both in the description of models and in data analysis there has occurred a good deal of 're-inventing the wheel' due to lack of appreciation of both similarities and differences in the behaviour of different systems. In principle all the problems I discuss depend on one concept: the law of mass action (see section 3.1). It will also be seen that every attempt is made to reduce the description of all models to a sum of exponentials (see section 4.2).

While an important reason for writing is to contribute to one's own education, one has to set some deadline which limits the range and depth of the author's knowledge. I would not claim like Benjamin Jowett (Master of Balliol College, Oxford from 1870 to 1893) 'I am Master of this college and what I do not know is not knowledge'. For instance I would have very much liked to have included more about noise analysis, correlation and the stochastic treatment of single channel behaviour. This would either have delayed this volume by years or it would have turned out to be superficial. The inclusion and exclusion of topics and of examples is very personal; it is based on my own interests and those of my immediate colleagues since I became involved in biochemical kinetics 40 years ago under the guidance of F. J. W. Roughton, Britton Chance and Quentin Gibson.

Problems arise in the presentation of a range of topics which can really only be understood in terms of mathematical derivations. The experience of potential readers with mathematical techniques will vary widely. I am afraid there will be sections in which the mathematically more competent will regard the algebraic detail unnecessarily lengthy and repetitive, while others will have to take some derivations on trust; though I hope the latter will not be too frequent. Some sections of algebraic treatments are placed in boxes. This indicates that they contain material which is explanatory and optional. The whole of chapter 2 is intended for readers who wish to improve their understanding or refresh their memory of algebra required for the design and analysis of kinetic experiments. Hopefully the references given satisfy those requiring further detail.

I have to thank many friends for their help and encouragement. In particular I want to refer to David Colquhoun. For a time I hoped he would be a co-author, which would have made an enormous contribution, especially to the treatment of channel kinetics by stochastic analysis and novel uses of matrix algebra. While it turned out that pressure on his time made it impossible for him to keep to anything like the schedule for completion, which I had set myself, discussions with him have influenced me in many ways. Inevitably phrases, or even paragraphs, may appear that should have been attributed to him. Mike Anson, Tom Barman, Tony Clarke, Julien Davis, John Eccleston, Mike Geeves, Heino Prinz and Frank Travers read the manuscript at various stages and made valuable contributions and revisions. John Corrie made constructive comments on section 8.4. David Trentham, my closest colleague for many years, worked particularly hard to correct many errors in the manuscript and made a number of useful suggestions for improvements. However, I am sure many careless mistakes remain in the final text and I bear the sole responsibility for this. My wife, Mary Gutfreund, not only tolerated my bad temper during the final months of the preparation of the manuscript, she also made a major contribution in the preparation of the references and many thorough revisions of the text.

It is a pleasure to express my appreciation to the US National Institutes of Health for the award of a scholarship which permitted me to spend 12 months over a period of 3 years at the Fogarty Center in Bethesda, Maryland. The facilities and intellectual environment in Bethesda enabled me to give this volume a good start. I also wish to thank my colleagues in the Department of Biochemistry at Bristol University for giving me continued hospitality and facilities.

H. Gutfreund

1

The time scales of nature: a historical survey

1.1 The history of the application of physical principles

Input into biophysics from physics and biology

It is of interest to consider the historical development of ideas on the formal description and interpretation of chemical kinetics and the contributions made to this field by those with an interest in biological problems. In recent years we have witnessed a return to a multi-disciplinary approach in which more physical scientists are becoming involved in the study of biology. However, it is worth noting that this has been a recurrent feature of this field throughout history. In the eighteenth and nineteenth centuries it was easier to be a polymath than it is today. An outstanding example was Hermann von Helmholtz (1821–1894) who started his professional life as an army medical officer and held in succession chairs of anatomy, physiology and physics. In 1871 he was offered, but declined, the world's most coveted post in physics, the Cavendish Professorship in Cambridge. He developed a most ingenious method for measuring the rates of muscle contraction and the conduction of signals in nerve fibres (Helmholtz, 1850). The design and detailed description of these experiments are well worthy of study 150 years later. At that time he was also involved in establishing that the laws of conservation of energy applied to metabolic processes and physiological functions. His further important contributions were in the physics of vision (Helmholtz, 1856) and of hearing (republished Helmholtz, 1954). Last, but not least, he was associated with the theoretical physicist J. Willard Gibbs (1839–1903) in formulating some of the most important laws in chemical thermodynamics and electrochemistry. These laws are used in the interpretation of the mechanism of the transmission of signals in nerve fibres and of other physiological processes. Thomas Young (1773–1829) and Sir George Stokes (1819–1903) contributed to our knowledge of colour

1

vision and respiratory pigments, respectively, through the development of spectroscopic techniques. Many of the principal characters involved in the birth of molecular biology were inspired by Schrödinger's (1944) influential little book *What is Life?*. They were trained as physicists (Delbruck, Crick and many others) and wartime (1939–1945) experiences with the development of radar benefited the research on nerve conduction (see Hodgkin, 1992).

Chemical kinetics and biology

In the context of the theme of this volume it is specially pertinent to refer to those who have been among the first to recognize the importance of chemical kinetics to the study of reactions of biological interest. This is not intended to be a comprehensive treatment of the history of this subject, but just a reminder of some of the seminal contributions. Arrhenius (1859–1927), whose work features in sections 7.1 and 7.2, as well as van't Hoff (1884–1911) and Hood (1885) have shown how systematic studies of the temperature dependence of rates and equilibria of chemical reactions can be represented by a logarithmic relation. The empirical plot of log(rate constant) against the reciprocal of the absolute temperature is still widely used to characterize chemical and biological processes. The theoretical significance and limitations of this treatment of experimental data are discussed in section 7.2. Arrhenius's contributions to the theory of electrolyte solutions (Nobel Prize, 1903) as well as to the theory of the influence of ions on the rates of reactions were of considerable importance for the quantitative study of physiological processes. He summarized his ideas in this field in a volume entitled *Quantitative Laws in Biological Chemistry* (Arrhenius, 1915). It is worth noting that textbooks of physiology often have more detailed discussions of the physical chemistry of electrolytes than all but specialized books on physical chemistry. Arrhenius was also interested in the origin of life on earth and was an early proponent of panspermia: the drift of microorganisms to earth from space.

　　Two major figures in the field of chemical kinetics, Hinshelwood (1897–1967) and Eyring (1901–1987), turned to the study of biological reactions during the later stages of their careers. The extensive investigations on the kinetics of bacterial growth and adaptation carried out by Hinshelwood and his colleagues are summarized in a monograph (Hinshelwood, 1946). These courageous studies preceded the discovery of the double helix, the definitive recognition of DNA as the transmitter of genetic information and the consequent development of molecular genetics. The reputation of

Hinshelwood's studies thus suffered from being overtaken by events. Eyring applied his ideas on the formulation of a theory of absolute reaction rates, through defined transition states, to a wide range of complex biological problems (Johnson, Eyring & Stover, 1974). Like a lot of controversial ideas, those of Eyring have stimulated many interesting developments. However, one does not have to be a vitalist to assume that a theory developed for simple reactions in the gaseous state will require much modification before it can be applied to complex molecules in aqueous solutions. Some of these modifications are discussed in sections 7.2 and 7.3 in connection with the effects of temperature and viscosity on reaction velocities. The theory of Kramer (1940), describing reactions in terms of a viscosity dependent diffusion over the energy barrier, has come into its own in the interpretation of the dynamics of protein molecules and protein–ligand interaction (see section 7.3).

Atkins (1982, p. 921) makes an interesting distinction between chemical kinetics and the field of chemical dynamics of individual steps. The former is descriptive in terms of rate equations defining the number of intermediates and their rates of interconversion. The latter is in the realm of chemical physics, but it is becoming of interest as modern techniques help to describe the dynamic behaviour of macromolecules. A similar distinction was made above between the descriptive and the mechanistic role of kinetic studies on systems at a more complex level. This emphasizes the important point that general questions about the mechanism of a reaction or a functional process are open ended. It is essential that specific questions are asked and methods are used which are appropriate to their resolution.

1.2 Kinetics and biological problems

Why measure the rates of biological processes?

The title of this chapter is a translation of one used by Manfred Eigen for a lecture given to the Max Planck Society (Eigen, 1966): 'Der Zeitmasstab der Natur'. It covers one of the aims of this volume, namely to provide information about the rates at which various biological processes proceed and how they form a temporal structure. This leads to the second aim, which is to show what kinetics can tell one about the mechanisms of complex functions. A third aim is the exploration of systems at different levels of organization, from the single step of a chemical reaction or the binding of a defined compound to a purified protein, to complex physiological events such as muscle contraction, or the electrical response of photoreceptor cells to the absorption of a photon by the visual pigment. Ogden (1988), when reviewing a new technique for the initiation of reactions in situ (see section 8.4), wrote 'many processes in cell physiology will be understood only when the rates and equilibria of the reaction steps involved are known'. In some cases the aim is principally the elucidation of the detailed chemical mechanism of action of a natural effector at its primary site of interaction; this can be used for the design of drugs. In other, more complex systems the resolution of the sequence of events will help to characterize individual steps, demonstrate their contribution to the overall rate and indicate which change in condition affects specific steps.

Kinetics as a theoretical discipline plays a similar, though some might say less exalted, heuristic role in biology to that of formal genetics. It is, however, of more specialized application and although such analogies must not be pushed too far, some examples will illustrate the point. Genetics is both a theoretical discipline, proposing models and analysing data and also an experimental science concerned with the testing of such models as well as with the discovery of new phenomena. It is applied to systems of widely different levels of organization from large populations of higher animals to chemically well-defined events in prokaryotes, viruses and even to replication in systems containing nucleic acids and enzymes in free solution (Mills, Peterson & Spiegelman, 1967; Spiegelman, 1971). Model building and testing is an important part of genetics and new discoveries about the existence of functions under the control of a single or of multiple genes are made by such studies. Genetic techniques applied to the study of physiological functions can provide information about the sequence of the involvement of different proteins (gene → products). For instance Hall (1982) presented an extensive review of the application of genetics to the analysis of the number of functional proteins and their role in phototransduction, chemosensory behaviour and membrane excitability.

A warning is necessary concerning the use of mathematical models in biology. These must be closely linked to the design of experiments which can test them. Examples will come up in the text which demonstrate the danger of models taking over from reality. It is also important to realize that a biological system does not necessarily work at optimum efficiency. Just as in private enterprise, to succeed one only has to be the most efficient in one's immediate environment.

Dynamics of populations

The application of kinetic equations to biological problems on the largest scale is found in modelling of the behaviour of populations. The use of kinetic as well as genetic methods in this field is expanding with the wide availability of computers. Kinetic models of the time dependence of population sizes were formulated more than a century before the application of population genetics. The derivation of exponential growth curves for populations also preceded the use of such equations for the description of the exponential time course of chemical reactions discussed in section 2.1. The similarity of the approach is illustrated by the fact that one often talks about the population of some molecular species or of a state. Hutchinson (1978) provides a very elegant introduction to the development of equations describing the growth of populations. Other classic surveys of the history of this field are given by d'Arcy Thompson (1948) and Lotka (1956). Exponential growth results in population explosions. We can use a microbial example, discussed in more detail in section 3.3. If one starts off with one cell at time $t = 0$, in 24 hours one would obtain (under ideal conditions) 2^{72} cells (i.e. 5×10^{21} descendants) weighing 10^6 kg. On Alexander Pope's advice 'the proper study of mankind is man' and the earliest applications were to human populations.

Demography is an application of quantitative population dynamics of importance for the planning of adequate resources for public services and for the correct assessment of insurance premiums. At the beginning of the nineteenth century Malthus drew attention to the dangers of unrestricted population growth and the resulting competition for resources. Equations for growth rates and survival rates were derived respectively by Verhulst (1804–1849) and Gompertz (1779–1865). Statistics about births and deaths had already been tabulated by Graunt (1620–1674), who discovered the slight excess of male over female births a continuing trend in human populations. Gompertz's (1825) paper 'On the nature of the function expressive of the law of human mortality and on a new mode of determining the value of life contingencies' is often referred to as the origin of the Gompertz curve. His aim was to deduce life contingencies, or as he put it 'the power to oppose destruction', from tabulated data. There are no plots or curves in his publication, only algebraic expressions derived to fit the data (see also comments on Arrhenius plots in section 7.2).

The Gompertz equation is often written in its modern form:

$$N(t) = N(0) \exp\{-\exp(a + kt)\} \tag{1.2.1}$$

(with constants a and k) for N survivors at time t. Some algebraic properties of equation (1.2.1), which is obtained on integration of $dN/dt = kN(a - \ln N)$, are discussed by Winsor (1932). An equation

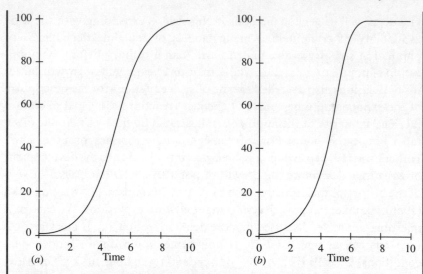

Figure 1.1   Simulation of (*a*) logistic (symmetrical) and (*b*) Gompertz (asymmetrical) growth curves. See equations (1.2.5) and (1.2.6), respectively.

for population growth proposed by Verhulst (1804–1849), which is, for unknown reasons, called the logistic equation, has the following mathematical foundation. If one starts from the premise that the rate of growth of a population is a function of its size (see the law of mass action, section 3.1), namely the number of its members N, then we can write

$$dN/dt = f(N)$$

This can be expanded by Taylor's theorem to

$$dN/dt = a + bN + cN^2 + dN^3 + \ldots \tag{1.2.2}$$

(with additional constants b, c and d). The simplest rate equation for growth is

$$dN/dt = bN \tag{1.2.3}$$

If we add the next term

$$dN/dt = bN + cN^2$$

on substituting for $c = -a/K$ we obtain (K is a constant)

$$dN/dt = aN(K - N)/K \tag{1.2.4}$$

which gives, on integration, one of the forms of the logistic equation

$$N(t) = K/\{1 + \exp(-bt)\} \tag{1.2.5}$$

For this growth curve the term $(K - N)/K$ in equation (1.2.4) is added as a feedback term to the equation for exponential growth $dN/dt = bN$. This provides an inverse linear dependence of growth on population density.

The logistic curve, like the Gompertz curve, is sigmoidal (see figure 1.1) but the former is symmetrical while the latter is asymmetrical. Both equations can be used for growth or decay by change of a sign in an exponent and they have been used to model many quite varied systems. Easton (1978), for instance, used the Gompertz equation to describe the sodium and potassium conductance changes in giant squid axons in the following form (using Easton's nomenclature):

$$Y = a \exp\{-b \exp(-kt)\} \tag{1.2.6}$$

where the conductance Y is related to the asymptotic conductance a and

$$b = \frac{\text{asymptotic conductance}}{\text{initial conductance}}$$

The logistic equation (1.2.4) can be used to analyse autocatalytic reactions like the proteolytic activation of proteases and the activation of protein phosphokinases by auto-phosphorylation. Many other differential equations developed for the description of population growth, predator–prey interactions and the spread of epidemics have similarities to equations applied to the analysis of the progress of chemical reactions and of control phenomena. References will be made to common applications in relevant sections and simple examples of population problems are found in Pielou (1969) and Maynard Smith (1974). The more general treatment of the problems mentioned is a topic for much active mathematical research on the solution of non-linear differential equations. Incidentally, models of population genetics, of population dynamics and the kinetics of reactions in organized systems all involve diffusion phenomena (see section 7.4). Population growth as well as biochemical systems can, under certain feedback conditions, go into oscillations. Examples from the biochemical literature are given by Segel (1984).

Kinetics and physiology

Helmholtz's classic experiments of kinetic measurements on nerve conduction and muscle contraction have already been referred to. Further examples of the application of kinetics, which are used to demonstrate various points of physiological interest in detail in different sections, come from the kinetic analysis of visual perception and their molecular components. Important information about its mechanism can be obtained from studies at several levels of organization. Attempts can be made to correlate the kinetics of psychophysical observations with electrophysiological measurements on the retina or its component cells. These, in turn, can be compared with the rates of bleaching of rhodopsin and of the consequent cascade of enzyme reactions. The last of these reactions can be studied in solutions of reconstituted systems of purified proteins (see section 4.2). Fuortes & Hodgkin (1964) followed the time course of the first electrical response in the retina after bleaching the visual pigment rhodopsin. The equations which they developed to describe their results were used to postulate the number of distinct events during this process. Further electrophysiological studies and the recent elucidation of the biochemistry of a sequence of reactions subsequent to photon capture have confirmed the complexity of the mechanism of transmission of the message (see Stryer, 1988; and section 4.2).

In general there is a set of criteria which can be used to demonstrate whether an observed step is, or can be, on the direct pathway of a complex process. The rate of the step and its dependence on changing conditions has to be compatible with the overall rate. It is of course easier to exclude a step from the direct pathway, because it is too slow, than it is to ascertain its inclusion. This applies in the exploration of enzyme reactions, as will be discussed in detail in section 5.1, as well as to physiological responses. An example of the latter, which will be used to illustrate points in different sections, is the relation between steps in the hydrolysis of ATP by myosin and structural and mechanical changes during muscle contraction. Similarly the time course of calcium release and removal has to be correlated with the stimulation and relaxation of contraction and other phenomena.

It will be apparent to the reader that the study of rate processes in biological systems has several related objectives. First it can be used in a purely descriptive manner. The physiologist wants to know how fast different muscle types can contract, how fast a nerve can conduct a signal or how fast a cell can respond to a chemical messenger. Each of these is quite a complex process, which can be resolved into elementary steps. A second objective would be, for instance, to discover which step or steps are responsible for the difference in the rate of contraction of fast and slow muscle. The use of a range of methods for kinetic investigations increases the resolution by separating complex phenomena into a sequence of well-defined ones. Different methods of perturbation, that is initiations of reactions, and different monitors to follow them provide not only rate data but also characterize the nature of the events. The effects of a change in conditions (temperature, composition of medium, etc.) help to provide information on controlling factors and optimum performance (see section 5.2 for examples).

In the present volume information obtained about the temporal resolution of complex processes will be used as examples to define the problems which can be solved by the application of kinetic reasoning and techniques. There are systems of biological importance for which the rates at which they function are relatively unimportant. For example, the actual rates of oxygen uptake and release by the respiratory proteins haemoglobin and myoglobin are of marginal significance compared to the physiological importance of their equilibrium oxygen binding properties. Yet, without doubt, more kinetic investigations have been carried out on the reactions of haemoglobin than on any other protein. These experiments with a wide range of techniques were designed to explore the interesting mechanism involved in the cooperativity of the interaction with oxygen and other ligands. While equilibrium binding studies provide essential physiological and thermodynamic information about the reactions of haem proteins with oxygen and other ligands (they define the problem), only the results of kinetic investigations can describe the mechanism in terms of the number of reaction intermediates and their rates of interconversion. It is also often only possible to obtain equilibrium constants from kinetic data (see chapter 6).

Kinetic investigations tend to expose open ended questions since one can continue to divide steps into ever increasing resolution. This will be found to be a repetitive theme in the present volume. The exploration of rates of enzyme reactions can move from an interest in metabolic control by changing substrate and effector levels to the exploration of the detail of the mechanism of a single reaction, which would be principally of interest to a physical organic chemist. There are also many different levels of interest in most biological problems. For this reason the design of each investigation should start by asking specific questions. The word 'mechanism' does not mean the same to a physiologist and a chemist. An individual step in muscle contraction, as defined from the point of view of a physiologist, is a complex series of events for the chemist. This kind of demarcation will be highlighted again in chapter 7 when physical factors controlling rates are discussed. The methods described in other chapters will be illustrated with examples which correspond to varying levels of molecular detail.

In some cases kinetic considerations provided the initiative for the search for a specific addition to the sequence of events during a process. For instance, the role of CO_2 in respiratory function requires a more rapid formation of bicarbonate than that achieved through the uncatalysed hydration

$$H_2O + CO_2 = HCO_3^- + H^+$$

with a half life of 20 seconds. This led to the search for and discovery of the enzyme carbonic anhydrase (for a review see Roughton, 1935). This reaction is sometimes quoted as evidence that all, even the simplest, chemical reactions in biological systems are enzyme catalysed. Similarly, for the role of Ca^{2+} in excitation contraction coupling in muscle, Hill's (1948) calculations showed that the rate of diffusion of the ion is not fast enough to be compatible with the overall speed of response. The discovery of the role of the sarcoplasmic reticulum in calcium secretion and removal helped to resolve this problem.

1.3 Time scales of nature and of measurement

From elementary chemical steps to physiological processes

We shall now turn to a discussion of the range of rates of reactions from elementary chemical steps to physiological functions. This will then be linked to a discussion of methods which cover different time scales and monitors which are suitable for recording different signals. Table 1.1 shows the time scales over which biological processes, and some of the chemical steps associated with them, occur. The photochemical initiation of the utilization of solar energy for biosynthesis and the earliest reactions of the photopigments involved in visual response or bacterial ion transport occur in picoseconds (or faster). Some rate limiting steps in enzyme reactions are 12 or more orders of magnitude slower. Response to control signals and the adaptation of functions, such as contractile processes, occur on three separate time scales. The effect of a messenger or effector binding can be noticed within 10^{-6} to 10^{-1} seconds, the stimulation of the synthesis of a new protein through gene activation will take 10^2 seconds while adaptation to a new environment through natural selection takes of the order of 10^8

Table 1.1a. *Physiological responses and time scales*

Response	Time scale (frequencies per second)
Vibration of fly wings	200
Hummingbird	75
Rattlesnake (on a hot day)	100
Fastest human muscle	50
Human eye	50
Blow fly detecting flicker	250
Protein synthesis	10^{-3}
Range of enzyme turnover	1 to 10^9
Rate of signal conduction in myelinated nerve fibres	
outer diameter 2 μM	10 m s^{-1}
outer diameter 20 μM	100 m s^{-1}

Table 1.1b. *Time scales of kinetic methods*

Method	Time constants (range in seconds)
Fluorescence decay	10^{-6}–10^{-10}
Ultrasonic absorption	10^{-4}–10^{-10}
Electron spin resonance	10^{-4}–10^{-9}
Nuclear magnetic resonance	1–10^{-5}
Temperature jump	1–10^{-6}
Electric field jump	1–10^{-7}
Pressure jump	down to 5×10^{-5}
Stopped flow	down to 10^{-3}
Phosphorescence	10–10^{-6}

seconds. The major interest in the present volume is focused on the time window from about 10^{-6} to 100 seconds, which covers the range of ligand binding to proteins and the subsequent conformational rearrangements and chemical transformations. According to molecular dynamics calculations on protein conformation changes the major observed events, with time constants ranging from 10^{-4} to 1 second, are in fact the results of picosecond structural fluctuations on a pathway searching for the new conformation. When considering the suitability of a kinetic technique for the study of individual steps of an enzyme reaction one is confronted with a wide range of turnover numbers which represent the slowest step. The fastest reported turnover rates (see section 7.4) are for superoxide dismutase and carbonic anhydrase. They are of the order of 10^8 and 10^6 s^{-1} respectively, while some digestive enzymes and ribulose bisphosphate carboxylase-oxygenase (probably the most abundant enzyme in nature) have turnover rates of seconds. The interesting developments in the application of kinetics to the study of enzymes and other functional proteins are in the design of new experimental methods and in the formulation of models suitable for algebraic description and, above all, for testing proposed models. The actual solution of the necessary equations by analytical or numerical methods does not involve any problems which are not, at least in principle, common to other fields using applied mathematics. The special cases do, however, prove to be of interest to mathematicians.

The approach to a correct model through kinetic investigations is an iterative process: preliminary experiments result in a proposed model. Simulation of this model allows predictions to be tested with further experiments. These experiments result in alterations or refinements of the model and so on (hopefully not ad infinitum). Methods for simulations of models and the least square analysis of experimental and simulated data, are discussed in chapter 2. Jencks (1989) gives the warning that models can take on a life of their own. He points out that mathematical models which explain all the data can be physically unrealistic. A bon mot attributed to Francis Crick is 'if a theory explains all the facts it is likely to be wrong because some of the facts will be wrong'.

There is an interesting relation between methods which have been developed for the study of rate processes and what actually occurs in biological systems. All techniques depend in some way on disturbing a system and observing the time course of the response to the disturbance as a new equilibrium or steady state is approached. Fluctuations about an equilibrium, which can be studied in 'small' systems containing relatively few molecules, can also be interpreted in terms of rates of transitions between different states. This approach to kinetic analysis is widely used in studies of ion channels of excitable membranes. Some very specialized methods are involved (Sakmann & Neher, 1983) and only their potentialities can be discussed in this volume. Similarly, perturbations due to changes in concentrations or in the physical environment occur in nature in response to stimuli, these in turn initiate reactions. The time course of these responses provides information about the temporal structure of the system and its function. This is analogous to studies of physical and structural properties of materials. These are usually carried out by perturbing the material to get information about elasticity, viscosity, stiffness, etc. from the way it responds to such perturbations. In research on the molecular mechanism of muscular contraction mechanical responses of fibres are correlated with the kinetics of molecular events (see for instance Hibberd & Trentham, 1986).

Technical developments and time scales

In this discussion of time scales some reference must be made to examples of the major technical achievements, which make it possible to obtain kinetic data over a wide time scale and from responses to different perturbations. These methods, which have been selected on account of their special features and application to biological problems, will be referred to in discussions of the interpretation of the kinetic behaviour of different systems. The explanations are schematic rather than in technical detail. Many features relating to the monitoring of changes in reactant concentrations via changes in some physical parameter are common to all methods. A discussion of monitors would require chapters on electronics and optics. Factors which determine the time resolution are discussed below.

There are two criteria for the time resolution of a technique for following the rate of a process. The first is the precision with which zero time is determined and the second is the overall response time of the sensor (transducer of any physical change to an electrical one) and the system used for amplification and recording the signal. Some aspects of frequency response, time constants and signal to noise ratio are closely related to kinetic theory. The classic physiological experiments of Helmholtz, referred to above, were among the first in a long history of the use of galvanometers and smoke drums for the sensitive recording of electrical signals with a millisecond response. Interest in the theory of the frequency response of galvanometers has been maintained until fairly recently (Hill, 1965).

Whenever possible the basic molecular events of physiological processes are studied in detail in solutions either in parallel with, or as a guide to, experiments on the organized system. For example, the study of the reactions of myosin and the associated proteins of the contractile cycle, by the methods used for the investigation of mechanisms of soluble enzymes (see section 5.1), has helped in the planning of experiments and in the interpretation of the events observed in muscle fibres. The most useful methods for kinetic studies on physiological functions are those which can be applied to systems at different levels of organization. We shall return to them at the end of this chapter.

The time scale of methods for the study of reactions in solutions was greatly extended by the rapid flow techniques initiated by Hartridge & Roughton (1923a, b). The stimulus for this development was their interest in the reactions of haemoglobin with oxygen and other ligands. Up to that time there was no general procedure available for the study of reactions in solution, which allowed a time resolution much better than, say, 30 seconds. The methods of Hartridge & Roughton relied on the rapid linear flow (about 10 m s^{-1}) of the reaction mixture along a capillary tube. The progress of the reaction was monitored, with a spectroscope or a thermo-couple, at intervals along the tube. The initiation of the reaction, at the top of the capillary, was carried out either by injecting two reactant solutions through a mixing chamber, or by photoactivation by a constant strong light source. The use of the length of a capillary tube as a time axis was similar to that employed by Rutherford (1897) for the study of the recombination of ions. In his experiments, gas flowing down the tube was subjected to ionizing radiation at the top of the tube and the extent of recombination was monitored with electrodes at points downstream. The procedure involving mixing chambers, which made it possible to define the start of the reaction with an accuracy somewhat better than one millisecond, had many further applications. Initially the method could only be used when large volumes of reactant solutions were available since it relied on many individual readings during constant flow of the reaction mixture. The obvious interest in the application of flow techniques to the study of intermediates in the reactions of enzymes and respiratory proteins, which were being purified in small quantities, encouraged Roughton, Millikan, Chance and Gibson (see Roughton & Chance, 1963) to design equipment which was more economical in materials. Chance (1943) was inspired by Keilin & Mann's (1937) spectroscopic observations on different com-pounds of the enzyme peroxidase to carry out time resolved studies. He used an accelerated flow device to observe the kinetics of enzyme–substrate reactions. Keilin's studies of cytochrome spectra (for review see Keilin, 1966) also provided Chance (1952) with the background for his investi-gations of the kinetics of the reactions during cellular respiration. Rapid flow techniques were applied to studies on suspensions of mitochondria and whole cells. It thus became possible to study chemical reactions under conditions approaching those appertaining in vivo.

The modern commercially available stopped-flow apparatus has become a standard piece of laboratory equipment. A short burst of flow through the mixing chamber fills an observation tube with reaction mixture less than 1 ms old. When flow is stopped the signal from the photomultiplier or other transducer is fed into standard amplifiers and digital recording equipment

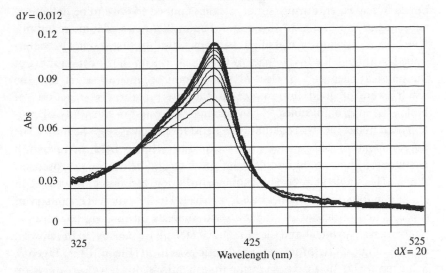

Figure 1.2   Repetitive spectra at 1.25 ms interval during the reaction of a
mutant of horse-radish peroxidase (Arg38 → Lys) compound I (1.25 μM) with
p-aminobenzoic acid (200 μM) at pH 7 and 25 °C. Experiments were carried out
with a HiTech Scientific Ltd SF-61 stopped-flow spectrophotometer with MG-
6000 rapid scanning system by Drs A. T. Smith and R. N. F. Thornley
(unpublished observations).

with adjustable response times. The development of these, as well as of most
other biophysical instrumentation, was the result of the availability of ever
improving electronic components, as well as of the mechanical and
electronic skills acquired by the pioneers in the field. Chance, for instance,
returned to biophysics in 1945 after making major contributions to the
development of amplifiers during his work on radar in the Radiation
Laboratory at the Massachusetts Institute of Technology. With modern
commercial stopped-flow equipment it is possible to record spectra (spanning 200 nm) repetitively at approximately 1 ms intervals (see figure 1.2).

   Another milestone in the history of kinetic studies on solutions of interest
to biologists was the application of chemical relaxation techniques (Eigen &
de Maeyer, 1963, and chapter 6). It is of interest to note that the initiation of
rapid biochemical reactions by the perturbation of equilibria through a step
change of intensive parameters, as well as through photochemical activation, was among the topics of a discussion at the Royal Society of London
in 1934 (Hill, 1934). Eigen's ideas, on the extraction of kinetic information
from the relaxation of a system after rapid perturbation, were stimulated by
observations on periodic perturbations. The discovery of sound absorption

bands in sea water, during 'sonar' studies connected with mine detection, subsequently suggested to Eigen and Tamm that rates of ionic reactions, which were previously classified as immeasurably fast, could be determined. Although the first impact of this work was on the understanding of the physical chemistry of electrolytes, it soon became apparent that the results were of great importance in the interpretation of reactions of biological macromolecules. Eigen became interested in the study of proteins and nucleic acids. Together with de Maeyer he was responsible for the development of temperature, pressure and electric field perturbation techniques, which could resolve reactions into and below the microsecond range. The single step equilibrium perturbation technique, discussed in chapter 6, found wide application to problems in the physical chemistry of proteins and nucleic acids. Eigen's own interests moved, via cooperative phenomena in proteins and the kinetics of DNA base pairing, to the slowest of all biological rate processes: pre-biotic evolution (Eigen, 1971; Eigen & Schuster, 1977, 1978). These latter investigations do, however, draw on information about molecular recognition obtained from studies of nucleotide base pairing with relaxation techniques.

Physiologists had investigated the time course of mechanical and electrical changes in vivo long before then. Several other techniques have been developed in recent years. These can be used for rate measurements on solutions and on, for instance, muscle fibres or excitable membranes. Pressure perturbation has been used successfully to study actin–myosin interaction in solution and in fibres (Geeves & Ranatunga, 1987; see also chapter 6). This is of particular historical interest since it was the first relaxation method, proposed by Cattell (1934, 1936) for the initiation of rapid reactions. Ultrasound as well as magnetic resonance techniques are used at all levels from detailed molecular dynamics to in vivo studies on whole animals, including *Homo sapiens*. The contribution of NMR to our potential knowledge about the structure and dynamics of complex biological molecules is so great that it can only be covered in a meaningful way by a comprehensive treatment in a separate volume. The photo-flash liberation of caged substrates is ideally suitable for studies at different levels of organization (McCray & Trentham, 1989; see also section 8.4). Correlation techniques and noise analysis, which can only be briefly referred to in this volume, are also suitable for the study of a wide range of phenomena (see Sakmann & Neher, 1983). Time resolved X-ray crystallographic techniques have been extensively developed recently. Their application to the study of structure changes during enzyme–substrate interaction has yielded exciting information (Moffat, 1989; Hajdu & Johnson, 1990; Schlichting *et al.*,

1990). The Laue techniques used for this purpose can give a picture of the whole changing structure, but the time resolution still restricts it to slow changes (seconds rather than milliseconds). A much faster resolution is obtained from records of single reflections in diffraction patterns of active muscle fibres after electrical stimulation (Huxley & Faruqi, 1983).

In concluding this introductory chapter we cannot resist the temptation to comment on the healthy competition between the two complementary fields of structure analysis and kinetic investigations. In about 1965 it became clear that deductions about the structures of active sites of some enzymes, studied by the techniques of protein chemistry and transient kinetics, would be faced with reality – or supposed reality – from X-ray crystallographic investigations. Some of us were sceptical about the idea that crystallographers were going to solve structure/function problems unaided. Gutfreund & Knowles (1967) wrote 'making a model of a horse from photographs does not tell us how fast it will run'. Incidentally, in 1882, Muybridge, of San Francisco, used a series of cameras with sequential 2 millisecond exposures to obtain a dynamic picture of the movement of a horse's leg (Porter, 1967). Karplus & McCammon (1983) and Frauenfelder, Parak & Young (1988) describe the gradual transition from the crystallographers' view of the 1960s and 1970s of a rigid structure to the new dynamic view of conformational mobility. None of this diminishes the important role of information about average structures in the elucidation of mechanisms of protein function. However, the complementarity of different approaches to our knowledge of reactions of proteins must be emphasized. Some aspects of this will be discussed in chapter 7.

A method which can be combined with kinetic and structure studies, to enhance the potentialities of both of them, is the replacement of specific amino acids in a protein. The change of selected amino acid residues in specific positions along the polypeptide chain, by site directed mutagenesis, is providing a very valuable tool for the study of mechanisms of catalysis and of protein folding (Matouschek *et al.*, 1989). Not only can the importance, within the mechanism, of a particular residue be assessed, but it is also possible to insert reporter groups (tryptophan) or cysteine as a reactive group for labelling in a chosen position. The kinetics of folding should get an impetus from this approach. The dogma that the three-dimensional structures of proteins are determined by their primary structures (the amino acid sequence) requires elaboration. During folding there must be a search for a potential energy trough, one of many. To find the trough for the native structure the conditions must be such that the pathway to the required trough has the lowest energy barrier; this is a kinetic problem discussed in section 7.2. Similar kinetic control phenomena of thermodynamic anomalies are encountered in the prevention of protein denaturation and freezing under conditions where this is expected to occur. Physical or chemical conditions are chosen which slow down considerably the inevitable final outcome determined by thermodynamic equilibrium.

2
Some mathematical introductions

2.1 The origins of exponential behaviour

Mathematics as the language for kinetics

Some topics covered in this volume are more easily described by mathematical equations than by words, while others can only be described in algebraic language. An attempt is made in this chapter to state the mathematical principles which are the foundation of kinetic behaviour and to present the methods used to derive the equations which describe this behaviour. It is not essential to understand the contents of this chapter to benefit from the rest of this volume, but it is difficult to avoid mistakes in kinetic investigations without it.

It will be seen that exponential curves of one sort or another crop up everywhere in the study of the rates of reactions. Not only is the change of concentration (or membrane current, etc.) frequently described by an exponential function of time (or by the sum of several such functions), but exponentials also appear in some of the more exotic ways of measuring rates, such as in the study of fluctuations and noise. Furthermore the study of individual molecules, as is possible for some sorts of ion channel, gives rise to probability distributions (e.g. the distribution of the length of time for which an individual channel stays open) that are also described by exponentials (Colquhoun & Hawkes, 1983). Unfortunately some of these topics can only be referred to in passing in the present volume. It is, however, important to realize that it is sometimes useful to distinguish between the lifetime of a state (or reaction intermediate) and the rate (or probability) of its decay.

It is no coincidence that all these different approaches to kinetics involve exponential curves; the fundamental reason is the random behaviour of individual molecules (as will be discussed below). However, this is not the only way of looking at the phenomenon of exponential behaviour, and indeed it may not be a helpful way when we are considering empirical descriptions of the behaviour of complex systems, rather than a molecular description of a relatively simple reaction mechanism. For example, exponentials are also common in mechanical and electrical systems and in empirical studies of the rate of growth of populations of bacteria (or humans), which were discussed in chapter 1. We shall therefore first consider exponential behaviour from the point of view of systems that contain large numbers of molecules, and then move on to the level of individual molecules.

Underlying principles of exponential behaviour

It is common for the *rate* of change in a quantity, at any particular moment, to be directly proportional to the *magnitude* of the quantity at that moment. Such behaviour is, for example, implicit in the *law of mass action* which states that the rate of change of the concentration of a reactant is directly proportional to the concentration of the reactant; the proportionality constant being the *reaction rate constant* in this case. In other words, the larger the concentration the faster the reaction, and vice versa. This means that the reactant will disappear rapidly at first, while its concentration is high, but as its concentration progressively falls (towards its equilibrium value) the rate of reaction also falls in proportion. The next section is devoted to a discussion of this proportionality.

If the rate of change of a variable (i.e. its first derivative) is equal, apart from a constant, to the value of the variable itself, then the way in which the variable changes with time must be described by a function that is the same after it has been differentiated as it was before. There is only one function (of time, t, in this example) that does not change when it is differentiated, namely the infinite series

$$f(t) = 1 + t + t^2/2! + t^3/3! + t^4/4! + \dots \tag{2.1.1}$$

where, for example, 4! denotes $4 \times 3 \times 2 \times 1$ (read as '*factorial* 4'). It is readily seen that on differentiating equation (2.1.1), term by term, it is unchanged. This property makes this function of the greatest importance and it is given a special name, *exponential function*, which may be written in either of the following forms

$$f(t) = e^t, \quad \text{or} \quad f(t) = \exp(t) \tag{2.1.2}$$

The constant denoted by 'e' here is a universal constant which is at least as important as π, even if not so well known. From (2.1.1) and (2.1.2) it is clear (put $t = 1$) that the value of the constant is

$$1 + 1 + 1/2! + 1/3! + 1/4! + \ldots = 2.718\,281\ldots \tag{2.1.3}$$

This is, like π, an infinite and non-recurring decimal; it can be evaluated to any required degree of accuracy by adding more terms to the series. This argument can be written, more formally, as

$$\frac{d(e^t)}{dt} = e^t$$

This implies that if we have an equation that states that some function varies in proportion to its own magnitude, so that $df(t)/dt = f(t)$ then the solution of this equation must be of the form e^t. Any reader who wishes to know more about these and other elementary principles of calculus should consult Silvanus P. Thompson's book (1914); any book still in print more than 80 years after first publication must be worth reading!

It is exactly this sort of relationship that has been used to describe the simplest sort of population growth in chapter 1. The rate at which a population produces children will be (in the absence of constraints produced by a limited supply of food or space) directly proportional to the number of individuals in the population. The more individuals there are to produce offspring, the more offspring there will be and an exponential growth of the population can be expected. This is discussed in chapter 1 with some examples in the context of the general application of kinetic arguments in relation to biological problems.

An example of a reaction mechanism

Consider, for example, the case of an enzyme–substrate complex, or a drug–receptor complex; say the ligand (substrate, or drug) is denoted A, and the enzyme (or receptor) is denoted R. The reaction can therefore be written as

$$AR \xrightarrow{k} A + R \tag{2.1.4}$$

where k denotes the dissociation rate constant for the complex. Suppose some complex is already present, and then, at $t = 0$, the ligand is suddenly removed. The complex will dissociate, and, according to the law of mass action (see above and section 3.1), the rate of dissociation will be directly proportional to the concentration of the complex, c_{AR}, the value of which, at time t, we write as $c_{AR}(t)$. Thus the rate of change of $c_{AR}(t)$ is

$$\frac{dc_{AR}(t)}{dt} = -k c_{AR}(t) \tag{2.1.5}$$

Apart from the constant k, this is just the same as the equation we have just been discussing. Its solution is therefore exponential in form. In fact it has the form $c_{AR}(t) = e^{-kt}$, a function that starts (at $t = 0$) at 1.0, and decays towards zero as t increases. The concentration at $t = 0$ will not necessarily be 1.0, but some specified value, $c_{AR}(0)$ say. The final solution is therefore

$$c_{AR}(t) = c_{AR}(0)\exp(-kt) \tag{2.1.6}$$

In this example it is assumed that the rate of reassociation is negligible, the derivation for a reversible reaction is given later. The shape of this exponentially decaying solution is shown in figure 2.1a. As $c_{AR}(t)$ gets smaller the slope gets shallower in direct proportion, as stated in equation (2.1.5).

At first sight it might seem rather perverse to define logarithms (known as *natural logarithms*) to the base 2.718 281 ..., rather than using some simpler number such as 10 (the base of common logarithms), but it is precisely because results like (2.1.7) are so common that natural logarithms are so useful. The natural log (denoted ln) of e^x, is, by definition of logarithms, simply x. Thus, for example, if we take natural logs of both sides of (2.1.6) we get

$$\ln(c_{AR}(t)) = \ln(c_{AR}(0)) - kt \tag{2.1.7}$$

which shows immediately that a plot of $\ln(c_{AR}(t))$ against time, t, will be a straight line, the intercept, at $t = 0$, being $c_{AR}(0)$ and the slope being $-k$. This plot has been widely used in the past as a test of whether experimental data are changing exponentially (as indicated by the linearity of the plot), and, if they are, as a method of estimating the value of the rate constant k. Some further discussion of such plots and variants of them, will be found in section 3.2, but they have now been largely superseded by least squares fitting methods for estimating parameters (see section 2.3), although they may still be useful for display purposes.

The solution of equation (2.1.5) would normally be written in a more formal (but less intuitively helpful) way as follows. The equation would be rearranged in the form

$$\frac{dc_{AR}(t)}{c_{AR}(t)} = -k dt$$

and then integrated

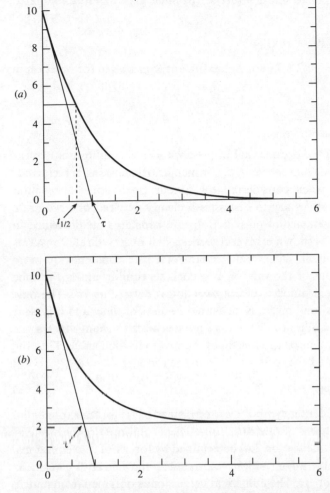

Figure 2.1 Illustration of the time constant (τ) and the half life ($t_{1/2}$) for an (a) irreversible and (b) reversible reaction. In (a) $k = 1 \text{ s}^{-1}$ and in (b) $k_{12} = 1 \text{ s}^{-1}$ and $k_{21} = 0.25 \text{ s}^{-1}$. It can be seen from the intercept with the baseline that $\tau^{-1} = k_{12} + k_{21}$.

$$\int_{c(0)}^{c(t)} \frac{\mathrm{d}c(t)}{c(t)} = -k \int_{t=0}^{t=t} \mathrm{d}t \tag{2.1.8}$$

Formally the solution can be found by noting that the right hand side is simply $-kt$ and by finding (e.g. in a table of indefinite integrals) that the integral on the left hand side is $\ln(c(t))$. The integral is therefore evaluated as

$$\ln(c_{AR}(t)) - \ln(c_{AR}(0)) = -kt \tag{2.1.9}$$

which is the same as (2.1.7), so, by taking antilogs we get the solution in (2.1.6).

Measures of the rate of decay

The rate at which the exponential curve approaches equilibrium is measured by an *observed rate constant*, k_{obs}, which has dimensions of reciprocal time. Note that, when considering the kinetic predictions of reaction mechanisms, it is necessary to distinguish clearly between observed rate constants, and fundamental mass action rate constants, as described in section 3.1. It will be shown in several sections that observed rate constants are usually a function of several fundamental rate constants. A more convenient measure of the rate of approach to equilibrium is the time constant, which is denoted τ (Greek tau) and is defined as $1/k$. The time constant or relaxation time, τ, is measured in units of time, and it is clear from equation (2.1.6) that it is the time for the value to change to $1/e$, i.e. 37% of its initial value, or to get 63% of the way to its final value. Thus the solution given in (2.1.6) is often written in the form

$$c_{AR}(t) = c_{AR}(0) \exp(-t/\tau) \tag{2.1.10}$$

In sections 4.2 and 5.1 the symbol λ is used instead of $1/\tau$ for the exponential coefficients in solutions of rate equations. These solutions are the eigenvalues of the matrix (see section 2.4) determined by the set of rate equations. The following problem can result in some confusion when authors use λ as equivalent to k_{obs} or $1/\tau$. The numerical values for λs obtained from matrix solutions are always negative while the values for k_{obs} and therefore of τ are always positive. Hence

$$c_{AR}(0) \exp(-t/\tau) = c_{AR}(0) \exp(\lambda t)$$

This problem is discussed in more detail in section 4.1.

The parameter τ is closely related to the more familiar, but less convenient, half time for the process. This is the time taken for the initial value to halve. If the half time is denoted as t_{50} then, in the notation of (2.1.6), $c_{AR}(t) = 0.5\,c_{AR}(0)$ when the time is $t = t_{50}$. Thus

$$0.5\,c_{AR}(0) = c_{AR}(0) \exp(-kt_{50})$$

and

$$t_{50} = \tau \ln(2) = 0.693 \tau \tag{2.1.11}$$

It is an important characteristic of single exponential processes that the time constant, or the half time, is indeed constant. At *whatever* point on the curve the measurement is started (the 'initial' value is measured), the time taken for the variable ($c_{AR}(t)$ in this example) to fall to $1/e$ of this initial value is always τ (and the time taken to fall to $1/2$ of it is always $t_{1/2}$). The positions of τ, and of $t_{1/2}$, are illustrated graphically in figure 2.1.

The term *time constant* is the generally agreed mathematical name for the parameter τ in (2.1.10). The same quantity is also referred to as a *relaxation time* (which is quite acceptable jargon and generally used for small perturbations, see chapter 6), or simply as *rise time*; however the use of *decay time* is ambiguous and therefore unacceptable.

An electronic example

Similar equations occur in many contexts. For example, in electronics the current that flows into a capacitor as it charges or discharges is directly proportional to the rate of change of the voltage across the capacitor, so

$$I_c(t) = C \, dV(t)/dt \tag{2.1.12}$$

where $I_c(t)$ is the current flowing into the capacitor at time t, $V(t)$ is the voltage across the capacitor at time t, and C is the capacitance (measured in farads) which is a constant. Consider the simple circuit in figure 2.2, which can be taken as a representation of the electrical properties of a cell membrane. It consists of a resistor (representing the resting resistance of the membrane) and a capacitor (representing the membrane capacitance) in parallel. The potential across the capacitor, which is the same as the potential across the cell membrane, is measured between the two terminals. From Ohm's law a current, $I_r(t) = V(t)/R$ flows through the resistor, and a current $I_c(t)$, given by (2.1.12), flows through the capacitor. If a current is passed through the membrane, charging up the capacitor, until the potential across the membrane attains some value denoted $V(0)$, and then at $t = 0$ the current is switched off, the membrane potential will decay back towards zero. Since the net membrane current, $I_r + I_c$, is zero we have $V(t)/R + C \, dV(t)/dt = 0$, i.e. $-dV(t)/dt = V(t)/(RC)$. This has exactly the same form as equation (2.1.5) and by the same arguments its solution is $V(t) = V(0)\exp(-t/RC)$; so the membrane potential decays back towards zero following an exponential time course with a time constant given by

Output Waveforms of *RC* Circuits as a Function of Circuit Time Constant τ for Square Wave Input

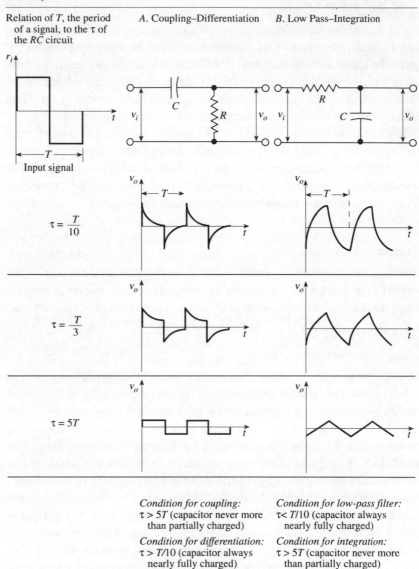

| Relation of *T*, the period of a signal, to the τ of the *RC* circuit | *A*. Coupling–Differentiation | *B*. Low Pass–Integration |

Condition for coupling:
τ > 5*T* (capacitor never more than partially charged)

Condition for differentiation:
τ > *T*/10 (capacitor always nearly fully charged)

Condition for low-pass filter:
τ < *T*/10 (capacitor always nearly fully charged)

Condition for integration:
τ > 5*T* (capacitor never more than partially charged)

Figure 2.2   Illustration of different exponential and other responses of electrical circuits.

$\tau = RC$. The larger the membrane capacitance, or the larger its resistance, the slower this decay will be. Measuring this time constant (by injecting rectangular current pulses) allows the membrane capacitance to be measured. From this an estimate of membrane area can be made since it has been found that almost all membranes have a capacitance close to $1\,\mu F\,cm^{-2}\,(0.01\,F\,m^{-2})$. It is, incidentally, clear from this argument that the membrane capacitance can result in problems when trying to measure the time course of currents that flow through a membrane when, for example, ion channels open; the time course of the membrane potential $V(t)$ will not follow accurately the time course of the underlying changes in membrane current. It is precisely for this reason that all experiments in which the time course of membrane currents has to be measured must be conducted at a *constant membrane potential* so, from equation (2.1.12), no current can flow into the membrane capacitance; the measured current will therefore represent accurately the opening and shutting of ion channels. Such experiments, and the apparatus used to do them, are described by the term 'voltage clamp'. An excellent treatment of these topics is presented by Hille (1992). Time constants introduced by electronic equipment used for all kinetic experiments have also got to be taken into consideration! (See section 2.3.)

Macroscopic and microscopic events

Consider again the case of the enzyme–substrate complex, or drug–receptor complex, as in scheme (2.1.4), but now consider a *single* enzyme or receptor molecule. In this case it makes no sense to talk about the *concentration* of the complex at time t, as we did above. At any particular moment the complex is either in existence, 100% occupancy, or it is not, 0% occupancy. We must instead speak of the *probability* that the receptor is occupied at time t. This, like $c(t)$, will decline exponentially with time in the experiment described above. What is happening at the level of the single molecule is that a ligand molecule from the solution, when it gets sufficiently close to the binding site, with a suitable orientation, will get bound and a complex will be formed. When the concentration of free ligand is high, successful collisions will be more frequent, so the durations of time for which the binding site is vacant will be, on average, shorter than at low concentrations of free ligand.

Once the complex has formed, it has 'no way of knowing' what the free ligand concentration is, so the length of time for which the complex remains in existence (the *mean lifetime* of the complex) should be quite independent of the ligand concentration. The ligand will eventually dissociate when the random thermal movements of the complex happen to produce such a big stretching of the bonds that hold the complex together that the bonds break and the ligand escapes. The length of time that elapses before this occurs will be randomly variable, but its average value, the mean lifetime of the complex, will depend on the amount of energy needed to break the bonds, and this is measured by the dissociation rate constant, k, defined above in the context of the law of mass action. The relationship is indeed very simple; the mean lifetime of the complex is simply $1/k$ – rapid dissociation on the macroscopic scale corresponds to short-lived states on the microscopic scale. In an experiment of the type discussed above, the free ligand is suddenly removed at $t = 0$. Therefore, after this time no more complexes can be formed and we are looking only at the break-up of complexes that happened to be already in existence at $t = 0$. These complexes will last on average for a length of time $1/k$ before breaking up. But the actual length of time is a random variable, and so must be described by a probability distribution. This distribution is not described by the familiar bell-shaped curve of the Gaussian distribution, but actually has the shape of a decaying exponential curve (so, in a sense, short lifetimes are more common than long ones). Some of the complexes that were in existence at $t = 0$ will last for a short time, others for much longer. The complexes that remain in existence at time t after removing the ligand will be precisely those that happen to have a lifetime greater than t. This proportion will, because the distribution of lifetimes is exponential, also follow an exponentially decaying curve, the time constant for this curve being simply τ, i.e. it is the mean lifetime. The way in which this random behaviour of single molecules sums to produce the observed macroscopic behaviour is derived in some detail by Colquhoun & Hawkes (1983). An interesting discussion on the statistics of *waiting times*, including some illuminating paradoxes, are found in Colquhoun (1971).

2.2 Modelling of kinetic mechanisms

Mathematical principles

Many books are available with detailed and clear accounts of numerical integration and data handling. The first of these topics can be explained relatively succinctly to a level which should help the reader to understand the procedures for numerical solutions of differential equations describing kinetic mechanisms. Most of the algorithms applied to kinetic problems are based on one mathematical principle, stepping methods which have been progressively refined. Other procedures, such as series expansion and Monte Carlo methods, will be found in texts on numerical analysis. The Monte Carlo method (see for instance Kleutsch & Frehland, 1991) has proved itself particularly useful for the simulation of the growth of individual myosin filaments (Davis, 1993) and microtubule lattices (Martin, Schilstra & Bayley, 1993). However, the methods available for data handling are numerous and involve a range of different principles. A brief description of these in the next section should avoid the blindfolded use of commercial computer programs for simulating kinetic models and for the interpretation of experimental results in terms of parameters fitted to such models.

The numerical solutions of differential equations

Kinetic mechanisms are generally described by differential equations of the type

$$\frac{dc_i(t)}{d} = f'(k_i, c_i(t), t) \tag{2.2.1}$$

(Equation (2.2.1) is simplified to correspond to a set of irreversible steps.)

In the simplest case with linear concentration terms (see section 2.4), as in $A \rightarrow B$, the differential equation

$$\frac{dc_A}{dt} = -k\,c_A(t) \tag{2.2.2a}$$

can be integrated to give

$$c_A(t) = c_A(0)\exp\left(-k_1 t\right) \tag{2.2.2b}$$

(see section 2.1).

The analytical solution in equation (2.2.2) can, of course, be graphically represented in terms of c_A against t. The matrix method described in section 2.4 and other procedures applied to examples in sections 4.2 and 5.1, can be used to obtain solutions for rate equations describing quite complex reactions, as long as they are linear in concentration terms. For equations with non-linear terms it is usually not possible to obtain analytical solutions. In such cases the mechanism can only be described by a set of $n-1$ differential equations for n reactants and one equation for conservation of mass. These have to be solved numerically.

The solution of a set of coupled differential equations (DEs) describing a reaction is an *initial value* problem, which means that all c_is are known at some starting value $c_i(t=0)$. The basic principle of numerical solutions of DEs can be illustrated using Euler's stepping method on the three state system:

$$A \longrightarrow B \quad \frac{dc_A(t)}{dt} = -k_{12}c_A(t)$$

$$B \longrightarrow C \quad \frac{dc_B(t)}{dt} = k_{12}c_A(t) - k_{23}c_B(t) \tag{2.2.3}$$

The differential equations (2.2.3) can be converted into finite difference equations by changing from the differential (infinitesimal) operator d to the finite small difference operator Δ:

$$\Delta c_A(t) = [-k_{12}c_A(t)]\Delta t$$

$$\Delta c_B(t) = [k_{12}c_A(t) - k_{23}c_B(t)]\Delta t \tag{2.2.4}$$

This procedure can be continued as follows:

$$c_A(t+\Delta t) = c_A(t) + \Delta c_A(t),$$

$$\Delta c_A(t+\Delta t) = [-k_{12}c_A(t+\Delta t)]\Delta t \dots$$

The corresponding values for c_A are used at every time interval for the calculation of a new value for c_B by the same procedure However, the solution to this problem is not quite as simple as it looks. The value for Δt has to be very small, about one-tenth of the shortest time constant. If the steps are too large one does not get an accurate solution, or even worse, the sequence shoots off at a tangent. Most people now use one of the further developments outlined below. However, it must be emphasized that PCs have now become so fast that the above simple Euler method can be very useful. If no other program is available and one has an elementary knowledge of programming *Quickbasic* or some other high level language, such an exercise does teach one to think about one's model! One of the problems of much of the folklore about programs and programming is that it grew up during a time when desktop machines had a small fraction of the power they have today and *Basic* had nothing like the facilities available in modern versions. It is now very powerful for mathematical work, but beware of *Visual Basic*; I would not recommend the latter for attempting such programs.

There are a number of ways of making stepping methods more efficient, that is faster and still accurate. One favoured in my laboratory, which is still simple enough to be programmed by a novice, is a version of the predictor corrector method. In this procedure one calculates the slopes at t using the concentration $c_A(t)$ and at $t + \Delta t$ using $c_A(t + \Delta t)$ obtained from the first predictor step; one then uses the average slope from the two calculations to obtain the *corrected* value for $c_A(t + \Delta t)$. This allows one to make Δt longer and the procedure is about three times as fast as the simple Euler method, for comparable accuracy. There is also a second order *predictor corrector* method, but for such further refinements a specialized volume (for instance Conte & de Boor, 1981) must be consulted.

Other popular methods for numerical solutions of DEs are the Runge–Kutta methods. They again come in forms of different order, depending on the number of selected points on each sub-interval for which the function is evaluated and averaged. The development of these methods includes quite sophisticated analyses of errors (deviations from the true solutions) which occur with functions of different properties. A major problem in the numerical integration of rate equations is *stiffness*. A differential equation is called stiff if, for instance, different steps in the process occur on widely different time scales. It is very inefficient to compute with time intervals suitable for the steepest part of the progress curve (see Press *et al.*, 1986, chapter 16).

The use of numerical integration for simulating reaction mechanisms makes a great contribution to the study of complex systems. It should, however, be applied with a scalpel rather than a sledge hammer. Whenever possible analytical equations should be obtained, at least for limiting cases, and compared with numerical solutions. When this is not possible it is advisable to write out the differential equations rather than to rely entirely on a symbolic processor to convert a description of the mechanism. Saving time is not the most important aspect of any procedure involved in research. Some laborious occupations help one to think about the problem.

2.3 Data collection, reduction and analysis

The revolution through digitizing and the use of PCs

During the last 20 years the increasing power of desktop computers available in every laboratory has completely changed the way data are collected, analysed and presented. In this section I wish to emphasize the advantages and the dangers of this development. In the acquisition of data, bias was gradually removed when instruments changed from analog (dials with parallax) to digital panels and finally to direct digital recording on some storage medium (tape, disk, or computer memory). When plotting data, whether for binding hyperbolas, exponential records or for the logarithmic evaluation of rate constants, human bias will inevitably be introduced. The paper by DeSa & Gibson (1969) set a new trend for dealing with the output of instruments used for biochemical kinetics.

The demands on the skills of a scientist are open ended. For optimal planning and interpretation of experiments an understanding of data collection and electronics as well as of statistics and computer programs is essential. Almost all information about the time course of reactions comes through an electrical signal. The output of a photomultiplier, an electrode, or some other sensor of a physical change goes to a recording device via an amplifier and intentional, as well as unintentional, filtering (introduction of a time constant).

Noise and filtering (time constants)

The signal to noise ratio is the greatest challenge encountered in the improvement of measurements. There are three theoretical limitations: random thermal noise (Johnson noise); shot noise due to the fluctuation of the number of electrons per unit time making an impact on a sensor, for instance on the collecting anode of a photomultiplier; and finally there is flicker noise, which depends on the use of (non-ideal) components in the circuits. This subject, like most other aspects of the design of electronic equipment is discussed in the admirable book *The Art of Electronics* by Horowitz & Hill (1980). Unfortunately the theoretical noise limitations are rarely of importance. The effects of pickup from other equipment through the air or mains electricity and the imperfect shielding and construction of the apparatus usually introduces much larger effects.

Apart from reducing the instrument noise which is introduced unnecessarily, the presence of random noise can be reduced by filtering and averaging. Noise which introduces transients with time constants of the same order of magnitude or slower than any expected from the real data (often fluctuations in the light source) can only be reduced by averaging a number of records. The noise is reduced by a factor of $\sqrt{n}$, where n is the number of traces averaged. This can, of course, be implemented in the form of a simple computer program. Fast random noise can be removed with a cutoff RC filter (see section 2.1). For the correct design of such devices Horowitz & Hill (1980) should be consulted. Serious mistakes have been made through the introduction of spurious relaxation times into the recorded data by *RC* filters. A classic example of this is described by Geraci & Gibson (1967).

An alternative and advantageous way of filtering noise is to do it numerically either by simple averaging, say over blocks of 10 data points, or by one of several more sophisticated filtering algorithms (see Press *et al.*, 1986). The sizes of memories of devices for storing digital data are, in most cases, large enough to record at least ten times as many points as are required for the evaluation of a single time constant. Thus a reduction of data points through averaging or filtering gives an adequate set for subsequent calculations and presentation. If data capture is planned with numerical filtering in view, then it is possible to reconsider the chosen time constant of the filter. This can be of importance if it is found, on subsequent review of the experiment, that the data may contain higher frequency components. Recording data on a logarithmic time scale is sometimes advantageous; for instance, when the system under investigation has a number of relaxation times orders of magnitude apart (Austin *et al.*, 1976; Walmsley & Bagshaw, 1989).

Colquhoun (1971) and others pointed out more than 20 years ago that the widely used linear derivations of the hyperbolic ligand binding equation do not give correct statistical weighting to the data. Similarly logarithmic plots for the evaluation of exponentials are not a good substitute for fitting data to the non-linear rate equations. In sections 2.3 and 2.4 some of the old graphical methods are reviewed to help the reader when consulting the substantial section of the literature where they are still in use. At that time, of course, non-linear least square analysis had to be carried out very laboriously with calculators, even for the simplest two parameter problem, or the data had to be sent off to a distant computer to be treated by anonymous programs.

However, there are pitfalls in the now common practice of transferring data straight from laboratory instruments, via some form of digital store, into a commercial computer program for evaluation. Note the dictum 'rubbish in = rubbish out' when the analysis is carried out by the 'user friendly' but incompletely understood programs provided by the instrument manufacturer or a software house. The operator may neither ask the right questions of the program nor understand the questions the program asks of the data and is, therefore, unlikely to appreciate the answers. To give the most blatant example, it is easy to get a good fit with two or more exponentials to a record of a single second order reaction, even if it is free of noise (see figure 2.3). In this example two exponentials gave a near perfect

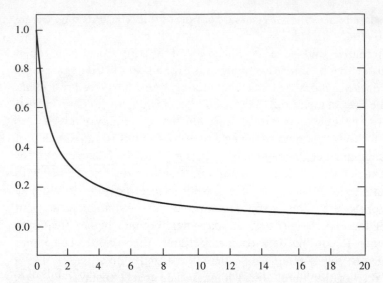

Figure 2.3   Simulation of a second order reaction ($k = 1$ M^{-1} s^{-1}) and a superimposed curve simulated as the sum of two exponentials with $1/\tau$ of 1.72 and 0.37 s^{-1} respectively.

fit. If the time scale of the simulated second order reaction were further extended a third exponential would improve the fit! Errors are also often introduced by incorrect fixing of the final baseline. Some programs will fit the wrong exponential plus a sloping baseline. The reader should try the following illuminating experiment. Plot the two curves:

$$y(t) = 0.41 \exp(-0.69t) + 0.58 \exp(-0.54t)$$
$$y(t) = 0.96 \exp(-0.59t) + 0.032 \exp(-1.24t)$$

The resulting curves should be superimposed and/or analysed.

A numerically specified function, that is a dataset, can almost always be made to fit one or more of the following: sum of exponentials, power series or Fourier series; regardless of its physical significance. Acton's (1970) chapter entitled 'What not to compute' is essential reading. As he points out 'most of us can more easily compute than think'. None the less, if correctly used, digital collection and computer analysis of data have brought many insoluble problems into the realm of the soluble.

Non-linear least square procedures

Fitting of data to a linear equation

$$y(x) = a_1 + a_2 x \tag{2.3.1}$$

by '*least square*' to obtain slope (a_2) and intercept (a_1) is such a standard procedure that it does not need any comment here. For non-linear equations several methods are in common use, in addition to least square. A recent volume of *Methods in Enzymology* (Brand & Johnson, 1992) provides an extensive overview of many different routines for fitting data to non-linear equations. This volume should be consulted for approaches other than the one recommended here. It requires a multi-author volume to do this, since most individuals are convinced that their particular method is, taking all facts into consideration, superior to any of the others. In my laboratory non-linear least square fitting of data to rate equations is carried out by the differential correction technique (McCalla, 1967; Edsall & Gutfreund, 1983). We have modified this with the Marquardt procedure. Both Bevington (1969), a volume which has kept its popularity as a guide for data handling and error analysis for 25 years, and Press *et al.* (1986) describe the Marquardt (1963) algorithm for a convenient gradient search for minimizing χ^2 (equation (2.3.2)). This is the objective of all least square procedures.

$$\chi^2 \equiv \sum_{i=1}^{n} \left[\frac{1}{\sigma_i^2} [y_i - y(x_i)]^2 \right] \tag{2.3.2}$$

where y_i is a data point and $y(x_i)$ is obtained from the equation with the correct or iteratively refined values for the parameters. The uncertainty, σ_i, for a data point is only obtained if sets of values for y_i are available for each x_i.

For equations with non-linear parameters a_i, for instance

$$y(t) = a_1 \exp(-a_2 t) \tag{2.3.3}$$

the parameters cannot be separated into different terms of a sum, as they can for the least square solution of the linear equation (2.3.1). The optimum values for n parameters a_j are obtained by minimizing the following function, with respect to each of them simultaneously:

$$\frac{\partial}{\partial a_j} \chi^2 = \frac{\partial}{\partial a_j} \sum_{i=1}^{n} \left[\frac{1}{\sigma^2} [y_i - y(t_i)]^2 \right]$$

There are a number of ways of iterative minimization of χ^2, which are illustrated in the references quoted above. They involve setting up a matrix of partial differentials of the function with respect to each parameter, first with estimated values and then with subsequent refinements. The secret of a good method is to reach a true minimum fast and to avoid stopping at a *local* minimum. There is no useful halfway house between this superficial description of the procedure used in commercial programs and the detail available in Bevington (1969) and Press *et al.* (1986).

If one is investigating a reaction with a very complex mechanism, that is with many parameters in the rate equation, fitting may not give a unique solution. In such a case a useful exercise is to overlay a simulated curve over the experimental record, using estimated values for the parameters in the rate equation. Successive changes in the values for the parameters will lead to a potential solution. This sounds more tedious than it is and, furthermore, it is a useful exercise because it will also show which parameters have the greatest influence on the behaviour of the function and it makes one think about the problem at the same time.

The above procedures are all concerned with obtaining the parameters for the best fit of an integrated rate equation to the experimental data. Many commercial programs are available, which give rapid solutions on most PCs and portable computers in present use (for instance the program *GraFit*, by Leatherbarrow, 1992). When the mechanism is very complex or non-linear (as defined in section 2.4) it is not possible to obtain a satisfactory analytical solution and (as discussed above) a set of differential equations, one for each state of the system, is the only available description of the mechanism. In such cases one can proceed either by trial and error (the above mentioned overlay method), using numerical simulation of the differential equations instead of the analytical equation, or by a procedure until recently only available for use on somewhat larger computers. This refers to programs like *FACSIMILE* (Flow And Chemistry SIMulator), which was gradually developed by Chance *et al.* (1977) from a numerical integration algorithm. This routine has been extended to facilitate non-linear least square fitting of data which can only be described by sets of differential equations. After every successive estimate of parameters a numerical solution has to be provided for the calculation of χ^2. Kinetic instrument manufacturers are, increasingly, making such programs commercially available.

Although the above minimization approach for the fitting of data to non-linear equations is the most commonly used, attention must be drawn to the availability of several other methods. The need for a multi-author volume (Brand & Johnson, 1992) on this subject is a testimony to their diversity.

Finally some comments are called for on judging the comparison of fitting different equations to the same experimental data. With some notable exceptions, such as the dissection of multiple spectral changes, the use of elaborate statistical diagnostics raises the suspicion of inadequate experimentation. However, the simple procedure of comparing the residuals when different equations – say the sums of one, two, or three exponentials – are used, is recommended. The eye is the best judge of systematic deviations along the time axis!

2.4 Guidelines for the derivation of equations used in the text

How much algebra does a kineticist need to know?

This section is not a substitute for one of the many good texts on mathematical methods written for scientists with different backgrounds. No one of these volumes will appeal to everybody, but I find Boas (1966) has the clearest and most comprehensive coverage of the mathematical problems arising in the present volume. It is intended that the brief summary of matrix algebra will help the reader to follow those sections of the book in which kinetic equations are derived. Specific examples of the derivation of rate equations by this method, including numerical evaluation of exponential coefficients and amplitudes, are found in sections 4.2 and 5.1.

Two other mathematical methods, the solution of second order differential equations and Laplace transforms, will be discussed in the relevant sections of chapter 5. These will be marked as non-essential reading.

An introduction to matrix algebra: matrices, vectors and determinants

A matrix is a table, and will be denoted by a bold capital symbol. It is rather like a spreadsheet, the entries in the table being defined by the row and column in which they occur. The entry in the ith row and the jth column of the matrix $\mathbf{A}$, an *element* of $\mathbf{A}$, will be denoted in lower case italic as a_{ij}. A matrix with n rows and m columns is said to be an $n + m$ matrix. Thus, for example, a 2×2 matrix can be written as

$$\mathbf{A} = \begin{bmatrix} a_{11} & a_{12} \\ a_{21} & a_{22} \end{bmatrix} \tag{2.4.1}$$

Clearly a 1×1 matrix is thus just an ordinary number (called a *scalar*). In the context of matrix algebra they are simply matrices that happen to have only one row (a *row vector*), or only one column (a *column vector*). They are manipulated just like any other matrix and will be denoted by a bold lower case symbol.

A determinant is a *number* (not a table) that can be calculated from the elements of a matrix. The determinant is usually written just like the matrix, except that it is enclosed by vertical lines rather than brackets (it therefore looks rather similar to the matrix, and it is important to remember that the whole symbol enclosed within the vertical lines represents a single number). For example, the determinant of the matrix in (2.4.1), denoted $\det(\mathbf{A})$, is represented as illustrated in (2.4.2), the right hand side of which shows how the number is calculated from the elements of $\mathbf{A}$.

$$\det(\mathbf{A}) \equiv \begin{vmatrix} a_{11} & a_{12} \\ a_{21} & a_{22} \end{vmatrix} = a_{11}a_{22} - a_{12}a_{21} \tag{2.4.2}$$

The general method for the evaluation of a determinant can be illustrated with the 3×3 matrix

$$\mathbf{B} = \begin{bmatrix} b_{11} & b_{12} & b_{13} \\ b_{21} & b_{22} & b_{23} \\ b_{31} & b_{32} & b_{33} \end{bmatrix}$$

This matrix $\mathbf{B}$ can be divided into cofactors and minors. The minors are obtained by deleting the row and column of the cofactor and for the cofactors b_{11} and b_{12} the minors are respectively

$$\begin{bmatrix} b_{22} & b_{23} \\ b_{32} & b_{33} \end{bmatrix} \quad \text{and} \quad \begin{bmatrix} b_{21} & b_{23} \\ b_{31} & b_{33} \end{bmatrix}$$

The value for the determinant of the 3×3 matrix is obtained by summing three products of the cofactors of a row or a column with their signed minors. The sign is obtained from the sum of the indices of the cofactors, even sums positive and odd sums negative, thus the determinant of $\mathbf{B}$ is

$$b_{11}(b_{22}b_{33} - b_{32}b_{23}) - b_{12}(b_{21}b_{33} - b_{31}b_{23}) + b_{13}(b_{21}b_{32} - b_{31}b_{22})$$

Clearly for larger matrices the situation becomes quite complicated and, while the principle ought to be understood, computer programs are available for the numerical as well as for the symbolic evaluation of large determinants. I find the programs *MapleV2 for Windows* (Waterloo Maple Software, 160 Columbia Street West, Waterloo, Ontario, Canada N2L

3G1) and *Mathcad Plus* 5.0 (MathSoft Inc., 101 Main Street, Cambridge, MA 02142, USA) very useful for the treatment of kinetic equations. Anything larger than a 3×3 matrix becomes very cumbersome to treat without an efficient program. Programs for specific mathematical operations, which one understands in principle, are less dangerous to use than black box programs which solve a whole problem of simulation or data analysis.

A matrix that has a determinant of zero, $\det(\mathbf{A}) = 0$, is said to be a *singular* matrix and is also of significance in the solution of sets of equations – see below.

Multiplication of matrices

The definition of the product of two matrices may, at first sight, seem a bit perverse, but it turns out to be exactly what is needed for the convenient representation of, for example, simultaneous equations. Multiplication goes 'row into column'. For example

$$\mathbf{C} = \mathbf{AB} = \begin{bmatrix} a_{11} & a_{12} \\ a_{21} & a_{22} \end{bmatrix} \begin{bmatrix} b_{11} & b_{12} \\ b_{21} & b_{22} \end{bmatrix}$$

$$= \begin{bmatrix} (a_{11}b_{11} + a_{12}b_{21}) & (a_{11}b_{12} + a_{12}b_{22}) \\ (a_{21}b_{11} + a_{22}b_{21}) & (a_{21}b_{12} + a_{22}b_{22}) \end{bmatrix}$$

Thus the element in the ith row and the jth column of the product, $\mathbf{C}$, is obtained by taking the ith row of $\mathbf{A}$, and the jth column of $\mathbf{B}$, multiplying the corresponding elements, and adding all these products. Clearly the number of columns in $\mathbf{A}$ must be the same as the number of rows in $\mathbf{B}$. If $\mathbf{A}$ is an $n \times m$ matrix, and $\mathbf{B}$ is $m \times k$, then the product, $\mathbf{C} = \mathbf{AB}$, will be an $n \times k$ matrix. The recurring examples of products of matrices or vectors are the representation of a set of simultaneous rate equations in terms of [matrix of coefficients (rate constants)] $\times$ [vector of occupation of states] (see for instance sections 4.2 and 5.1).

It is important to note that, contrary to the case with ordinary (scalar) numbers, it is *not* generally true that $\mathbf{AB}$ and $\mathbf{BA}$ are the same; that is they are not commutative. We must therefore distinguish between *pre-multiplication* and *post-multiplication*. In the product $\mathbf{AB}$, it is said that $\mathbf{A}$ pre-multiplies $\mathbf{B}$ (or that $\mathbf{B}$ post-multiplies $\mathbf{A}$).

If a matrix is multiplied by an ordinary (scalar) number, x say, this means simply that every element of the matrix is multiplied by x. Thus $x\mathbf{A}$ is a matrix with elements xa_{ij}.

The identity matrix and the division of matrices

The matrix equivalent of the number 1 is the identity matrix, denoted I, for which all the diagonal elements are 1, and all others are 0. Therefore for the case of a 2×2 matrix

$$I = \begin{bmatrix} 1 & 0 \\ 0 & 1 \end{bmatrix} \tag{2.4.3}$$

The definition of matrix multiplication shows that multiplication of any matrix by I does not change it. Thus

$$AI = IA = A \tag{2.4.4}$$

The importance of the identity matrix is discussed below in connection with the determination of eigenvalues of matrices of kinetic coefficients.

If we define $C = AB$, then what is B? In ordinary scalar arithmetic we would simply divide both sides by A to get the answer. When A and B are matrices the method is analogous. We require some matrix analogue of what, for ordinary numbers, would be called the reciprocal of A. Suppose that some matrix exists, which we shall denote A^{-1}, that behaves like the reciprocal of A in the sense that, by analogy with ordinary numbers,

$$A^{-1}A = AA^{-1} = I \tag{2.4.5}$$

where I is the identity matrix defined in (2.4.3). The matrix A^{-1} is called the *inverse of* A. If we post-multiply both sides of (2.4.5) by A we get simply $A = A$.

Thus the solution to the problem posed initially can be found by pre-multiplying both sides of $C = AB$ by A^{-1} to give the result as $B = A^{-1}C$. In the case of a 2×2 matrix the inverse can be written explicitly as

$$A^{-1} = \begin{bmatrix} a_{22}/\det(A) & -a_{12}/\det(A) \\ -a_{21}/\det(A) & a_{11}/\det(A) \end{bmatrix} \tag{2.4.6}$$

where each element is the cofactor (see p. 38) of that element in the original matrix divided by $\det(A)$ (the number defined in (2.4.6)). It can easily be checked that this result is correct by multiplying it by A. The result is the identity matrix, I.

It is apparent from (2.4.6) that the inverse cannot be calculated if $\det(A) = 0$; this would involve division by zero. Thus singular matrices (see above) cannot be inverted. It is also clear, for example from (2.4.6), that only square matrices (number of rows = number of columns) can be inverted.

The inverse of a product can be found as

$$(\mathbf{AB})^{-1} = \mathbf{B}^{-1}\mathbf{A}^{-1} \tag{2.4.7}$$

To write down explicitly the elements of the inverse of a matrix for larger matrices we need the definitions of cofactors and minors given above in connection with the evaluation of determinants. Each position in $\mathbf{A}^{-1}$ is occupied by the signed minor of the cofactor for that position, divided by the determinant of the matrix and then transposed. The explicit form rapidly becomes very cumbersome, and accurate numerical calculation of the inverse of matrices larger than, say 3×3, requires special techniques that are available in computer programs such as *Maple* (see above). Problems arise in numerical work if a matrix is *ill conditioned*, that is a matrix with a determinant which is very small compared to its elements.

Cramer's rule

A set of n simultaneous equations with n unknowns and non-zero constant terms, a_{ij} and c_i can be solved using matrix algebra. The two equations

$$\begin{aligned} a_{11}x_1 + a_{12}x_2 &= c_1 \\ a_{21}x_1 + a_{22}x_2 &= c_2 \end{aligned} \tag{2.4.8}$$

can be solved by multiplying the first equation by a_{22} and the second equation by a_{12}, subtracting the result and solving for x_1

$$x_1 = \frac{c_1 a_{22} - c_2 a_{12}}{a_{11}a_{22} - a_{21}a_{12}}$$

solving for x_2 in a similar way we obtain

$$x_2 = \frac{a_{11}c_2 - a_{21}c_1}{a_{11}a_{22} - a_{21}a_{12}}$$

The set of equations (2.4.8) can be written in the matrix form

$$\begin{bmatrix} a_{11} & a_{12} \\ a_{12} & a_{22} \end{bmatrix} \begin{bmatrix} x_1 \\ x_2 \end{bmatrix} = \begin{bmatrix} c_1 \\ c_2 \end{bmatrix} \tag{2.4.9}$$

and it can be seen that the denominator of the solutions is the determinant of the coefficients a_{ij} and the numerator for the solution for each unknown

x_i is the determinant of these coefficients with the vector of the constant coefficients replacing the ith column of $\mathbf{A}$. This method for solving a set of linear equations is called Cramer's rule and can be applied to a large system using computer programs to evaluate the determinants.

Homogeneous equations

The term *homogeneous equation* can be defined using the example of the matrix equation (2.4.9); it means that all elements of vector $\mathbf{c}$ are zero. Then, if $\det(\mathbf{A}) \neq 0$, all elements in vector $\mathbf{x}$ must be zero. Hence there is only one solution to the equations, namely zero; the equations are linearly independent. If, however, $\det(\mathbf{A}) = 0$ then there are an infinite number of solutions and only the ratio of the values for the unknowns can be determined. This is done by the following procedure.

Using the example

$$a_{11}x_1 + a_{12}x_2 + a_{13}x_3 = 0$$
$$a_{21}x_1 + a_{22}x_2 + a_{23}x_3 = 0 \qquad\qquad (2.4.10)$$
$$a_{31}x_1 + a_{32}x_2 + a_{33}x_3 = 0$$

and calculating the determinant of a minor of $\mathbf{A}$ which is not zero, for instance $a_{11}a_{22} - a_{21}a_{12}$ and substituting an arbitrary constant for $x_3 = t$ we can solve the first two equations

$$a_{11}x_1 + a_{12}x_2 = -a_{13}t$$
$$a_{21}x_1 + a_{22}x_2 = -a_{23}t$$

using Cramer's rule described above:

$$x_1 = \frac{\begin{vmatrix} -a_{13} & a_{12} \\ -a_{23} & a_{22} \end{vmatrix}}{\begin{vmatrix} a_{11} & a_{12} \\ a_{21} & a_{22} \end{vmatrix}} t \qquad x_2 = \frac{\begin{vmatrix} a_{11} & -a_{13} \\ a_{21} & -a_{23} \end{vmatrix}}{\begin{vmatrix} a_{11} & a_{12} \\ a_{21} & a_{22} \end{vmatrix}} t$$

A solution determined in this way will satisfy all three equations for any value of t and the ratio of all unknowns is obtained.

Linear and non-linear equations

In many places throughout this volume algebraic solutions can only be achieved if the system can be described in terms of *linear equations*. A homogeneous equation is linear if, and only if, the sum of two solutions is also a solution. For example, if in the equation $a_1x + a_2y = h$ we substitute $a_1 = -0.4$, $a_2 = 0.2$, $h = 2$ we obtain: $y = 10 + 2x$. This equation has the solutions:

(1) $x_1 = 3$, $y_1 = 16$
(2) $x_2 = 5$, $y_2 = 20$

as well as the solution

(3) $x_3 = 8$, $y_3 = 26$

The solution of the third equation is the sum of the solution of the first two minus 10, the factor which has to be subtracted to convert all the equations into homogeneous ones.

In the context of kinetic equations linearity is most easily defined by the following example:

$$\frac{dc(t)}{dt} = -k_1c + k_2c^2 + k_3c^3 \ldots$$

If $k_i = 0$ for $i > 1$ then

$$\frac{dc_1(t)}{dt} + \frac{dc_2(t)}{dt} = \frac{d(c_1 + c_2)}{dt} = -k_1(c_1 + c_2)$$

Therefore, with no higher terms in the series, the equation is linear. However, if $k_2 \neq 0$ then we obtain the inequality

$$\frac{d(c_1 + c_2)}{dt} \neq -k_1(c_1 + c_2) + k_2(c_1 + c_2)^2$$

because $k_2(c_1 + c_2)^2 \neq k_2c_1^2 + k_2c_2^2$. The introduction of higher order terms result in a non-linear equation in which two solutions cannot be added to give a valid solution.

The importance of the definition of linearity lies in the fact that it is a condition for making kinetic problems soluble without recourse to numerical integration. It is a precondition for the application of matrix algebra for the derivation of the parameters of a kinetic equation in terms of eigenvalues and eigenvectors as shown below.

Eigenvalues and eigenvectors of a matrix

The eigenvalues, λ_i, of an $n \times n$ matrix are the n roots of the *characteristic polynomial* of the matrix. This polynomial is defined by subtracting λ from each diagonal element, and equating the determinant of the result to zero. Expansion of the determinant produces a polynomial that contains powers of λ up to λ^n. Thus the eigenvalues of a matrix X are the roots of the polynomial

$$|X - \lambda I| = 0 \tag{2.4.11}$$

We may note that the sum of all the eigenvalues of a matrix (for instance the matrix of the coefficients of a set of rate equations; see for instance equation (4.2.17) is always equal to the sum of all the rate constants in the underlying mechanism. Another check on setting up the correct matrix is that the sum of each column must be zero.

Denote the eigenvalues as λ_i ($i = 1, 2, \ldots, n$); then we can define a corresponding *column eigenvector*, denoted m_i, for each eigenvalue (an $n \times 1$ matrix) which is defined as the solution of the equation

$$(X - \lambda_i I)m_i = 0 \tag{2.4.12}$$

Only the relative values of the elements of the eigenvector are defined; any constant times the eigenvector is also a solution of (2.4.12) so the scaling of the eigenvectors is arbitrary. There is one eigenvector for each eigenvalue, so it does not matter which order we put the eigenvalues in, as long as the eigenvectors are kept in the same order. It will be seen that the eigenvectors, m_i, provide the information about the amplitudes A_i corresponding to the exponentials terms λ_i in $\Sigma \exp(\lambda_i t)$. How these operations are carried out in the solution of specific sets of linear differential equations is demonstrated with examples in sections 4.2 and 5.1.

This operational definition of the derivation of eigenvalues and eigenvectors, like the rest of this section, should be sufficient to follow the algebraic procedures applied in this volume. Many good texts are available to develop a deeper understanding of the algebraic methods described above.

3
Elementary kinetic equations: ground rules

3.1 The law of mass action, rates and equilibria

The law of mass action

The rates of chemical reactions almost always obey the *law of mass action*. Although the direct proportionality to concentration is sometimes modified, this law states that

> the rate of reaction is directly proportional to the product of the concentrations of the reactants.

The proportionality constant is known as the *rate constant* for the reaction in question. For particular sorts of reaction this name may be given a rather more descriptive name, for example, the *association rate constant* for a reaction involving association of two molecules, the *dissociation rate constant* for the reverse reaction, or the *isomerization rate constant* for a reaction that involves isomerization of one form of a molecule to another. The rate constant is a measure of how fast a reaction takes place (for a specified concentration), or, more precisely, it indicates how *frequently* the reaction occurs (hence it has the units s^{-1}). At the level of single molecules the rate constant is a measure of the probability (per unit time) that the reaction will happen in the next time interval.

How constant are rate constants?

It will be supposed in all that follows that rate 'constants' are indeed constant in the sense that they change neither with time nor with reactant concentration. The value of a rate constant may, and usually will, depend on temperature, on pressure, on electric field (e.g. on membrane potential) and suchlike intensive variables (see sections 6.1 and 7.2). The assumption that the rate constant is really constant throughout the reaction therefore depends critically on keeping conditions constant throughout the observation period (this is another reason why a voltage clamp method is needed in experiments on ion channels to keep the membrane potential constant). If any condition is changed in order to perturb the equilibrium, it must be changed in a stepwise fashion at $t = 0$ and thereafter held constant. This is done in temperature jump, pressure jump and concentration jump experiments, described in chapters 6 and 8, as well as in voltage jump experiments, which are a specialty of electrophysiologists (Hille, 1992).

It is important to mention in this context that in many reactions changes in ionization result in the uptake or release of protons (see sections 3.4 and 6.4). Unless the effect of proton concentration is being considered explicitly as part of the postulated reaction mechanism, it is also essential to keep pH constant during the observation period. The importance of protons, and hence of pH, was recognized by Sorensen (1912) when he introduced the concept of buffered solutions. The effects of changes in pH on the reactivity of enzymes and other proteins are discussed in sections 6.4 and 3.4.

The use of the law of mass action to describe a simple binding reaction

Consider again the case of a simple binding reaction, as in section 2.1. The ligand A (e.g. a substrate or drug molecule) can bind reversibly to a macromolecule R (e.g. an enzyme or receptor) to form a complex AR, thus

$$A + R \underset{k_{21}}{\overset{k_{12}}{\rightleftharpoons}} AR \tag{3.1.1}$$

The arrows in each direction have been labelled with the mass-action rate constants, which are denoted k_{12} for the forward (association) reaction, so k_{12} is the association rate constant, and k_{21} for the backward (dissociation) reaction, so k_{21} is the dissociation rate constant. The two reactants for the association reaction are the free ligand, the concentration of which will be denoted c_A, and the vacant enzyme/receptor molecule the concentration of which can be denoted c_R. The rate of the forward association reaction is thus given by the law of mass action as the product of these two concentrations, multiplied by the rate constant for this process, k_{12}. The rate of the reaction can be defined in terms of the rate of change of the concentration of any one of the three reactants. The change of the concentration of the complex with time is dc_{AR}/dt (apart from the sign this is the same as the rate of change of the concentration of vacant enzyme/receptor with time). Thus the rate of the reaction is

$$\text{rate of association} = \left[\frac{dc_{AR}}{dt}\right]_{assoc} = k_{12}c_A c_R \tag{3.1.2}$$

The dissociation reaction has only one reactant, namely the complex AR, the concentration of which may be denoted c_{AR}, so the rate of dissociation is this concentration multiplied by the dissociation rate constant, k_{21}. Dissociation results in a decrease in the concentration of AR so the sign is negative, thus

$$\text{rate of dissociation} = -\left[\frac{dc_{AR}}{dt}\right]_{dissoc} = k_{21}c_{AR} \tag{3.1.3}$$

The overall rate of change of concentration of a complex with time is the difference between the rate at which a complex is formed by the association reaction and the rate at which it breaks up by the dissociation reaction. The final result of the application of the law of mass action to the rate of change of c_{AR} for the mechanism in (3.1.1) is therefore

$$\left[\frac{dc_{AR}}{dt}\right] = k_{12}c_R c_A - k_{21}c_{AR} \tag{3.1.4}$$

An alternative notation

The meaning of 'concentration of free ligand' is obvious, and, in the case of an enzyme in solution, the meaning of 'concentration of complex' is equally obvious. However, in the case of an enzyme or receptor that is bound to a membrane, the meaning of the word concentration is not so clear because the molecules are not evenly distributed throughout the solution. In such cases it is common to refer not to the 'concentration' of the complex but to the 'fraction of molecules that are bound', or the 'fraction (of binding sites) that are occupied', or simply to the 'occupancy'. This amounts to no more than dividing the above equations by a constant, namely the total concentration of the macromolecule, $c_R(\text{total})$ say. The macromolecule exists in two different states, vacant R and complexed (or occupied) AR. The concentrations of one or both of these may vary with time, but at all times they must add up to the total concentration of macromolecule present, i.e.

$$c_R + c_{AR} = c_R(\text{total}) \tag{3.1.5}$$

If we define the *proportion* (hence the symbol p) of macromolecules in the complexed form as $p_{AR} = c_{AR}/c_R$(total) (alternatively known as the fraction occupied, or the occupancy) and the fraction that is *not* occupied as $p_R = c_R/c_R$(total), then dividing (3.1.5) by c_R(total) gives

$$p_R + p_{AR} = 1 \tag{3.1.6}$$

The fractional occupation, p, is dimensionless and is always within the range 0 to 1. As seen above, we denote the molar concentrations of reactants, say free ligand, by c_A rather than by the often used symbol [A]. When, for example, one wishes to emphasize that this concentration is varying with time one can simply write $c_A(t)$, which is a clearer notation than writing $[A](t)$, or $[A]_t$. Dividing (3.1.4) by c_R(total) gives the rate of change of the concentration of complex (of the occupancy), in this notation the rate of change of $p_{AR}(t)$ is

$$\frac{dp_{AR}(t)}{dt} = k_{12}c_A(t)p_R(t) - k_{21}p_{AR}(t) \tag{3.1.7}$$

The order of reactions and the dimensions of rate constants

A reaction that involves only one molecular species, such as the dissociation reaction in equation (3.1.3) is referred to as a *first order* reaction. It is clear from inspection of equation (3.1.3) that in such a case the rate constant, k_{12}, must have the dimensions of reciprocal time (s^{-1}) (i.e. of frequency). A reaction that involves the encounter of two reactant molecules, such as the association reaction in equation (3.1.2), is referred to as a *second order* reaction. Inspection of (3.1.2) shows that the rate constant in this case must have the dimensions of $1/(\text{concentration} \times \text{time})$, e.g. it will be measured in units of $M^{-1}s^{-1}$ when concentration is expressed in molar (M) units. Genuine reactions with order greater than two are rare and will not be dealt with here; they would require ternary collisions, which are likely to have very low probabilities.

A second order reaction may appear to behave like a first order reaction under certain circumstances. The concentration of free ligand may vary with time (this was indicated explicitly in (3.1.7) and implicitly in (3.1.4)). In general it must do so because as complex is formed, free ligand is removed from solution. However it is quite common for experiments to be done under conditions in which the concentration of free ligand remains, to a sufficiently close approximation, the same throughout the observation period. In the case of an enzyme reaction in solution this can be achieved by adding a large excess of the substrate. In the case of a receptor or enzyme bound to a cell surface it can be achieved by rapidly perfusing the cell with a solution containing a fixed concentration of the free ligand. Under such circumstances the rate of association can be written as $k_{12}c_A c_R$, or as $k_{12}c_A p_R(t)$ where c_A is now a constant. This is known as a *pseudo first order* reaction. The practical consequences of evaluating second order reactions under different conditions of ratios of reactant concentrations are discussed in detail in section 3.2.

Small perturbations

In the literature of chemical kinetics it is common to read of the *linearization of the rate equation*, a technique that can be applied when there are only *small perturbations* (see the definition of linear equations in section 2.4). These expressions mean, virtually always, that the concentration of free ligand can, to a good approximation, be regarded as constant (it is perturbed only slightly during the course of the reaction) in the way just described. In other words the system appears to be first order, and exponential behaviour is to be expected. This approach to simplify rate equations is discussed in detail in chapter 6.

Solutions of rate equations

Equations like (3.1.4) or (3.1.7) are simple to solve in the case where the free ligand concentration is constant (not varying with time), that is under pseudo first order conditions. This requires the solution of a first order differential equation with *constant coefficients*, which was published by A.V. Hill (Hill, 1909) in the biological context, 10 years before Langmuir (1918) used the same equations to describe adsorption of gases to light bulb filaments. The notation of equation (3.1.7) will be used now, because it makes it clearer which things are varying with time. If we substitute $p_R(t) = 1 - p_{AR}(t)$ in (3.1.7) we obtain

$$\frac{dp_{AR}(t)}{dt} = k_{12}c_A[1 - p_{AR}(t)] - k_{21}p_{AR}(t)$$

$$= -(k_{12}c_A + k_{21})p_{AR}(t) + k_{12}c_A \tag{3.1.8}$$

This, apart from the added constant $k_{12}c_A$, has the same form as equation (2.1.5). Its solution is

$$p_{AR}(t) = p_{AR}(\infty) + [p_{AR}(0) - p_{AR}(\infty)]e^{-t/\tau} \tag{3.1.9}$$

where $p_{AR}(0)$ is the occupancy (or concentration) at $t=0$, $p_{AR}(\infty)$ is the occupancy (or concentration) after an infinite time (i.e. at equilibrium) and the time constant, τ, for the transition from $p_{AR}(0)$ to $p_{AR}(\infty)$ is given by

$$\tau = 1/(k_{12}c_A + k_{21}) \tag{3.1.10}$$

The above elementary principles of the derivations of rate constants will be treated more generally and in greater detail in section 3.2 using the more familiar form with *concentration*, as in equation (2.1.10), rather than *occupancy*. The important problem of solutions of rate equations for second order reactions under different conditions of ligand concentrations will receive particular attention there.

The equilibrium condition

The equilibrium solution can be found directly from equations (3.1.4) or (3.1.7) by equating the rate of change to zero (or, equivalently, by equating the association and dissociation rates in (3.1.2) and (3.1.3)). This gives, in terms of equation (3.1.7),

$$k_{12}c_A(t)(1 - p_{AR}(\infty)) + k_{21}p_{AR}(\infty) = 0$$

so

$$p_{AR}(\infty) = \frac{k_{12}c_A}{k_{12}c_A + k_{21}} = \frac{c_A}{c_A + K} \tag{3.1.11}$$

where we define the *equilibrium dissociation constant*, K, as the ratio of the two rate constants

$$K = k_{21}/k_{12} \tag{3.1.12}$$

so that K has the dimensions of concentration. Equation (3.1.11) is often referred to as the Langmuir equation, though it should more properly be called the Hill–Langmuir equation. Inspection of it shows that K is the concentration at which half the enzyme/receptor binding sites are occupied. The shape of this relationship, which is illustrated in figure 3.1 is often referred to as being *hyperbolic*, because (3.1.11) describes (in the region of interest where concentrations are non-negative) part of one limb of a rectangular hyperbola.

The dynamic nature of equilibrium

When equilibrium is attained, the concentrations of all reactants become constant (they do not vary with time), but this does not, of course, mean

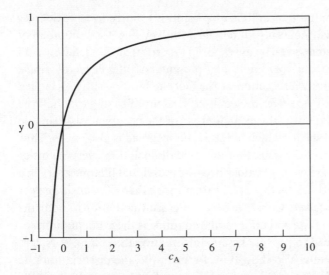

Figure 3.1 Relation between the equation for the rectangular hyperbola $(xy=a)$ and the binding equation $y=c_{AR}/[c_{AR}+c_R]=c_A/[c_A+K]$, where c_R, c_A, c_{AR} are respectively the concentrations of binding sites, ligand and complex and K is the dissociation constant. The simulated curve is based on $K=1$ and $c_R+c_{AR}=1$.

that nothing is happening, which can be imagined by considering the fate of a single molecule at equilibrium. From (3.1.3) the rate of dissociation at equilibrium is $k_{21}c_{AR}(\infty)$, where $c_{AR}(\infty)$ denotes the equilibrium concentration of the complex, so the molar concentration of complex is decreasing by this amount per second. This is, at equilibrium, exactly equal to the rate of association, which, from equation (3.1.2), is $k_{21}c_Ac_R(\infty)$, M s^{-1}, so the net concentration of the complex remains constant. Dividing by the total concentration of the macromolecule, c_R(total), gives, in terms of occupancy, the dissociation rate at equilibrium as $k_{21}p_{AR}(\infty)$, so this is the frequency (s^{-1}) of transitions from liganded to unliganded state per molecule. At equilibrium this is equal to $k_{12}c_Ap_R(\infty)$, the frequency of associations.

As a numerical example, suppose, as above, that $k_{12}=1\times10^8$ M^{-1}s^{-1}, and $c_A=1$ μM, then $k_{12}c_A=100$ s^{-1}. If the dissociation rate constant is taken as $k_{21}=100$ s^{-1} (so the mean lifetime of the complex is $1/k_{21}=10$ milliseconds) then 50% of the molecules will be in the complexed form at equilibrium since $K=k_{21}/k_{12}=1$ μM, so $p_{AR}(\infty)=0.5$. Thus if the total concentration of R, c_R(total), is 1 nM then there is $c_R(\infty)=0.5$ nM of complex present at equilibrium. The dissociation, and association, rate would therefore be 50 nM s^{-1}. Multiplying by Avogadro's number there are, therefore, on average, 3×10^{13} dissociations per second, and 3×10^{13}

associations per second (per millilitre of solution); fairly frenetic activity even at these lowish concentrations. Dividing by the total number of molecules in the system gives, on average, $k_{21}p_{AR}(\infty) = 50$ dissociations and associations per second per molecule present. At the level of single molecules this is simply a reflection of the fact the binding site is occupied for 50% of the time (the single-molecule equivalent of saying that half the sites are occupied), and that the mean lifetime of the occupancy is 10 ms, so the mean length of time for which the site is unoccupied is also 10 ms. Thus there is, on average, one association and one dissociation every 20 ms at equilibrium. These are average values because individual lifetimes vary in a random way; sometimes there will be rather more than 50 transitions per second, sometimes fewer. Consequently there are fluctuations about the average equilibrium state and under some conditions these fluctuations, or *noise*, can be analysed to provide kinetic information (Sakmann & Neher, 1983). These arguments also show that, for example, the dissociation rate constant, k_{21}, can also be interpreted as the mean frequency of dissociation transitions for one molecule per unit of time spent in the complexed state; in single channel recordings it is sometimes possible to measure such transition frequencies directly in order to estimate the value of rate constants.

Equilibrium versus steady state

In the discussion above the value of $c_{AR}(t)$, or of $p_{AR}(t)$, attained after a sufficiently long period, was referred to as its equilibrium value and K was referred to as an equilibrium constant. In systems such as that represented by scheme (3.1.1), a genuine equilibrium can be attained (this may be defined as a state of zero entropy production (see for example Denbigh, 1951). However the derivation of (3.1.11) did not, strictly speaking, require the assumption of equilibrium, but only the weaker assumption of a *steady state* which means only that concentrations are not changing with time. It is quite possible for a system to reach a steady state while not being genuinely at equilibrium, for example if there is some process that removes one of the reactants at a constant rate. The classical Briggs–Haldane treatment of enzyme catalysis (see section 3.3) supposes that the enzyme system is approximately in a steady state for most of the time that it is working, though in fact the substrate is being progressively used up. In the case of a steady state that is *not* a genuine equilibrium, the solution will have the form given in (3.1.11), but the constant K cannot be interpreted as an equilibrium constant (it will be some more complex combination of rate constants, as exemplified by the Michaelis constant of enzyme kinetics (see p. 82). The problem of steady states in general and of enzyme reactions in particular is developed in section 3.3.

Observed rate constants and fundamental rate constants

In the example just discussed, re-equilibration of the system is predicted to proceed with an exponential time course, the time constant τ for this process being given in equation 3.1.10 as $\tau = 1/(k_{12}c_A + k_{21})$. The observed rate constant for the exponential process is therefore $1/\tau = k_{12}c_A + k_{21}$. It can not be emphasized often enough that this *observed rate constant* (which is what can be measured experimentally) is actually a combination of two different *fundamental* or *mass action rate constants*, namely k_{21} and k_{12}. There is clearly scope for confusion when the term 'rate constant' is used without qualification. It is for this reason that the term 'rate constant' will be used in this book to mean the fundamental constants, defined by the law of mass action to describe the frequencies of transitions, whereas the rate at which observed exponentials decay will be described in terms of the observed *time constants* for the process, as was done in equations (3.1.9). and (3.1.10). One of the most frequent sources of confusion in discussions of kinetic problems is the one between rates (dc_x/dt) and the mass law (or fundamental rate constants).

The fundamental rate constants will (almost) always be denoted by the symbol k, with subscripts to indicate different rate constants: for example it is common to use k_+ and k_- to indicate the forward and backward steps of an individual transition, or to use k_{ij} for the rate of the transition from the ith state to the jth state. As the reader will have noticed, the latter form is used in this volume. An equilibrium constant, the ratio of two rate constants, will normally be denoted by an upper case K, with suitable subscripts. The measured time constants, τ, will be quoted with a subscript where necessary. The relation between k_{obs}, τ and λ, the eigenvalues of the matrix of rate constants, was discussed in section 2.1 and will be illustrated in sections 4.2 and 5.1.

3.2 The evaluation of first and second order rate constants

Introduction

In this section we shall review the common practices for the presentation and analysis of rate constants for single steps of reactions. The theoretical basis of the problem has already been developed in sections 2.1 and 3.1 in terms of exponential behaviour and the law of mass action. Here the emphasis is on simple practical criteria for and evaluation of first and second order rate constants. Repetition of certain topics, from a different angle, should be helpful to some readers.

First order rate constants

Irreversible first order transitions:

$$A \longrightarrow B$$

have been described (p. 22) by the differential equations

$$dc_A/dt = -kc_A(t) \quad \text{or} \quad dc_B/dt = k\{c_A(0) - c_B(t)\} \tag{3.2.1}$$

with the single rate constant k, the initial concentration $c_A(0)$ and the concentrations of $c_A(t)$ and $c_B(t)$ at time t. Integration of equation (3.2.1) gives, as discussed in section 2.1,

$$\ln c_A(t) + C = -kt$$

The integration constant C is determined from the initial conditions, $c_A(t) = c_A(0)$ at $t = 0$, when both sides of the equation equal zero:

$$\ln c_A(t) - \ln c_A(0) = -kt$$

and

$$c_A(t)/c_A(0) = \exp(-kt) \tag{3.2.2}$$

The properties of this exponential function and its application to many processes have been discussed in section 2.1 and chapter 1 respectively.

The form in which the above expressions are used for analysing data will depend on whether the record of the reaction shows the disappearance of A (equation (3.2.2)) or the appearance of B. The latter is described by

$$c_B(t)/c_B(\infty) = 1 - \exp(-kt) \tag{3.2.3}$$

The classical way to evaluate a first order rate constant from an exponential record, is to plot the logarithm (base e) of the reactant or product concentration against time to obtain the slope $-k$. The fact that the units of first order rate constants are s^{-1} and do not contain a concentration term, has simplifying consequences. The quantity on the Y ordinate of the plot need not be the logarithm of concentration of one of the reactants, but can be any measurement on the logarithmic scale (absorbance, fluorescence, conductivity, etc.) which is directly proportional to it.

Zero time can be taken anywhere along the reaction record as long as the maximum or minimum signal at $t = \infty$ is known. Under practical conditions of noisy traces and instrumental drift, uncertainty of the correct baseline at $t = \infty$ is the greatest source of error in estimating first order rate constants. If the end point of a record is taken too early, the derived first order rate constant will be greater than the true value and second order reactions terminated too early can give misleadingly acceptable exponential curves suggesting first order behaviour (see section 2.3).

Guggenheim (1926) proposed a method for the evaluation of first order rate constants from data which leave the end point (baseline) ill defined. An extension of this procedure by Gutfreund & Sturtevant (1956) allows for a constant drift. This latter procedure has proved very useful in the determination of first order rate constants for the approach to a steady state (see section 5.1) and is illustrated in detail elsewhere (Gutfreund, 1972). These secondary plots are now largely of historical interest, but can be found useful when only hand calculators are available for data analysis. Apart from rough estimates obtained from log plots or half lives (see section 2.1), rate constants are now generally evaluated by fitting exponentials with non-linear least square methods using desktop computers or programmable calculators (see section 2.3).

Reversible reactions: rate of approach to equilibrium

Reversible reactions, first order in both directions, are defined by two rate constants, k_{12} and k_{21}, which are related to the equilibrium constant K. The present derivation is intended to add practical details to the basic theory in sections 2.1 and 3.1. For the reaction

$$A \underset{k_{21}}{\overset{k_{12}}{\rightleftharpoons}} B$$

we can write

$$k_{12}/k_{21} = c_B(\infty)/c_A(\infty) \tag{3.2.4}$$

with equilibrium concentrations at $t = \infty$. The principle of reversible ligand binding and the expression for the observed rate constant for such processes are discussed in section 3.1. Let us derive this rate constant another way. Approaching equilibrium from the initial conditions at $t = 0$, $c_A(t) = c_A(0)$ and $c_B(t) = 0$ the rate of disappearance of A is given by

$$dc_A/dt = -k_{12}c_A(t) + k_{21}\{c_A(0) - c_A(t)\} \tag{3.2.5}$$

Using the equilibrium condition

$$k_{12}c_A(\infty) = k_{21}\{c_A(0) - c_A(\infty)\} \qquad (3.2.6)$$

to eliminate k_{12}, we obtain

$$dc_A/dt = \{k_{21}c_A(0)/c_A(\infty)\}\{c_A(\infty) - c_A(t)\} \qquad (3.2.7)$$

Using the expression which can be obtained from (3.2.6)

$$c_A(0)/c_A(\infty) = (k_{12} + k_{21})/k_{12}$$

to substitute in (3.2.7), we can write the differential equation in a form familiar for first order reactions

$$dc_A/dt = (k_{12} + k_{21})\{c_A(\infty) - c_A(t)\} \qquad (3.2.8)$$

The sum of the two rate constants appears in place of the single one of equation (3.2.1). Integration provides again a logarithmic or an exponential equation. Using the same reasoning for the identification of the integration constant from the initial conditions we can express the equation for a first order reversible reaction in the form

$$c_A(t) - c_A(\infty) = \{c_A(0) - c_A(\infty)\}\exp\{-(k_{12} + k_{21})t\} \qquad (3.2.9)$$

Records of such reactions are analysed in the same way as first order reactions and the resulting reciprocal time constant, τ^{-1}, is $k_{obs} = k_{12} + k_{21}$.

It is not intuitively obvious why the observed rate constant of approach to equilibrium is the sum of the rate constants for the forward and the reverse reaction. However, this point is illustrated diagrammatically in terms of time constants in section 2.1 (figure 2.1). This behaviour of reversible reactions has often been misinterpreted in the literature (see for instance the example in section 5.3).

Isolated true first order reactions can, by their very nature, only be studied through perturbation of their equilibrium. It is necessary to produce (or change the concentration of) one of the reactants of reaction (3.2.4) by some process which is faster than the subsequent reaction. For the case of an essentially irreversible reaction some chemical stimulus or catalyst, like a change in pH or the addition of a metal ion, can serve to start the reaction. Photochemical initiations of reactions (see p. 256 and section 8.1) also come within that category. Readily reversible reactions can also be studied by following the chemical relaxation after a rapid perturbation of the equilibrium by changes in temperature or pressure. This approach is discussed in detail in chapter 6. Most first order reactions discussed in this volume are part of a sequence of events initiated by a rapid second order reaction. In some cases a first order equilibration within a sequence of reactions can be studied in isolation. This is possible when everything that happens before it is very much faster and the following step is much slower than the equilibration. An example of this is the interconversion of metarhodopsin-I to metarhodopsin-II after a flash of light, which is discussed in section 4.2. In such cases the reaction step of interest is 'uncoupled' from the rest of the sequence. Frequently, however, the exponential time course, idealized in the present section, is perturbed by prior and by subsequent steps. Chapters 4 and 5 are devoted to the analysis of a variety of consecutive processes in enzyme catalysis, muscle contraction and signal transduction.

Parallel reactions

The importance of checking amplitudes for an assurance that the correct mechanism is assigned to a reaction will be frequently emphasized. This has already been alluded to with reference to the relation between observed rate constants and equilibrium amplitudes (see p. 56). If a side or parallel reaction occurs, the amplitude of the observed changes in product concentrations will not be equal to those of the reactant concentrations. In a system represented by the scheme:

$$A \underset{}{\overset{k_{12}}{\rightleftharpoons}} B$$
$$\Big\downarrow k_{13}$$
$$C$$

the rate of disappearance of A is given by

$$c_A(t) = c_A(0)\exp\{-(k_{12} + k_{13})t\} \tag{3.2.10}$$

That means that the observed rate constant is $(k_{12} + k_{13})$. The rate of appearance of B can be derived by integration of

$$dc_B/dt = k_{12}c_A(t) = k_{12}\,c_A(0)\exp\{-(k_{12} + k_{13})t\}$$

with the initial condition $c_B(0) = 0$

$$c_B(t) = \{k_{12}/(k_{12} + k_{13})\}c_A(0)\{1 - \exp(k_{12} + k_{13})t\} \tag{3.2.11}$$

similarly for the formation of C

$$c_C(t) = \{k_{13}/(k_{12} + k_{13})\}c_A(0)\{1 - \exp(k_{12} + k_{13})t\} \tag{3.2.12}$$

The pre-exponential factors of equations (3.2.11) and (3.2.12) determine the amplitudes for the product concentrations. For parallel reactions with reversible steps the matrix method described in section 4.2 should be used to derive the rate equations.

Second order reactions

Reversible binding of a ligand to a specific site on a protein or nucleic acid molecule is, with the exception of photochemical responses, the universal mode of initiation of biological processes in vivo. Ligand binding and recognition are the essential events in catalysis, signal and energy transduction, cross membrane transport as well as in cellular defence mechanisms. A number of different techniques with a range of time scales and optical or electrochemical monitors are used for their study, depending on the character of the signals obtained from particular systems. In subsequent chapters appropriate methods are referred to for specific examples discussed. However, it will be emphasized repeatedly that several experimental techniques should be applied to follow the time course of each process. So far we have only discussed first order reactions, which are sometimes called unimolecular transitions in the chemical literature to emphasize the stoichiometry of the reaction. The rest of this section is concerned with the derivation of rate constants for second order reactions and the consequences of their coupling to a subsequent transition.

Some aspects of nomenclature must be referred to in these introductory remarks on ligand binding. First, the term ligand is used differently in the biochemical and the chemical literature. In the field of metallochemistry the ligand is the chelator or host (for instance EGTA for Ca^{2+}), that is the metal binding site. Biologists almost universally use the term ligand for metals, substrates, effectors and transmitters which bind to the receptor or host site. To add to the confusion, the word substrate can be found in use to describe a layer of material onto which something is bound. This terminology is found in the literature when reactions on surfaces are discussed.

Association and dissociation constants: a problem of nomenclature

Another problem of nomenclature concerns the distinction between association and dissociation equilibrium constants. Since one is usually interested in the affinity (strength of binding), which is proportional to the association constant, the latter is used in some of the literature on ligand binding. There are, however, two reasons why the dissociation constant for ligand A from the complex AR

$$K_D = k_{21}/k_{12}, \qquad AR \underset{k_{12}}{\overset{k_{21}}{\rightleftharpoons}} A + R \qquad\qquad (3.2.13)$$

is in frequent use. First the Michaelis constant, K_M (see section 3.3), is expressed as the enzyme substrate dissociation constant and is used in many fields apart from enzymology (see discussion in section 3.1). Second the dissociation constant is expressed in units of molarity and, provided that $c_A(0) \gg c_R(0)$, it corresponds very conveniently to the ligand concentration at half occupancy of binding sites in the absence of cooperative interaction (see the end of this section). Then the binding curve (plot of occupancy against free ligand concentration) is represented by a rectangular hyperbola, which is described by the Hill–Langmuir equation derived in section 3.1. An extensive discussion of binding curves is given by Edsall & Gutfreund (1983). In the presentation of kinetic equations describing a sequence of reactions it is, however, often clearer if the equilibrium constants all point in the same direction and for the three step reaction one should write as follows:

$$A + R \underset{k_{21}}{\overset{k_{12}}{\rightleftharpoons}} AR_1 \underset{k_{32}}{\overset{k_{23}}{\rightleftharpoons}} AR_2 \underset{k_{43}}{\overset{k_{34}}{\rightleftharpoons}} AR_3$$

$$K_1 = k_{12}/k_{21}, \qquad K_2 = k_{23}/k_{32} \qquad K_3 = k_{34}/k_{43}, \text{ etc.}$$

We shall adopt the nomenclature of using K_D or K_M for dissociation constants and designate association constants, like those above, as K_1 and subsequent equilibrium constants for the ith step as K_i.

The rate equation for an irreversible second order reaction of the type $A + B \rightarrow X$ is

$$dc_X/dt = k\,c_A(t)c_B(t) \qquad\qquad (3.2.14)$$

The units of k in equation (3.2.14) have already been defined as $M^{-1}s^{-1}$. The significance of the inclusion of concentration units in second order as distinct from first order rate constants, will be seen to be important in the correct evaluation of rate constants from experimental records.

Choice of experimental conditions

In the different biochemical reactions to be discussed the relative concentrations of the reactants determine which of three solutions of equation (3.2.14) can be applied to the interpretation of data:

(a) $c_A(0) = c_B(0)$

(b) $c_A(0) \gg c_B(0)$ or $c_A(0) \ll c_B(0)$

(c) The general case when the two reactant concentrations are not equal but neither exceeds the other by a large factor.

'How large is large?' will be discussed below. Under condition (a)

$$\mathrm{d}c_A/\mathrm{d}t = -k\,c_A(t)\,c_B(t) = -k\,c_A^2(t) \tag{3.2.15}$$

The special case of dimerization reactions $A + A \to B$ will be considered separately. From equation (3.2.15) one can write

$$k\,\mathrm{d}t = -\mathrm{d}c_A/c_A^2(t)$$

which integrates, using the initial condition $c_A = c_A(0)$ at $t=0$, to give

$$c_A^{-1}(t) - c_A^{-1}(0) = kt \tag{3.2.16}$$

The second order rate constant can be evaluated from a plot of $c_A^{-1}(t)$ against time. Since the units of second order rate constants contain a concentration term, absolute values for concentration have to be plotted on the ordinate. It is also important to remember that the progress curves of second order reactions have very long tails. Care has to be taken to obtain true end points.

A second order reaction does not follow exponential decay and one cannot talk about time constants, except when first order conditions are imitated (see below for solution under condition (b)). We have to revert to the term half life, which differs by a factor of ln2 from the time constant of a first order reaction. For a first order decay the half life (note its relation to the time constant, see section 2.1), which is defined as the time taken for half completion of a reaction, is independent of the starting point. For a second order decay the half life is inversely proportional to the reactant concentration under the condition $c_A(0) = c_B(0)$.

Condition (b) is in fact the one with most frequent application to the study of biochemical reactions since it results in linear rate equations for the kinetic behaviour of systems. When the concentration of one of the reactants, A, is sufficiently in excess of the other, R, it remains essentially constant during the whole of the reaction $A + R \to AR$. Then $kc_A(0)$ is a pseudo first order rate constant and

$$\mathrm{d}c_A/\mathrm{d}t = -kc_A(0)c_R(t)$$

can be integrated to give

$$c_R(t) = c_R(0)\,\exp\{-kc_A(0)t\} \tag{3.2.17}$$

A plot of $\ln c_R(t)$ against time or a least square fit to the exponential will

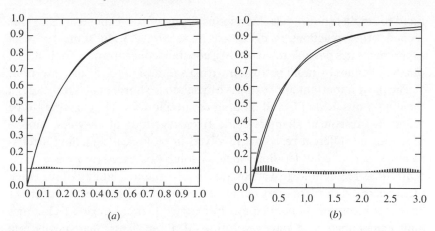

Figure 3.2   Simulated pseudo second order reactions $A + R \rightarrow AR$: $k = 1$, $c_R = 1$ and in (a) $c_A = 5$ and in (b) $c_A = 2$. The exponentials fitted for the two cases give $1/\tau$ of 4.710 s^{-1} and 1.712 s^{-1}. The deviations from true pseudo second order are discussed in the text. Residuals between the simulated reaction and the fitted exponential are shown across the graph.

give the value of $c_A(0)k$. If a series of experiments is carried out, at different concentrations of A, the resulting set of the observed pseudo first order rate constants can be plotted against $c_A(0)$; the slope of this plot is the second order rate constant. A straight line assures that the interpretation is correct.

The multiplication of the second order rate constant k(M^{-1}s^{-1}) by the fixed concentration $c_A(0)$ results in a pseudo first order constant with dimension s^{-1}. For example, if the rate constant for the association of A with R is 10^8 M^{-1}s^{-1} and $c_A(0) = 1$ μM, then the observed rate constant is 100 s^{-1}.

Appreciable errors are introduced in the calculation of pseudo first order rate constants if the excess of one of the reactants over the other is less than tenfold. This can be illustrated by simulating a series of second order reactions, keeping the rate constant and one of the initial reactant concentrations constant, while varying the initial concentration for the other reactant for each case. If the resulting computer generated curves are then analysed in terms of single exponentials, it is important to note that the time constants (τ) obtained, when the excess of one of the reagents $c_A > c_R$ is less than tenfold, depend critically on the proportion of the reaction analysed. When the analysis is extended over 97% of the theoretical amplitude the following deviations from the theoretical values for kc_A are obtained: for $c_A(0)/c_R(0)$ equal to 10, 5, 3 and 2 respectively the values for $1/\tau$ obtained were 3, 5.8, 10.7 and 14.5% smaller than the real value. Figure 3.2 illustrates a comparison of simulated reactions with twofold and fivefold excess ligand fitted to single exponentials. Erroneous fitting of two exponentials to true second order reactions is discussed in section 2.3. In some types of experiments it is possible to maintain a constant concentration, for instance of a free ligand, by rapid perfusion with a solution of fixed concentration.

The most important applications of pseudo first order conditions are in the ubiquitous field of ligand binding to specific sites. The examples to be found in subsequent chapters come from reactions of enzymes, muscle proteins and different receptors involved in biological signalling mechanisms. Practical considerations such as solubility, excessive rates at concentrations required to give adequate signals, and availability of reagents, often make it difficult to carry out experiments at very large excess of either ligand or receptor. As pointed out in chapter 1, rapid mixing techniques limit experiments to a time resolution of about 1 ms; that means rate constants larger than about 700 s^{-1} can not be determined by that method. It is in this connection that the illustration of errors introduced if the excess reactant concentration is only between 3 and 10 times greater than the other (figure 3.2), is of great importance. Other methods for the study of the kinetics of ligand binding are discussed in chapter 6 (relaxation methods). The stochastic methods (Colquhoun & Hawkes, 1983) developed for membrane receptors will find wider application as the observation of single molecules becomes technically feasible. The special case of pseudo first order kinetics due to rapid second order pre-equilibration prior to an observed first order process is discussed below.

For the analysis of second order reactions carried out under condition (c), when the concentrations of the two reactants are not equal but neither is significantly in excess of the other, the following general equation has to be solved:

$$\mathrm{d}c_X/\mathrm{d}t = k\{c_A(0) - c_X(t)\}\{c_B(0) - c_X(t)\}$$

or

$$kt = \int \mathrm{d}c_X\{c_A(0) - c_X t)\}\{c_B(0) - c_X t)\} \tag{3.2.18}$$

If $c_A(0)$ is not equal to $c_B(0)$ equation (3.2.18) can be transformed to simplify integration by partial fractions:

$$\frac{\mathrm{d}c_X}{\{c_A(0) - c_X(t)\}\{c_B(0) - c_X(t)\}} = \left[\frac{1}{\{c_B(0) - c_A(0)\}} \frac{1}{\{c_X(t) - c_A(0)\}} \frac{1}{\{c_X(t) - c_B(0)\}} \right]$$

Hence we can write equation (3.2.18) in the form

$$\mathrm{d}c_X\{[1/\{c_X(t) - c_B(0)\}] - [1/\{c_X(t) - c_A(0)\}]\} = k\{c_B(0) - c_A(0)\} \, \mathrm{d}t$$

which can be integrated to give

$$\ln\left[\frac{c_X(t)-c_B(0)}{c_X(t)-c_A(0)}\right]=kt\{c_A(0)-c_B(0)\}+\text{constant} \qquad (3.2.19)$$

Several algebraic problems have to be noted. When $c_A(0)>c_B(0)$ then $c_X(\infty)\rightarrow c_B(0)$ and when $c_A(0)<c_B(0)$ then $c_X(\infty)\rightarrow c_A(0)$.

The signs of the two sides of the equation and hence the fraction under the logarithm of equation, can be changed. The integrated equation (3.2.19) can be rearranged by using the substitutions

$$c_X(t)=c_A(0)-c_A(t)=c_B(0)-c_B(t)$$

and the form in which the equation is often used is

$$\ln\left[\{c_A(t)c_B(0)\}/\{c_B(t)c_A(0)\}\right]=kt\{c_A(0)-c_B(0)\},$$

but this can be rearranged for different initial conditions and for the observation of the time course of changes in concentration of A, B or X.

The relative complexity of the expressions obtained for the time course of second order reactions with generalized initial conditions [condition (c)] will convince the reader that conditions (a) or (b) should be used whenever possible.

Assembly reactions

Some comments are required on dimerization reactions of the type:

$$A+A\longrightarrow X$$

which can be self associations with, or without, chemical transformations. The differential equation describing the disappearance of A is

$$dc_A/dt=-2kc_X(t)^2=-2k\{c_A(0)-c_X(t)\}^2 \qquad (3.2.20)$$

This can be integrated for the initial conditions $c_A(t)=c_A(0)$ and $c_X(0)=0$

$$[1/\{c_A(0)-c_X(t)\}]-1/c_A(0)=2kt \qquad (3.2.21)$$

By convention the rate constant is written as $2k$ since two molecules of A disappear on each successful collision. A plot of $1/c_A(t)$ against time results in a slope equal to k. When self associations of more than two monomers occur, statistical effects as well as the likely pathway have to be taken into account (see below). It is unlikely that polymers are formed via third, or higher, order collisions. Sequential reactions of the type

monomer$\longrightarrow$dimer$\longrightarrow$tetramer (or trimer) are to be expected. If the dimerization is reversible the differential equation for the rate of disappearance of monomer is

$$dc_A/dt = -2k_{12}c_A(t)^2 + 2k_{21}c_X(t)$$

(note that two molecules of A are produced from each molecule of X).

The above equation can be written with all concentrations expressed in terms of monomer or in terms of dimer. Correspondingly different forms of integrated equations will be obtained.

The general equation for reversible second order reactions becomes quite complex and difficult to fit to data. The same is true of dimerization equilibria. The most suitable methods for the study of reversible second order reactions, as well as of dimerization and polymerization processes, are the relaxation methods. This approach is discussed in detail in chapter 6. Some equations for complex reversible reactions are given here for the record. More detail of their derivation and different limiting conditions are given by Capellos & Bielski (1980) and Moore & Pearson (1981). For the special case of the reaction

$$A + B \underset{k_{21}}{\overset{k_{12}}{\rightleftharpoons}} X \quad \text{with } c_A(0) = c_B(0),$$

and using the equilibrium conditions

$$k_{12} = k_{21}c_X(\infty)/\{c_A(0) - c_X(\infty)\}^2$$

to substitute for k_{21} the integrated rate equation is

$$k_{12}t = [c_X(\infty)/\{c_A^2(t) - c_X^2(0)\}]$$
$$\ln[c_X(\infty)\{c_A^2(t) - c_X(t)c_X(\infty)\}/c_A^2(t)\{c_X(\infty) - c_X(t)\}] \qquad (3.2.22)$$

The other rate constant, k_{21}, can be obtained from k_{12} and the equilibrium constant. It can be seen that when the equilibrium is largely in favour of X, this equation becomes very sensitive to the precise equilibrium condition.

A special case: binding to cell surfaces

The derivation of another equation for the analysis of reversible second order reactions, which was obtained by Erickson *et al.* (1987), can be useful if the choice of reactant concentrations is limited. The treatment was specifically developed for the investigation of the rates of ligand binding to cell surfaces with multiple receptor sites. In this connection a relation is derived between k_{12} and the density of receptor sites on cell surfaces. The same group (Goldstein *et al.*, 1989) also studied ligand dissociation in such systems. Using the simple model (one site per receptor model) with $c_R(0)$ total binding sites, $c_A(0)$ total ligand concentration,

$$A + R \underset{k_{21}}{\overset{k_{12}}{\rightleftharpoons}} AR$$

the rate of change is

$$-dc_R/dt = k_{12}c_A(t)c_R(t) - k_{21}c_{AR}(t)$$

The concentration terms for free A and for AR are eliminated by using conservation of mass

$$c_A(0) = c_A(t) + c_{AR}(t) \quad \text{and} \quad c_R(0) = c_R(t) + c_{AR}(t)$$

and we obtain the differential equation

$$-dc_R/dt = -k_{21}c_R(0) + [k_{21} + k_{12}\{c_A(0) - c_R(0)\}]c_R(t) + k_{12}c_R^2(t)$$
(3.2.23)

At this stage one has the choice between fitting the record for the disappearance of binding sites to the differential equation (3.2.23) or to the rather complex integrated equation. Erickson *et al.* (1987) obtained the following equation by integration from $c_R(0)$ to $c_R(t)$, from zero time to time t:

$$\frac{c_R(t)}{c_R(0)} = \frac{b[d\exp(te) - 1] + e[d\exp(te) + 1]}{2c[1 - d\exp(te)]c_R(0)}$$
(3.2.24)

where

$$a = k_{21}c_R(0), \qquad b = k_{12}\{c_R(0) - c_A(0) - k_{12}\}, \qquad c = -k_{12},$$
$$d = \{-2k_{12}c_R(0) + b - e\}/\{-2k_{12}c_R(0) + b + e\} \quad e = \sqrt{b^2 - 4ac}$$

Rate of approach to equilibrium

For the case of a pseudo first order rate constant for approach to the equilibrium $A + R \rightleftharpoons AR$, under conditions of a defined dissociation constant and constant (excess) ligand concentration ($c_A(0) \gg c_R(0)$), we can derive the full expression, which includes the expected amplitude of complex formation. At equilibrium when $c_{AR}(t) > c_{AR}(\infty)$

$$c_{AR}(\infty) = c_R(0)\{c_A(0)/[c_A(0) + k_{21}/k_{12}]\}$$

and

$$c_R(0)\{c_A(0)/[c_A(0)+k_{21}/k_{12}]\}-c_{AR}(t)$$
$$=c_{AR}(0)[c_A(0)/\{c_A(0)+k_{21}/k_{12}\}]\exp[-\{k_{12}c_A(0)+k_{21}\}t]$$

or in terms of the fraction of sites in the liganded state

$$c_{AR}(t)/c_R(0)=\{c_A(0)/[c_A(0)+k_{21}/k_{12}]\}\{1-\exp[-\{k_{12}c_A(0)+k_{21}\}t]\}$$

The record of the time course of the appearance of the complex will be represented by an exponential corresponding to the observed rate constant $k_{obs}=k_{12}c_A(0)+k_{21}$. If the simple one step mechanism applies, a plot of k_{obs} against $c_A(0)$ will result in a straight line with slope k_{12} and ordinate intercept k_{21}. A plot of amplitude A against $c_A(0)$ will be a hyperbola, described by equation (3.2.25), where K_D is the dissociation constant, A is the amplitude at a given concentration of ligand and A_{max} is the maximum amplitude when $c_{AR}(t)=c_R(0)$.

$$A=A_{max}\, c_A(0)/\{c_A(0)+K_D\} \tag{3.2.25}$$

Another test for the adequacy of this model (a single reversible binding step) is that the ratio of the rate constants obtained from the plot of k_{obs} against $c_A(0)$ must agree with the dissociation constant obtained from the plot of amplitude against $c_A(0)$. A practical problem arises in the design of such an experiment in the case of small dissociation constants. The wider the range of ligand concentration the better is the test; it should vary over, at least, an order of magnitude on either side of the dissociation constant. However, this condition, combined with the requirement of excess ligand over binding site concentration may limit the latter and thus the size of the signal. In some cases pseudo first order conditions can be achieved by using constant ligand concentration and varied excess binding site concentration.

If either or both of the above criteria for a single step second order association are not fulfilled, then the simplest alternative model is the following

$$A+R\underset{k_{21}}{\overset{k_{12}}{\rightleftharpoons}}AR\underset{k_{32}}{\overset{k_{23}}{\rightleftharpoons}}X \qquad c_A(0)\gg c_R(0) \tag{3.2.26}$$

Provided a ligand binding reaction is studied in sufficient detail, by varying conditions and methods of observation, this is in fact always the minimum complexity likely to be found. If the equilibration of the first step is rapid compared to the formation of X, the concentration of AR, at any time during the reaction, can be described as a fraction of the total of $c_R(0)$ $=c_R(t)+c_{AR}(t)$ available for conversion to X

$$c_{AR}(t)/\{c_R(t)+c_{AR}(t)\}=c_A(0)/\{c_A(0)+K_D\} \tag{3.2.27}$$

The result of this rapid equilibration of the first step is that the rate of the reaction $AR \rightleftharpoons X$ is controlled at any time by the fraction $\{c_{AR}/(c_{AR} + c_R)\}$ at equilibrium. The observed rate constant for the formation of X is

$$k_{obs} = k_{23}\{c_A(0)/[c_A(0) + K_D]\} + k_{32} \tag{3.2.28}$$

where $K_D = k_{21}/k_{12}$. At high concentrations of ligand, when $c_A(0) \gg K_D$, the observed rate constant is $k_{23} + k_{32}$. However, when $c_A(0)$ is not in excess of K_D the observed rate constant is dependent on ligand concentration and K_D. Thus one obtains apparent second order kinetics for a first order reaction which is preceded by a rapid equilibrium of ligand binding.

The two criteria, described above for a simple second order reaction, would fail for the reaction of scheme (3.2.26). The plot of k_{obs} against $c_A(0)$ for a series of experiments over a wide concentration range, from $c_A(0) \ll K$ to $c_A(0) \gg K$, will have an initial slope $k_{12}K$. This can be derived by differentiation of equation (3.2.28) with respect to $c_A(0)$ and using the condition $c_A(0) = 0$. At $c_A(0) \gg K$ the plot reaches the value $k_{23} + k_{32}$. The method of global analysis (Beechem, 1992) can be used to fit all the data of rate profiles, obtained under different conditions, in a single operation. The equilibrium constant for the formation of X from A and R is also composite; it is the product of the equilibrium constants for the two steps. Here we come up against the disadvantage of using the dissociation constant for describing binding equilibria. The steps subsequent to ligand binding are always described by forward equilibrium constants, as for instance for the second step of equation (3.2.26):

$$AR \overset{K_2}{\rightleftharpoons} X \quad K_2 - k_{23}/k_{32}$$

The question as to which part of the reaction is characterized by the overall equilibrium constant depends on which reactant is monitored by the signal used.

Multiple ligand binding sites

Many host molecules (enzymes, membrane receptors, respiratory proteins, etc.) have multiple binding sites for one particular ligand molecule. An important distinction has to be made in such cases between

(a) identical, non-interacting (independent) sites
(b) non-identical (chemically distinguishable) sites
(c) interacting identical sites (positive or negative cooperativity)
(d) non-identical and interacting sites

The last situation, (d), presents great complexities and will only be usefully described in exceptional cases. One such case is haemoglobin with two of each α and β chains binding ligands. The large signals obtained from kinetic experiments with this system permit data analysis in terms of both cooperativity and non-identity of sites.

Data obtained from equilibrium binding experiments can be analysed by a variety of algebraic procedures to distinguish between some of these cases. These are discussed in detail by Edsall & Gutfreund (1983, chapter 5) in conjunction with their thermodynamic analysis; site interaction is, by definition, a thermodynamic problem.

For identical, non-interacting sites the overall kinetic behaviour of the reaction of a ligand with a receptor with n binding sites is the same as that of the reaction on n separate receptor molecules, with one binding site each. However, sometimes one wishes to correlate the function of a protein with multiple sites with the proportion of sites occupied on an individual molecule. We shall proceed from the specific to the general, outlining the procedure for calculating the fraction of the total number of protein molecules (with four binding sites each) with none, one, two, three, four sites occupied during the reaction:

$$A + R \underset{k_-}{\overset{k_+}{\rightleftharpoons}} A_1R \rightleftharpoons A_2R \rightleftharpoons A_3R \rightleftharpoons A_4R \tag{3.2.29}$$

For case (a) above the intrinsic (or microscopic) rate constants are, by definition, identical for each step. However, the statistical factors defined below have to be included in the rate equations:

$$
\begin{aligned}
\mathrm{d}c_{AR}/\mathrm{d}t &= 4k_+ c_R c_A - (k_- + 3k_+ c_A)c_{AR} \\
\mathrm{d}c_{AR_2}/\mathrm{d}t &= 3k_+ c_{AR} c_A - (2k_- + 2k_+ c_A)c_{AR_2} \\
\mathrm{d}c_{AR_3}/\mathrm{d}t &= 2k_+ c_{AR_2} c_A - (3k_- + k_+ c_A)c_{AR_3} \\
\mathrm{d}c_{AR_4}/\mathrm{d}t &= k_+ c_{AR_3} c_A - 4k_- c_{AR_4}
\end{aligned}
\tag{3.2.30}
$$

Under the above conditions with c_A constant, this set of equations can be solved for each intermediate, as shown in the discussion of consecutive first order reactions in section 3.1. The statistical factors attached to the rate constants are dependent on the number of free sites (for k_+) or occupied sites (for k_-) available on a molecule. For instance, there are four ways of occupying the first site, while there is only one way to dissociate. However, the fourth site can be occupied in one way only, while there are four ways in which dissociation can occur. This is discussed in more detail for binding equilibria by Edsall & Gutfreund (1983). The general rate equation for liganding the i's site on a molecule with a total of n sites is

$$dc_{A_iR}/dt = (n-i+1)k_+c_A(0)c_{A_{i-1}R} - \{ik_+c_A + (n-i)k_-\}c_{A_iR} \qquad (3.2.31)$$

When a sequence of reversible ligand association is followed by some physical signal proportional to the number of sites occupied, then, if all sites on the molecule are identical and independent, no intermediates are detected. In some cases the resolution of the concentrations of intermediates can be derived by an analysis of the degree of fluorescence quenching of protein by bound ligand during the progress of the reaction (see section 8.2). There are cases in which the reversal of binding is sufficiently slow to allow for independent determination of different liganded species. If multiple sites on receptor molecules are independent but not kinetically identical, analysis of the time course of ligand binding should reveal a number of time constants.

Kinetics of cooperativity

Kinetic investigations of cooperative binding processes (interacting sites) are few and far between apart from the extensive studies on haemoglobin. The simplest model, which we call the gating process, is illustrated by a tetramer in which all four (chemically identical) sites are in equilibrium between two states in which they are simultaneously either open (state R) or closed (state T). In a case of complete cooperativity all sites are locked in the open state if one of them is occupied:

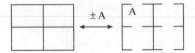

Ligand only has access to the binding sites when a receptor molecule is in the open state. If the fraction of the receptor molecules in the open state is given by

$$p_O = c_O/(c_C + c_O)$$

and the intrinsic rate constant is k_+, then the four rate constants in equation (3.2.29) become $p_O 4k_+$, $3k_+$, $2k_+$, k_+ respectively. The first of these is controlled by the proportion of sites in the open state (see pre-equilibria described by equations (3.2.27) and (3.2.28). The progress of such a cooperative process is compared with independent action in figure 3.3. Two features should be noted. The plot of total sites occupied against time shows a lag phase and the concentrations of reaction intermediates (the di- and tri-liganded states) are much reduced.

A similar series of first order reactions with a cooperative initiation is

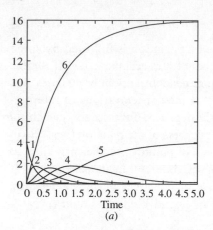 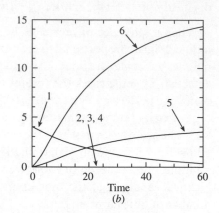

Figure 3.3   Kinetics of four site ligand binding: curves (1) to (5) represent the zero-, mono-, di-, tri- and fully liganded tetramers. Curve (6) represents the total concentration of liganded sites. (*a*) is based on the reaction of identical and independent sites with statistical multiplication factors (see p. 68) given to the intrinsic rate constant $k = 1$ s^{-1}. (*b*) is based on a cooperative system (identical sites) with intrinsic rate constants of 0.04 for the first step and 1.0 for the subsequent three steps of ligand binding. In this graph lines 2, 3 and 4 are omitted because they all straddle the baseline.

observed in the formation of α-helices from polypeptides. The steric requirement, and hence the entropy of activation, reduces the rate constant of the formation of the first intrapeptide hydrogen bond ($-C{=}O\ldots H\ldots N-$), compared to those of the subsequent helix formation, by a nucleation factor of about 3×10^{-3}. Hydrogen bond formation and rupture after initiation occur with rate constants of about 10^9 s^{-1}. The equilbrium constants for individual bond formation along the helix are, on average, only marginally larger than unity. Relatively stable helices are, therefore, only obtained with polypeptides containing many residues (see Zimm & Bragg, 1959).

A number of extensions of this simple model are obvious. The literature dealing with both equilibrium and kinetic aspects of cooperativity mostly follows the nomenclature of Monod, Wyman & Changeux (1965). Their model is defined in terms of two states, a weak binding T state and a high affinity R state. The ratio of the intrinsic dissociation constants $K_R/K_T = c$ and L_O, L_1, ..., etc. are the ratios of the concentrations of the oligomer (c_T/c_R) in the T and R states at different degrees of liganding. Cooperativity need not be, and usually is not complete; the equilibrium constants for the transitions between the two states can change with each successive liganding process. This model has been described in detail by Wyman & Gill

(1990). The cooperative transition from the T to the R states are generally thought to be due to changes in the quaternary structure (see Perutz, 1989). However, these transitions in the quaternary structure are triggered by ligand induced changes in the monomers, which could influence affinities even prior to the transition. These comments illustrate the potential compexities of kinetic profiles. Another interesting model is that derived by Geeves & Halsall (1987) for the sevenfold cooperativity of the interaction between myosin heads and actin filaments. They postulate that the heads can bind to open and closed filaments consisting of seven actin monomer units, but only the open complex can undergo the force generating structure transition. It is easy enough to write down a system of rate equations describing the sequence of the formation of the different liganded states for any cooperative model. The method described in section 4.2 can be used to derive solutions of these equations in terms of the time course of the fractional occupation of the states. These solutions become very complex for many models postulated in the literature. Only in exceptional cases, when experimental data of unusual signal and time resolution can be obtained, will it be possible to distinguish between different proposed models. For distinction between models in terms of binding isotherms the detailed analysis of Wyman & Gill (1990) should be consulted.

It is of interest to consider the interpretation of signals obtained during the time course of ligand binding for the case of the simple gating model. If the signal is proportional to the degree of liganding, then the form of the record will depend on the rate and equilibrium of the cooperative transition and the observed pseudo first order rate constant (the concentration dependent observed rate constant for binding, $k_+ c_A$). If the transition (opening) is slow, then rapid binding to the open sites will be followed by binding at the rate of opening. If the equilibrium is rapid, all sites will bind with an apparent rate constant ($Kk_+ c_A$). If the signal is due to the transition from the closed to the open state, then the rate of this transition can be directly evaluated. If binding occurs to both the T and the R states and the transition between these states is slow, acceleration of the reaction can be observed.

Further discussions of kinetic investigations into cooperative transitions will be found throughout this volume. The dynamic complexity of cooperative systems call for the application of a wide range of techniques. Observations following small perturbations of binding equilibria (see chapter 6) can yield information about ligand binding to different states as well as about the transition between these states (Kirschner *et al.*, 1966). Flash photolysis techniques (section 8.1) have been used to evaluate the rates of transitions between T and R states of haemoglobin at different degree of ligending (Eaton, Henry & Hofrichter, 1991). Some other investigations with rapid flow techniques will be discussed in section 5.1.

3.3 An introduction to practical steady state kinetics

Applications of steady state kinetics

The concept of a steady state or a stationary state is of interest for the description and simulation of many biological processes, as well as for approximations (the so-called steady state assumption) in the derivation of rate equations. The dynamic behaviour or the sequence of chemical reactions in a cell has many features which are more familiar to the chemical engineer than to those studying these processes in the usual laboratory environment. It is of interest to note that one of the classic papers on steady state theory by Denbigh, Hicks & Page (1948), who are chemical engineers by profession, uses examples from biology to illustrate the behaviour of open systems. Intracellular metabolic pathways and chemical production lines are both open systems with an approximately constant input and output, while in a laboratory synthesis is, usually, carried out in a closed system. Up to about fifty years ago the pricipal pre-occupation of those interested in physical aspects of biochemistry was with equilibrium thermodynamics. The postulates of classical thermodynamics do provide an essential guide to the limits of possible reactions, but they do not result in a realistic description of cellular processes in action; equilibria are necessarily end states. There is now considerable interest in the behaviour of steady state or non-equilibrium systems as the essential backbone to an understanding of energy transduction in membrane phenomena and muscle contraction (Hill, 1977). Our main concern in the present volume is the interpretation of the kinetic events during the transient approach to and the decay of the steady state. This involves the most interesting phenomena such as triggering and transmission of signals. However, in this section the principles of steady state kinetics are presented as a basis for subsequent discussions of transient events. The principal examples of steady state kinetics will be taken from enzymology, but first some more general comments are in place.

Definition of a steady state

Steady states or steady state approximations, are used to describe mechanisms and to solve equations for a great variety of phenomena such as population growth, chemical reaction sequences and heat conduction. The principle is illustrated by the general scheme:

$$\text{source} \to X_1 \to X_2 \to \ldots \to X_n \to \text{sink}$$

where the concentrations of X_1 to X_n remain constant during flow from source to sink. This steady or stationary state is created during an initial transient phase of the reaction which involves the adjustment of the n intermediate concentrations to a level at which the rates of formation of the i's intermediate, $k_i c_{x_{i-1}}$, are equal to their rates of removal, $k_{i+1} c_{x_i}$. The apparent rate constant for the steady state flux through such a sequence of irreversible reactions from a constant pool, is defined by a zero order rate constant k (see p. 80)

$$1/k_{\text{obs}} = \sum_{1}^{n} 1/k_i \tag{3.3.1}$$

The steady state behaviour of *systems* (see for instance von Bertalanffy, 1971, who discusses systems theory and feedback control) is of particular interest to students of cellular metabolism and other homeostatic functions. In subsequent chapters we shall be much concerned with the transients during the establishment and decay of steady states in a variety of functions. Although the analysis of systems in terms of the stability or instability of steady states, leading to the consideration of oscillations and dissipative phenomena, is also of special interest in biology, it is beyond the scope of this volume. A good introductory discussion is given by Segel (1984). The dynamics and stability of systems do have very wide applications and references to these can be found in Prigogine & Stengers (1984) and von Bertalanffy (1971).

The behaviour of chemostats

In section 1.1 feedback terms were introduced into the equations for the rates of growth of populations, and the resulting limits and steady states of population size were discussed. Related problems arise in the control of the synthesis of biological products in chemostats, which have become important for the production of genetically engineered proteins. The early development of fermenters, working under controlled conditions of composition of growth medium and continuous production, was due to Monod (1950) and Novick & Szilard (1950). They were interested in the natural rates of mutations in microorganisms during adaptation to specific nutrients under conditions of (prolonged) continuous (steady state) population growth.

In a culture of microorganisms exponential growth will occur, for a period, under certain optimum circumstances. From the data given by Stent & Callender (1978) for the growth of *Escherichia coli* under ideal conditions, we can derive a number of parameters which will serve as examples. From the turbidity of the culture, which is proportional to the number of organisms per unit volume, it was found that the concentration doubled every 1200 s. This means, that under such idealized conditions, if one started off with 1 cell at $t=0$, one would obtain, in 24 hrs, 2^{72} (i.e. 5×10^{21}) descendants weighing about 10^6 kg. However, the exhaustion of nutrients and the production of inhibitors would slow down the growth after a short time. In the microbiological literature the time taken to double the number of cells is called the generation time. We can calculate the rate constant for the above example of cell culture as follows:

$\ln 2 = kt$; $0.69 = 1200k$; hence $k = 5.75 \times 10^{-4} \, \text{s}^{-1}$

The rate constant in this equation corresponds to $1/\tau$, as derived in section 3.1. For a culture grown in an undisturbed medium the real growth curve would be S shaped. Such functions for growth curves and population limits in demographic or ecological systems have been developed on the basis of many different models and have already been referred to in section 1.1.

We shall now describe a simple realistic model for the growth of cultures of microorganisms in a chemostat. The rate of replication of microorganisms depends on the concentration of nutrients in the medium. Let us assume, for simplicity, that the rate of growth is limited by one substrate. The response, R, of many systems to a concentration, c_S, of a substrate can be described by

$$R_{observed}/R_{maximum} = c_S/[c_S + K] \qquad (3.3.2)$$

where K is a constant which is a measure of the affinity of the system (actually K is inversely proportional to the affinity, see p. 79) for the substrate which limits population growth. This form of *saturation kinetics* will be encountered in a number of phenomena discussed in this volume. Enzyme kinetics, membrane transport and many other processes involve ligand binding and the response of a receptor. At high concentrations of nutrient, $c_S \gg K$, growth proceeds at R_{max} and at low concentrations, $c_S \ll K$, growth is directly proportional to c_S. A characteristic property of K is that, at the point when $K = c_S$, $R_{obs} = \frac{1}{2} R_{max}$. Combining exponential rate of growth with the effect of the depletion of the nutrient we obtain a modified form of equation

$$\frac{\mathrm{d}N}{\mathrm{d}t} = \frac{kNc_S(t)}{c_S(t) + K} \tag{3.3.3}$$

where N is the number of organisms.

For many practical purposes large scale bacterial production is carried out in a chemostat. This device, a further development of the method used by Novick & Szilard (1950) and Monod (1950), allows continuous growth by constant removal of organisms and addition of nutrients, as well as the neutralization of inhibitory products, for instance by keeping pH constant. The design of the process is described by Herbert, Elsworth & Telling (1956) and a brief summary of the theory will be discussed as an example of steady state behaviour.

The kinetic problems involved in the continuous production of large quantities of microorganisms can be defined by the following model for a chemostat. This also serves as an example for the setting up of differential equations to describe a dynamic system. In such a system the volume of the reaction medium is kept constant by the continuous withdrawal of fluid containing cells and metabolic products and addition, at the same rate, of solution containing essential nutrients. The system is well stirred, so that diffusion, which can be an important factor in cellular processes, does not have to be taken into account. The chemostat is seeded to initiate growth and we make the simplifying assumption that only one nutrient, S, has to be added to maintain production of the organisms. The initial conditions at $t = 0$ are defined by the concentration of the critical nutrient $c_S(0)$ and the number of organisms $N(0)$ in the volume, V, of the reactor. The fraction, F, of the volume added and removed is v/V and the concentration of the nutrient solution added is $c_S(i)$.

We require two differential equations to define the time course of changes in the nutrient concentration $c_S(t)$ and the number of organisms $N(t)$ in the reactor. The exponential growth function is modified by its dependence on the nutrient concentration and the removal of organisms

$$\mathrm{d}N/\mathrm{d}t = [k(c_S) - F]\, N(t) \tag{3.3.4}$$

where $k(c_S)$ is the reciprocal for the time constant of exponential growth at a specified substrate concentration. The change in nutrient concentration depends on the addition and removal of the volume fraction as well as on the uptake of nutrient by the organisms. The latter is determined by a function $V(c_S)$ and the number of organisms. If we assume that $V(c_S)$ is related to $k(c_S)$ by a constant C, then we obtain

$$k(c_S) = CV(c_S) \quad \text{and} \quad dc_S/dt = F\{c_S(i) - c_S(t)\} - Nk(c_S)/CV$$

The nutrient concentration dependent growth rate constant can be defined by the hyperbolic function

$$k(c_S) = k_{max}c_S(t)/[K + c_S(t)] \tag{3.3.5}$$

where k_{max} is the value for $k(c_S)$ when c_S is much larger than K, the affinity of the microorganisms for nutrient. Taking all these factors into account we can proceed to an equation for population growth:

$$dN/dt = k_{max}N(c_S)/\{K + c_S(t)\} - FN \tag{3.3.6}$$

and for change in nutrient concentration:

$$dc_S/dt = \frac{k_{max}Nc_S(t)}{C\{K + c_S(t)\}V} + F\{c_S(i) - c_S(t)\} \tag{3.3.7}$$

This equation fulfils the limiting conditions set out above for $c_S \to \infty$.

The above description for the behaviour of a chemostat follows the arguments developed by Segel (1984) who in turn follows Rubinow (1975). Different numerical values for initial conditions and parameters can be inserted into the steady state equation:

$$dc_S/dt = 0 = \frac{k_{max}N(\infty)c_S(\infty)}{C\{K + c_S(t)\}V} + F\{c_S(i) - c_S(t)\} \tag{3.3.8}$$

When $V < k_{max}$ positive values will be obtained for the steady state concentrations, N and c_S; when $V > k_{max}$ the cells will all be washed out. A selection of experimental results will be found in the papers by Novick (1955) and Herbert *et al.* (1956). Other aspects of setting up and solving steady state equations will be illustrated with examples from enzyme systems.

Clearly other features, which must occur in a real system, could be added. The products of metabolism can act as feedback inhibitors or activators and, under some conditions, oscillations can occur. Such additions to the kinetic equations can be made by the reader with interests in different models. Feedback control can also be studied in the so-called turbidostat, which controls the flow through the chemostat to keep constant turbidity in the reactor to a specified level of concentration of microorganisms.

The role of steady states in enzyme kinetics

Steady state kinetics has played a major role in the study of reactions of isolated enzymes and of metabolic systems including many sequential and branch chain reactions. This approach to enzyme kinetics is invaluable for the interpretation of intermediary metabolism. However, for the resolution of enzyme mechanisms, methods for the identification of reaction intermediates, described in later chapters, are more informative. The large number of specialized volumes dealing with the steady state behaviour of enzyme reactions makes it superfluous to devote as much space to this topic as some biochemists might expect. Much of the material found in classic texts on enzyme kinetics will be progressively less used in practice. Discussion of the extensive literature on the application of derivative linearized forms of rate equations (Dixon & Webb, 1964; Segel, 1975) is restricted to a minimum since numerical fitting to the hyperbolic equations is more reliable and easy, thanks to several simple computer programs. In general, direct observations of reaction intermediates should be substituted for the often ambiguous conclusions drawn from steady state observations. It is, however, only fair to emphasize that some isotope exchange techniques applied to steady states have provided important conclusions specifically for chemical mechanisms (see for instance Britton, Carreras & Grisolia, 1972; Britton & Clarke, 1972; Albery & Knowles, 1976). In this volume it is more appropriate to discuss specialized kinetic problems of enzymology which are not treated extensively elsewhere. However, the basic arguments of steady state kinetics, which are often not properly understood, will be presented in this section to illustrate the methods, which can, in principle, be used to describe all enzyme reactions. The aspects of steady state kinetics treated here are a necessary background to transient kinetics of enzyme reactions, the method of choice for the elucidation of mechanisms. The results of transient kinetic investigations (the subject matter of section 5.1) must predict the correct *Michaelis parameters* (see below). It will be noted that the nomenclature of enzyme kinetics, especially that of saturation kinetics, has penetrated into the language and models of other areas of ligand receptor interaction. As mentioned before, it is one principal purpose of the present volume to cross fertilize the treatment of kinetic problems involved in the study of different biological systems.

The fact that the kinetics of enzyme action was of such major interest to physical biochemists resulted in the development of techniques and theoretical analyses which have subsequently found valuable applications to the study of many other proteins. The experiences gained in the study of the physical chemistry of enzymes have contributed a great deal to investigations on proteins involved in muscle contraction, membrane transport and the transmission of signals.

The development of enzyme kinetics

Different approaches to the algebraic treatment of steady state rate equations each have their own protagonists. To give a fair presentation of their relative advantages would require a large monograph. The basic principles of steady state kinetics of enzyme action were first clearly summarized by Haldane (1930) in a volume still worthy of detailed study. For a compendium of equations one can consult Segel (1975) and a balanced account of the information which can be obtained from steady state kinetics is given by Dalziel (1975), who also indicates the complementary value of transient kinetic studies. The latter are discussed in detail in section 5.1.

From a practical point of view the major difference between the two approaches to enzyme kinetics, steady state and transient rate measurements, is in the concentrations of enzyme used. Steady state experiments are carried out with catalytic amounts of enzyme at concentrations negligible compared to those of the substrates or products. The rationale of transient kinetic experiments, discussed in section 5.1, will be seen to rest on the observation of complexes of enzymes with substrates and products. The importance of the direct observation and characterization of reaction intermediates for an understanding of mechanisms will be illustrated in that section. This requires enzyme concentrations sufficiently high for detection of intermediates by spectroscopic or other physical monitors. There are a number of interesting systems in vivo with enzyme and substrates at comparable concentrations and the potential kinetic consequences of such situations will be discussed in sections 5.2 and 5.3. Jencks (1989) comments in connection with a review of the transient kinetic behaviour and mechanism of one such system, the calcium pump of the sarcoplasmic reticulum, that steady state kinetics could make no contribution to an understanding of its ATPase linked reaction. The same can be said of the mechanism of myosin-ATPase, which has been elucidated in detail by transient kinetic studies (see section 5.1).

Highly specialized steady state studies have been carried out on selected enzymes to explore their reaction mechanisms (see for instance Dalziel, 1975, and the isotope exchange experiments mentioned above). However, the major contribution of such investigations is to the definition of those kinetic characteristics of enzymes which are of importance for an understanding of their activity and control in metabolic pathways. It is essential to have information about the dependence of the rates of individual enzyme reactions on many concentration variables (substrates, inhibitors, activators and ionic composition of the medium). The mode of inhibition and activation by drugs is another important aspect of steady state studies on enzyme reactions. The steady state parameters are also an important diagnostic for the correct interpretation of information about mechanisms which has been obtained from other approaches.

An account of the history of the investigations which led to the present views on enzyme action via a series of intermediates can start with the oversimplified scheme

$$E + S \rightleftharpoons ES \rightarrow E + P \tag{3.3.9}$$

proposed by Michaelis & Menten (1913). Earlier proposals for the participation of enzyme–substrate complexes have been reviewed by Haldane (1930) and by Segal (1959). Michaelis & Menten's lasting contribution to enzymology was their proposal that such reactions can be described by two constants, the Michaelis parameters V_{max} and K_M. The Michaelis constant K_M was called the equilibrium constant for the first step. It was defined as a dissociation constant of the enzyme–substrate complex:

$$ES \rightleftharpoons E + S \quad K_M = c_S c_E / c_{ES} \tag{3.3.10}$$

This anomaly of nomenclature for a binding process has stayed with us and most biological binding processes are expressed in this way; as pointed out in section 3.2, this has some practical advantages. As will be seen, the Michaelis constant as well as the maximum velocity V_{max} (see equation (3.3.11)) determined by steady state methods are now recognized to represent complex functions of many rate constants. In the Michaelis scheme a single rate constant was assigned to the decomposition of the initial enzyme–substrate complex to free enzyme and product. An essential feature of equation (3.3.11) is that, as the substrate concentration is increased, a limiting maximum velocity is reached. Such saturation kinetics has been discussed above to indicate its wide application beyond enzymology to reactions stimulated by ligand binding to a specific site. The principle is the same as that of adsorption of gases on metal surfaces prior to catalysis, ligand binding to channels, active transport sites and oxygen binding to haemoglobin (see the discussion of the Hill–Langmuir equation in section 3.1).

If the maximum velocity, as $c_S \rightarrow \infty$, is designated V_{max} and the velocity,

dc_P/dt, at a given substrate concentration is v, then the general rate equation is

$$v/V_{max} = c_S/\{K_M + c_S\} \tag{3.3.11}$$

This equation should be compared with that derived for maximum occupancy of ligand binding sites (see section 3.2 and Edsall & Gutfreund, 1983). A plot of v/V_{max} against c_S corresponds to the rectangular hyperbola of the Langmuir adsorption isotherm and to the equations for ligand binding in solution, which have been referred to before as the Hill–Langmuir equation (see figure 3.1). However, in all these cases the experimental data only fit the hyperbolic equation if all binding sites are identical and independent (no site–site interaction). The unit of K_M is that of a dissociation constant, M, and the units of V_{max} are defined as M s^{-1}. This is often called a zero order rate constant; at the beginning of the reaction, while substrate remains approximately constant and there is no reverse reaction, product versus time is a straight line – the so called *initial rate*. Another measure for the rate of an enzyme reaction is the turnover number, which has the units of a first order reaction. It is obtained by dividing V_{max} by the enzyme active site concentration. In defining enzyme concentration care has to be taken to distinguish between that of the often polymeric protein molecule and the concentration of active sites. The latter is in more general use as the concentration variable for detailed kinetic analysis and is sometimes referred to as *normal* instead of *molar* concentration.

Briggs & Haldane (1925) removed the restrictive assumption that the enzyme–substrate complex is in equilibrium with free enzyme and substrate and introduced the steady state model, which gives the Michaelis parameters, K_M and V_{max}, a more complex meaning. The principles of steady state kinetics of enzyme reactions can be demonstrated with the more realistic, though still oversimplified, model of Haldane (1930). This contains the minimum number of intermediates, namely enzyme–substrate and enzyme–product complexes:

$$\text{E} + \text{S} \underset{k_{21}}{\overset{k_{12}}{\rightleftharpoons}} \text{ES} \underset{k_{32}}{\overset{k_{23}}{\rightleftharpoons}} \text{EP} \underset{k_{13}}{\overset{k_{31}}{\rightleftharpoons}} \text{E} + \text{P} \tag{3.3.12}$$

If the reaction is initiated by the addition of substrate to a solution of enzyme such that $c_E \ll c_S$, the rate of product formation will go through three phases. For a short transient period, while the concentrations of ES and EP build up to a steady state concentration, dc_P/dt will increase. This phase, which is too short to be observed with conventional methods (less than a second; see below, p. 85), is discussed in section 5.1. After this transient phase an approximately constant velocity of product formation will be observed, as long as the change in substrate concentration compared to total substrate, as well as the amount of product formed, is negligible. This is the steady state phase of the reaction. During the subsequent third phase dc_P/dt decreases as substrate is used up and the product concentration increases. In this section we are concerned with the second of the above three phases, although some comments will be made on reactions at very low substrate concentrations. This involves two related concepts of enzyme kinetics: the steady state and initial rate measurements. The principle of the steady state will be illustrated in relation to enzyme reactions as an extension of the discussion presented earlier in this chapter. Although the necessary condition for steady state enzyme kinetics is usually stated as essentially constant substrate concentration, the principal condition is that the concentration of the intermediates remain essentially constant during the course of the measurement. The term intermediates is used for the different states of the enzyme and the term reactants applies to the free substrates and products.

The following derivation of the kinetic equations for the scheme given in equation (3.3.12) depends on two simplifications. The first of these, essentially constant substrate concentration, has already been stipulated as a condition for maintaining a steady state. The second simplification is that the product concentration is essentially zero. The latter removes the necessity to consider the reversal of the reaction. The consequences of the relaxation of this latter condition will be shown later. With these assumptions we can write the rate equations for the two intermediates:

$$dc_{ES}/dt = k_{12}c_E(t)c_S(t) - \{k_{21} + k_{23}\}c_{ES}(t) + k_{32}c_{EP}(t) \tag{3.3.13}$$

$$dc_{EP}/dt = k_{23}c_{ES}(t) - \{k_{32} + k_{31}\}c_{EP}(t) \tag{3.3.14}$$

In equation (3.3.14) we substitute c_{ES} in terms of c_{EP} from equation (3.3.13) and eliminate the free enzyme concentration using the constant total enzyme concentration, $c_E(0)$, and the conservation equation

$$c_E(t) = c_E(0) - c_{ES}(t) - c_{EP}(t)$$

Thus we obtain an expression for the enzyme–product complex

$$c_{EP}(t) = \frac{k_{12}k_{23}c_S(0)c_E(0) - dc_{ES}/dt}{k_{12}c_S(0)\{k_{32} + k_{23} + k_{31}\} + \{k_{21}k_{32} + k_{21}k_{31} + k_{23}k_{31}\}} \tag{3.3.15}$$

If we substitute for $c_{EP}(t)$ using the expression

$$dc_P/dt = v = k_{31}c_{EP}(t) \quad \text{hence} \quad c_{EP}(t) = v/k_{31}$$

and use the steady state condition $dc_{ES}/dt = 0$, then we obtain

$$\frac{v}{c_E(0)} = \frac{k_{12}k_{23}k_{31}c_S(0)}{k_{12}c_S(0)\{k_{23}+k_{32}+k_{31}\}+\{k_{21}k_{32}+k_{21}k_{31}+k_{23}k_{31}\}} \tag{3.3.16}$$

The above equation can be rearranged into a form corresponding to equation (3.3.11)

$$\frac{v}{c_E(0)} = \frac{k_{23}k_{31}c_S(0)/\{k_{23}+k_{32}+k_{31}\}}{[\{k_{21}k_{32}+k_{21}k_{31}+k_{23}k_{31}\}/k_{12}\{k_{23}+k_{32}+k_{31}\}]+c_S(0)} \tag{3.3.17}$$

From this expression we can extract the following Michaelis parameters:

$$V_{max}/c_E(0) = k_{23}k_{31}/\{k_{23}+k_{32}+k_{31}\}$$

and $\qquad\qquad\qquad\qquad\qquad\qquad\qquad\qquad\qquad\qquad\qquad$ (3.3.18)

$$K_M = \{k_{21}k_{32}+k_{21}k_{31}+k_{23}k_{31}\}/k_{12}\{k_{23}+k_{32}+k_{31}\}$$

One of the most important things to learn about deriving any kind of equation is how to get it into a most useful form. The present problem serves as a good example. The proper substitutions were made to get an equation which expresses the rate of an enzyme reaction in terms of a set of rate constants and the known concentrations of substrate and total enzyme sites.

If the chemical transformation (ES → EP) is rate limiting and essentially irreversible, that is $k_{31} \gg k_{23}$ and $k_{32} \to 0$, then equation (3.3.18) simplifies to the Michaelis scheme as modified by Briggs & Haldane (1925):

$$K_M = \{k_{21}+k_{23}\}/k_{12} \quad \text{and} \quad V_{max}/c_E(0) = k_{23} \tag{3.3.19}$$

Michaelis assumed that $K_M = k_{21}/k_{12}$ is an equilibrium dissociation constant. The assumption that free enzyme and substrate are in true equilibrium with their complex holds only in some cases and should not be accepted as a generality. The real significance of K_M will be discussed below. However, the definition of K_M as the substrate concentration at which half the maximum velocity is observed will apply regardless of the complexity of the mechanism in terms of reaction intermediates, provided (a) all the active sites of an oligomeric enzyme are identical and independent (i.e. do not interact in a cooperative sense) and (b) substrates and/or products do not also act as inhibitors or activators. In the present discussion we shall use the term Michaelis kinetics for reactions which obey conditions (a) and (b) and which therefore respond to substrate concentration according to the general equation for a rectangular hyperbola illustrated in figure 3.1.

One can demonstrate that the physical meaning of K_M depends on the number of reaction intermediates and the position of the rate limiting step in the sequence of events. The condition that the rate limiting chemical transformation occurs immediately after the initial complex formation was first shown to be too restrictive by Theorell & Chance (1951). They demonstrated that in the reaction catalysed by horse liver alcohol dehydrogenase the dissociation of one of the products, NADH, was the slowest step of the pathway. Since that time the detailed exploration of enzyme mechanisms has demonstrated that conformation changes of the protein, either after substrate binding or before product dissociation, control the overall rates of many reactions. This will be discussed in more detail in section 5.1. The effect of a change in rate limiting step on K_M, by some altered condition, can be readily shown by a slightly simplified form of equation (3.3.18). If we consider the second as well as the third step as irreversible, i.e. $k_{32} = 0$ and $k_{13} = 0$, then

$$K_M = \{k_{31}k_{21} + k_{23}\}/k_{12}\{k_{23} + k_{31}\} \quad \text{and} \quad V_{max}/c_E(0) = k_{23}k_{31}/\{k_{23} + k_{31}\}$$

Now we can see what happens to K_M when the rate limiting step is changed between those controlled by k_{23} to k_{31}, that is from the chemical transformation to product dissociation. When $k_{31} \gg k_{23}$ then we obtain the Michaelis parameters of equation (3.3.17). However, if $k_{23} \gg k_{31}$ then we obtain

$$K_M = \{k_{21} + k_{23}\}k_{31}/k_{12}k_{23} \quad \text{and} \quad V_{max}/c_E(0) = k_{31}$$

In this latter case the Michaelis constant can be considerably smaller than the dissociation constant for the enzyme substrate complex, k_{21}/k_{12}. The Michaelis constant defines the substrate concentration when half the enzyme, in the steady state, is in the form preceding the rate limiting step. Of course the situation can be less well defined when there are several steps with similar rates! These comments are intended to underline the warning that Michaelis constants should not be used as measures for substrate affinities. A good example is the comparison of the K_M values for the chymotrypsin catalysed hydrolysis of acetyl-L-tyrosine ethyl ester and the analogous amide. A hundredfold decrease in k_{23} results in an equivalent increase in K_M.

Some further comments are necessary to define the units and dimensions of the Michaelis parameters. As stated above, the K_M has the units of a dissociation constant (concentration) and this remains so, regardless of its complexity in terms of many rate constants; V_{max} and v have the dimensions of a zero order reaction, concentration/time. A great many different units are in use in connection with assays of enzyme activity. However, it is recommended that moles of product per mole of enzyme sites per second should be used for any serious kinetic investigations. The term k_{cat} (or turnover number), which has the dimensions of a first order rate constant (units of seconds), is now widely used for $V_{max}/c_E(0)$. It is also common practice to use the term K_S for the substrate dissociation constant, when it can be determined independently from K_M. A common enzyme *unit* in use for enzyme assays and in commercial catalogues is µmoles per minute per milligram of enzyme. This can be converted to k_{cat} or k_0 (terms used for the turnover number) by multiplication as follows:

units × molecular weight/site × 1.667×10^{-5}

The validity of the steady state assumption

Criteria for the validity of the steady state assumption can be obtained operationally by consideration of how to obtain correct values for initial rates and their use in the evaluation of the Michaelis parameters. Before going into these practical considerations some reference should be made to theoretical treatments of steady state conditions (see Segel, 1975; Wong, 1975). We use the simple form, equation (3.3.19), for saturation kinetics, differentiating to obtain a relation between the change in the steady state intermediate $c_{ES}(t)$ and in $c_S(t)$ with time.

$$\frac{dc_{ES}}{dt} = \frac{k_{12}\{k_{21}+k_{23}\}c_E(0)}{\{k_{12}c_S(t)+k_{21}+k_{23}\}^2} \frac{dc_S}{dt} \tag{3.3.20}$$

This equation demonstrates that it is not an essential condition that $dc_{ES}/dt = 0$, but that it should be small compared to the rate of substrate utilization. This point is elaborated by Wong (1975). Another algebraic diversion gives some further insight into the approach to and the behaviour of the steady state. We shall use the simple minimum model with three states of the enzyme (E, ES and EP), with the interconversion ES → EP being rate limiting and irreversible. Although the transients involved in the approach to the steady state are discussed in detail in section 5.1, it is of interest in the present context to use this simple example to demonstrate the time scale involved. The rate of formation of ES is given by

$$dc_{ES}/dt = k_{12}\{c_E(0)-c_{ES}(t)\}c_S(0)-\{k_{21}+k_{23}\}c_{ES}(t)$$
$$= k_{12}c_S(0)c_E(0)-c_{ES}(t)\{k_{12}c_S(0)+k_{21}+k_{23}\} \tag{3.3.21}$$

The steady state concentration of the enzyme–substrate complex, $c_E(\infty)$, can be defined by

$$\{c_E(0) - c_{ES}(\infty)\}c_S(0)/c_E(\infty) = \{k_{21} + k_{23}\}/k_{12}$$

from which we can derive

$$c_{ES}(\infty) = k_{12}c_S(0)c_E(0)/\{k_{21} + k_{23} + k_{12}c_S(0)\} \tag{3.3.22}$$

Integration of equation (3.3.21) and use of the boundary conditions $c_{ES}(0) = 0$ result in the equation for the approach to the steady state:

$$c_{ES}(t) = c_{ES}(\infty)\{1 - \exp[-\{k_{12}c_S(0) + k_{21} + k_{23}\}]\}$$

A plausible example would be an enzyme reaction with

$$K_M = (k_{21} + k_{23})/k_{12} = 10^{-4} \text{ M} \tag{3.3.23}$$
$$k_{12} = 10^7 \text{ M}^{-1}\text{s}^{-1}, k_{23} = 900 \text{ s}^{-1} \quad \text{and} \quad k_{31} = 100 \text{ s}^{-1}$$

If the substrate concentration is assumed to be 10^{-4} M, substitution of these values in equation (3.3.23) gives a time constant for the transient, $\tau = 0.5$ ms. Such a delay in reaching the steady state can only be seen with the special techniques discussed in subsequent chapters. This simplified derivation of the rate of approach to the steady state takes account of only one intermediate to demonstrate the time scale of approach to the steady state. If we consider the realistic reaction mechanisms with two or more intermediates at significant concentrations in the steady state, then their rates of appearance can, in principle, be used to evaluate the rate constants of individual steps of the reaction sequence. The coupled reactions of all the intermediates will determine the rate of approach to the steady state (see section 5.1).

Traditionally the most common way to determine K_M and V_{max} was to measure the initial rate over a small range of the reaction for which the change in substrate concentration is negligible compared to the total substrate available. Such values for the initial rates, at different concentrations $c_S(0)$, can be fitted either in one of the linearized forms of rate equations illustrated in figure 3.4, or directly to the hyperbola by one of the non-linear fitting methods referred to in section 2.3. As demonstrated by Colquhoun (1971) and Wilkinson (1961), the non-linear methods are to be preferred to any of the commonly used linear plots, for the estimation of Michaelis parameters. Most scientists now have suitable desktop computers and fitting programs at their disposal and the time has come to abandon the various methods for the estimation of the parameters from linearized equations. Whatever method is used, it is essential that a theoretical hyperbola should be obtained from the calculated Michaelis parameters and the raw data points plotted over it. The eye is a good judge of systematic deviations of experimental points from the statistically fitted

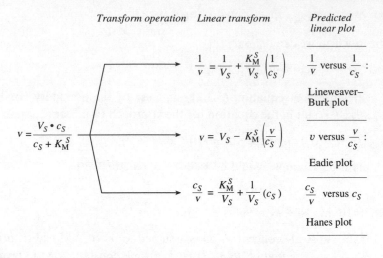

Transform operation	*Linear transform*	*Predicted linear plot*

$$\frac{1}{v} = \frac{1}{V_S} + \frac{K_M^S}{V_S}\left(\frac{1}{c_S}\right)$$

$\dfrac{1}{v}$ versus $\dfrac{1}{c_S}$:

Lineweaver–Burk plot

$$v = \frac{V_S \cdot c_S}{c_S + K_M^S}$$

$$v = V_S - K_M^S\left(\frac{v}{c_S}\right)$$

v versus $\dfrac{v}{c_S}$:

Eadie plot

$$\frac{c_S}{v} = \frac{K_M^S}{V_S} + \frac{1}{V_S}(c_S)$$

$\dfrac{c_S}{v}$ versus c_S

Hanes plot

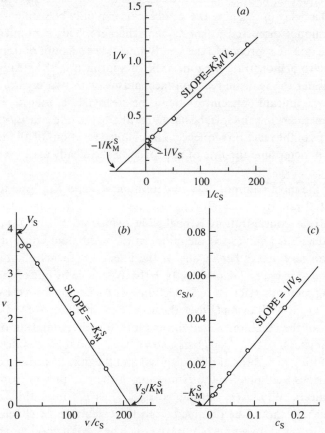

Figure 3.4   Illustration of three commonly used linear transformations of the Michaelis–Menten equation.

curve. The more sophisticated statistical methods for goodness of fit may not be very useful for relatively few points. Both linear and non-linear estimates of the Michaelis parameters are independent of the complexity of the mechanism provided the true initial rates are obtained and the saturation law applies under the conditions stated on p. 82. It is, of course, possible to fit the data for the whole reaction, from $c_S(0)$ to $c_S(t) = 0$, to the so-called integrated Michaelis equation. From

$$dc_P/dt = V_{max}\{c_S(0) - c_P(t)\}/\{K_M + c_S(0) - c_P(t)\}$$

we obtain on integration

$$\left. \begin{array}{l} V_{max}t = c_P(t) + K_M \ln\{c_S(0)/[c_S(0) - c_P(t)]\} + \text{constant} \\ \text{or} \\ \ln\{c_S(0)/[c_S(0) - c_P(t)]\}/t = V_{max}/K_M - c_P(t)/tK_M \end{array} \right\} \quad (3.3.24)$$

However, the above equations will only give reasonable values for the parameters in the unlikely case when the reaction is both essentially irreversible and not subject to product inhibition. Much more complex equations with many more parameters have to be used for a fit of the whole time course of the reaction to a realistic mechanism. There are hazards, too, in the determination of true initial rates. It is necessary to check that the change in substrate concentration during the period of initial rate measurement is sufficiently small so that the observed rate is the correct one for each specified value of $c_S(0)$. After a preliminary estimate for K_M and V_{max} one can readily calculate whether these parameters would cause a significant deviation from true linearity of initial rates, that is adherence to the steady state condition because of the difference in substrate concentration at the beginning and end of the measurement. This must not be left to the appearance of the initial slope. It is also important, for the correct evaluation of the Michaelis parameters, to heed the advice of Dowd & Riggs (1965) that the range of $c_S(0)$ must straddle both sides of K_M by a factor of 10.

Calculation of the steady state distribution by the determinant method

Many methods and minor variants of methods, have appeared in the literature for the derivation of steady state equations (see for instance King & Altman, 1956; Wong, 1975; Huang, 1979). They all depend on the solution of a set of linear equations. Matrix and other methods for deriving rate equations for the approach to the steady state are presented in section 5.1 and the principle used in equation (3.3.25) of writing rate equations in terms of a matrix of coefficients (the matrix of rate constants) and a vector of intermediate concentrations is presented in section 4.2. The general method for obtaining the relative concentrations of three intermediates in the steady state can be illustrated using the scheme for a fully reversible reaction:

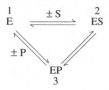

This involves the solution of the three simultaneous equations by the determinant method. For the present example we write the rate equations for the three states of the enzyme in the following form

$$
\begin{aligned}
dc_E/dt &= -\{k_{12}c_S(\infty)+k_{13}c_P\}c_E + &k_{21}c_{ES} &+ &k_{31}c_{EP} &= 0\\
dc_{ES}/dt &= &k_{12}c_S(\infty)c_E &-\{k_{21}+k_{23}\}c_{ES} + &k_{32}c_{EP} &= 0\\
dc_{EP}/dt &= &k_{13}c_Pc_E &+ k_{23}c_{ES} &-(k_{32}+k_{31})c_{EP} &= 0
\end{aligned}
$$

and writing the three simultaneous equations in the matrix form we obtain:

$$
\begin{bmatrix}
-\{k_{12}c_S(\infty)+k_{13}c_P\} & k_{21} & k_{31}\\
k_{12}c_S(\infty) & -\{k_{21}+k_{23}\} & k_{32}\\
k_{13}c_P & k_{23} & -\{k_{32}+k_{31}\}
\end{bmatrix}
\begin{bmatrix}
c_E\\
c_{ES}\\
c_{EP}
\end{bmatrix} = 0 \qquad (3.3.25)
$$

A mechanism with n intermediates is determined by $n-1$ equations when conservation of mass is taken into account. The determinant, or the distribution terms for the three states of the enzyme, can be calculated from the coefficients, for example for c_E, by omitting the column for c_E:

$$
c_E = \begin{bmatrix}
k_{21} & k_{31}\\
-\{k_{21}+k_{23}\} & k_{32}
\end{bmatrix} = k_{21}k_{32}+k_{31}\{k_{21}+k_{23}\}
$$

One writes similarly for c_{ES}

$$
c_{ES} = \begin{bmatrix}
-k_{12}c_S(\infty)+k_{13}c_P & k_{31}\\
k_{13}c_P & -\{k_{32}+k_{31}\}
\end{bmatrix}
$$
$$
= \{k_{32}+k_{31}\}\{k_{12}c_S(\infty)+k_{13}c_P\}-k_{31}k_{13}c_P
$$

and for c_{EP}

$$
c_{EP} = \begin{bmatrix}
k_{12}c_S(\infty) & -\{k_{21}+k_{23}\}\\
k_{13}c_P & k_{23}
\end{bmatrix} = k_{23}k_{12}c_S(\infty)+k_{13}c_P\{k_{21}+k_{31}\}
$$

Note that deletion of different equations can result in different amounts of algebraic manipulations.

The expressions for the steady state concentrations of the intermediates, obtained from the above determinants, can be substituted into the equations for the forward reactions:

$$v/c_E(0) = \{k_{23}c_{ES}(\infty) - k_{32}c_{EP}(\infty)\}/\{c_E(\infty) + c_{ES}(\infty) + c_{EP}(\infty)\}$$

or for the reverse reactions

$$v/c_E(0) = \{k_{32}c_{EP}(\infty) - k_{23}c_{ES}(\infty)\}/\{c_E(\infty) + c_{ES}(\infty) + c_{EP}(\infty)\}$$

$$dc_P/dt = v/c_E(0)$$

$$= \frac{k_{12}k_{23}k_{31}c_S(\infty) - k_{21}k_{32}k_{13}c_P(\infty)}{k_{21}k_{32} + k_{21}k_{31} + k_{23}k_{31} + k_{12}c_S(\infty)\{k_{23} + k_{32} + k_{31}\} + k_{32}c_P(\infty)\{k_{21}k_{23} + k_{32}\}}$$

$$(3.3.26)$$

At equilibrium all intermediate steps are at equilibrium and equilibrium concentrations are also designated at time $\rightarrow \infty$

$$dc/dt = dc_S/dt = 0 \quad \text{and} \quad c_{EP}(\infty)/c_{EP}(\infty) = k_{23}/k_{32}$$

If the Michaelis parameters for the forward reaction under conditions $c_P(t) = 0$ and for the reverse reaction under conditions $c_S(0) = 0$ are substituted in equation (3.3.26), the following useful expression is obtained

$$v = \frac{V_{max}^S K_M^P c_S(\infty) - V_{max}^P K_M^S c_P(\infty)}{K_M^S K_M^P + K_M^P c_S(\infty) + K_M^S c_P(\infty)} \tag{3.3.27}$$

(V_{max}^S, V_{max}^P, K^S, K^P are respectively V_{max} and K_M for the reactions with either substrate and product alone).

Equilibria and the Haldane equation

At equilibrium $v = 0$ and hence the numerator on the right hand side is zero. It follows that

$$V_{max}^S K_M^P c_S(\infty) = V_{max}^P K_M^S c_P(\infty) \tag{3.3.28}$$

and the equilibrium constant can be expressed by

$$K = c_P(\infty)/c_S(\infty) = V_{max}^S K_M^P / V_{max}^P K_M^S = k_{12}k_{23}k_{31}/k_{21}k_{32}k_{13} \tag{3.3.29}$$

Equations (3.3.27) and (3.3.28) are referred to as the Haldane relations in recognition of the original derivation by Haldane (1930). It can readily be shown that the overall equilibrium constant of a consecutive reaction is the product of all the equilibrium constants for the individual steps. The law of microscopic reversibility states that, at equilibrium, all steps of a consecutive or cyclic process have to be at equilibrium. For the sequence:

$$A \underset{}{\overset{K_1}{\rightleftharpoons}} B \underset{}{\overset{K_2}{\rightleftharpoons}} C \underset{}{\overset{K_3}{\rightleftharpoons}} D$$

$$K_1 = c_B(\infty)/c_A(\infty), \quad K_2 = c_C(\infty)/c_B(\infty), \quad K_3 = c_D(\infty)/c_C(\infty)$$
and $\quad K = c_D/c_A = K_1 K_2 K_3$

This is exactly what the Haldane relations demonstrate for the sequence of intermediates in enzyme reactions. They are useful criteria when changing Michaelis parameters are compared to the equilibrium constant for an enzyme catalysed reaction. We shall return to this problem in section 5.1, when we discuss how transient kinetic analysis can be used to determine the equilibrium constants of individual steps. In this connection the equations which express the concentrations of the intermediates in terms of the fraction of total amount of enzyme in the reaction mixture will turn out to be useful. Many enzyme reactions can be studied in both directions and the two sets of parameters for the reactions starting on either side (with S or P as the substrate) – V_{max}^S, K_M^S, and V_{max}^P, K_M^P – give further insight.

Some examples

In chapter 5 the phenomenon of 'on enzyme equilibria' is discussed with examples. This refers to the fact that the equilibrium between enzyme–substrate and enzyme–product complexes is often near unity, even if the overall equilibrium constant for the interconversion of free substrate to free product is a large number. This does not contradict the statement that enzymes (or catalysts in general) do not affect equilibrium constants of reactions. It has to be remembered that this definition of catalysis only applies to the equilibrium between free substrates and products. An example, which illustrates this in terms of the Haldane relation, is heart lactate dehydrogenase. By the methods discussed in section 5.1 it was shown that the equilibrium constant for the two complexes

```
NAD⁺        NADH
 |           |
 |   ±H⁺     |
E ⇌ EH⁺
 |           |
 |           |
Lactate    Pyruvate
```

is near unity, that is four orders of magnitude greater than the overall equilibrium constant of the reaction. This is explained, in terms of the Haldane relation, by the facts that the association constant for NAD^+ is $1/100$ that for NADH and similarly the association constant for lactate is $1/100$ that for pyruvate (see Gutfreund & Trentham, 1975, and section 5.1 for more examples and detailed calculations).

One further derivation, which finds wide application, must be discussed before more complex enzyme systems are briefly considered. The equation

$$v/c_E(0) = \{k_{cat}c_S(0)\}/\{K_M + c_S(0)\} \tag{3.3.30}$$

is used to describe saturation kinetics when the definition $V_{max}/c_E(0) = k_{cat}$ is taken into account. At very low substrate concentration, $c_S(0) \ll K_M$ equation (3.3.30) simplifies to

$$v/c_E(0) = \{k_{cat}/K_M\}c_S(t) \quad \text{as} \quad c_S(t) \to 0$$

This describes a first order reaction with the observed rate constant k_{cat}/K_M. Remembering that K_M has the dimensions of a dissociation constant, we can compare the above equation to that derived for two step ligand binding, when the second step is preceded by a rapid pre-equilibrium (see p. 66). The constant k_{cat}/K_M can be determined from the analysis of the record as a first order reaction. Although, even at such low initial substrate concentrations, the effects of reversibility or product inhibition may perturb the later parts of the reaction. The more usual procedure found in the literature is to plot $v/c_E(0)$ against $c_S(0)$ from a set of initial rate measurements and to take k_{cat}/K_M as the initial slope. As pointed out in section 3.2, one obtains an apparent second order rate constant.

Taking the expressions for the Michaelis parameters derived for the three step mechanism, with only the first step reversible (p. 83), we obtain

$$k_{cat}/K_M = k_{12}k_{23}/\{k_{21} + k_{23}\}$$

The right hand side of this equation can be rearranged to expose the relation of one or other rate constant to the rest. It is, however, readily ascertained that k_{cat}/K_M gives a minimum value for k_{12}. This has been widely used as an estimate for the rate of substrate binding from steady state kinetic investigations. During the detailed discussion of the rate of collision complex formation (section 7.4), criteria will be discussed which help to decide how close k_{cat}/K_M is, in different systems, to the true second order rate constant characteristic of the first step of enzyme–substrate complex formation.

In principle the methods described above and in the references quoted can be used to derive the parameters and steady state concentrations of intermediates for any mechanism following Michaelis type saturation kinetics. Many pages of algebra are found in the literature, which describe theoretical reaction pathways. They do not contribute anything useful, unless the resulting equations can be applied either to the interpretation of existing experimental data or the planning of realistic further experiments. The combinations and permutations of possible pathways become much more numerous with multiple substrates and the consideration of cooperative interactions of sites on oligomeric enzymes. Some comments on cooperativity are found in section 3.1. The small amount of kinetic information available will be referred to with examples in later chapters. Some confusion does arise when the term allosteric (often wrongly taken to be interchangeable with cooperative) is used to describe the almost universal phenomenon of substrate induced conformation change. Cooperative phenomena so far discovered are all concerned with substrate or effector binding and not with the subsequent catalysis. Substrate induced conformation changes may or may not be cooperative between the subunits of oligomeric enzymes. The problem is that the anomalous is usually regarded as more interesting than the usual. Hence, data are often 'fitted' to the former. A typical example is the associated phantom phenomenon of 'half of site activity'. Although it now has received the fate it deserved (except for some specially complex systems), it is a good example of what happens if the wrong interpretation of results is more interesting than the correct one. The usual errors of these particular investigations were reviewed by Gutfreund (1975).

We have emphasized before that only some pointers to the solution of problems in steady state kinetics will be given in this volume. To conclude this summary some aspects of enzyme reactions involving two substrates will be discussed. First, two conventions must be mentioned. In many reactions involving H_2O, H^+, or OH^- the concentrations of water and its ions are not considered stoichiometrically. The ionic concentrations are taken into account in terms of rapid equilibria (see sections 3.4 and 6.4). The distinction between substrates, coenzymes and prosthetic groups may not always be a sharp one. We shall treat coenzymes (NAD^+, NADH, ATP, etc.) as a second substrate. The term arises from the fact that, unlike other substrates, coenzymes are continuously recycled. The difference between coenzymes and prosthetic groups (biotin, riboflavin, pyridoxal phosphate, etc.) is that the latter are more or less firmly attached to the active site of the enzyme; the lifetime of the complex is very long compared to a single turnover of the enzyme reaction. The concentration of a prosthetic group is considered to be equivalent to that of the active site concentration. One major pastime of practitioners of steady state kinetics is to categorize multiple substrate reactions into three principal groups: ordered pathways, random pathways and substituted enzyme mechanisms. These are the most descriptive among other sets of names given to these schemes. The three types of mechanisms are best defined by examples. Substituted enzyme mechanisms, many of which involve prosthetic groups, can be identified unambiguously.

During the transamination

glutamate + oxaloacetate → α-oxoglutarate + aspartate

the amino group is transiently attached to the prosthetic group of the enzyme; pyridoxal phosphate is converted to pyridoxamine phosphate. Similarly the flavine groups of many oxidoreductases are transiently reduced during hydrogen transfer. Such mechanisms can be proved by separate partial reactions of donor or receptor with the enzyme or substituted enzyme. Details of the biochemical aspects of many reactions of prosthetic groups are found in standard texts, for instance Stryer (1988).

The distinction between ordered and random binding of substrates during ternary complex formation is more difficult and rarely absolute. For example the reaction of lactate dehydrogenase from pig heart muscle is classified as an enzyme reaction with an ordered mechanism in the reversible reaction

$$\text{lactate} + \text{NAD}^+ \rightleftharpoons \text{pyruvate} + \text{NADH} + \text{H}^+$$

It is found to a reasonable approximation that NAD^+ has to bind before lactate can do so and NADH has to bind prior to pyruvate. The converse must be and is also true; the nucleotide will only dissociate from its complex with the enzyme after the other substrate has done so. The effect of one bound ligand on the affinity for the other can be observed in many other systems apart from enzymes. Examples will be found in subsequent sections. There are two principal ways of using steady state data to distinguish between different mechanisms. Either one can use the diagnostic primary and secondary plots described by Dalziel (1975) and references given by him, or one derives the equations for different mechanisms and compares the fit to the primary data directly by non-linear analysis. There are also a number of isotope exchange techniques, which provide good criteria for distinction between different reaction sequences.

We have devoted much space to the most elementary concepts of enzyme kinetics. These are required as a baseline for an understanding of the transient kinetic techniques discussed in section 5.1. The latter are the only ones which can, but do not always, give unambiguous descriptions of mechanisms. Steady state kinetics has had very limited success in the elucidation of molecular mechanisms, as distinct from its importance in providing metabolic information. Enzyme mechanisms in a wider sense (the reactions of myosin, transport proteins, etc.) have been explored more usefully by transient kinetic techniques. These latter reactions will be used to describe the transient kinetic approach to the study of mechanisms.

3.4 Steady state kinetics of reversible effectors and ionic equilibria

Practical reasons for studying inhibitors

Kinetic studies of the effects of reversible inhibitors and activators on reactions of substrates and transmitters with binding sites, have important applications in the exploration of structural specificity. They contribute to our understanding of molecular recognition and to both the systematic and empirical search for compounds of pharmacological interest. The discussion of the interaction with ions (H^+, Ca^{2+}, etc.), which influence binding as well as catalysis, is sometimes treated separately from compounds which depend on their steric properties for their effects. As will be seen in the present discussion, the effects of ions are often activating over one part of the concentration range under investigation and inhibiting over another. The distinction between the effects of simple ions and other compounds is, at times, somewhat artificial and the theoretical interpretation of both types of effects is often the same. Only experiments with compounds which form non-covalent rapidly reversible interactions with the functional protein under investigation can be interpreted with the steady state equations derived in the present section. This excludes effects resulting from chemical modification and irreversible structure changes. Transient kinetic studies of other phenomena involving effectors will be discussed in the following chapters.

Classification of reversible inhibitors

As noted in a number of sections of this volume, the nomenclature of enzymology has been adopted for the description of reactions of receptors in general. Competitive, non-competitive and uncompetitive inhibitors are classified in terms of their effects on the Michaelis parameters K_M and V_{max}

respectively. Even for mechanisms which result in K_M being a complex function of rate constants (see p. 89), direct competition of an inhibitor with the substrate for the specific binding site can be interpreted in terms of the Michaelis equilibrium derivation. For example, even in the minimum mechanism:

significant concentrations of the four states of the enzyme can occur in the steady state and all six rate constants of the reaction can contribute to K_M (see section 2.4). Using the determinant method to obtain the distribution terms for the four intermediates and substituting into

$$\frac{v}{c_E(0)} = \frac{c_{EP}(t)k_{31}}{c_E(t) + c_{FI}(t) + c_{ES}(t) + c_{EP}(t)}$$

the following steady state rate equation results

$$\frac{v}{c_E(0)} = \frac{\dfrac{k_{12}k_{23}k_{31}}{k_{12}(k_{23} + k_{31} + k_{32})}}{c_S(0) + \left[\dfrac{k_{-i}}{k_i} c_I(0) + 1\right] \dfrac{k_{21}k_{23} + k_{21}k_{31} + k_{23}k_{31}}{k_{12}(k_{23} + k_{32} + k_{31})}} \tag{3.4.1}$$

Comparing this equation with equation (3.3.11) (where V_{max} is substituted for $k_{cat}c_E$) one can write

$$\frac{v}{c_E(0)} = \frac{k_{cat}c_S(0)}{c_S(0) + [\{c(0)/K_i\} + 1]K_M} \tag{3.4.2}$$

where $K_i = k_{-i}/k_i$, the dissociation constant for the inhibitor. At any fixed concentration $c_I(0)$ the dependence of the steady state rate on the substrate concentration can be used to evaluate the apparent K_M

$$K_{M\,obs} = K_M[1 + (c_I(0)/K_i)] \tag{3.4.3}$$

The other Michaelis parameter, $V_{max} = k_{cat}c_E(0)$, is not affected by a competitive inhibitor. The clear distinction of competitive inhibition from any other is that, on substrate saturation, the resulting maximum velocity is always the same, regardless of inhibitor concentration.

In contrast to the above scheme for competitive inhibition, non-competitive inhibition affects both V_{max} and K_M and cannot be compensated for by increasing substrate concentration. The distinction between non-competitive and uncompetitive inhibition, which is made in some texts on enzymes, is not very informative. Some authors may have a semantic disagreement with these definitions, but, as reaction mechanisms are explored in greater detail, it is found that most inhibitors do not conform absolutely to any of the terms. Non-competitive inhibition will be discussed below principally in terms of ionic equilibria, which often affect both Michaelis parameters. For problems involving substrate and product inhibition specialized texts on enzymology should be consulted (Dixon & Webb, 1964; Segel, 1975).

Inhibitors and metabolic control (feedback and allosteric phenomena)

Some comments are required on effectors which are of importance in metabolic and other regulation phenomena. The term allosteric inhibition was introduced by Monod, Changeux & Jacob (1963) to explain the discovery of feedback inhibitors, which are competitive with the substrate but not closely structurally related to it. Such inhibitors were called 'allosteric' in contrast to the usual competitive inhibitors which are 'isosteric' with the substrate to fit competitively into the same binding site. Allosteric competitive inhibition occurs at a site (allosteric site) which is separate from the substrate binding sites. Communication between these separate sites about respective occupancy occurs through conformation changes and these are transmitted through subunit interfaces. The term was extended to the related control phenomenon of allosteric activation, which occurs when a metabolite binds to an allosteric site and decreases the K_M for a substrate. Although the allosteric model covers effectors which can change the maximum velocity and/or the K_M, most examples found are for so called K effectors of K_M. Feedback and feed-forward effects (inhibition and activation) have been found to occur, generally, in oligomeric proteins with accompanying cooperative interactions between sites on the separate functional units (see p. 69 and for a review see Perutz, 1989). Allosteric effects can, in principle, occur in monomeric as well as in oligomeric proteins and even in oligomers allosteric effects need not be associated with cooperativity. This multiplicity of information transfer through tertiary and quaternary conformation changes, has resulted in some confused nomenclature. In the biological literature the term allosteric, which should be restricted to the interaction between sites on a single functional unit, which may, but need not, result in quaternary structure changes, is often indiscriminately applied to the well-known physical phenomenon of cooperative transition. The latter term should be restricted to interactions between different functional units (see also the discussion of cooperative binding; section 3.2). Allosteric phenomena, as defined above, must also be distinguished from the almost universal local conformation change resulting from the binding of a ligand to a specific site. This is usually referred to as substrate induced conformation change or induced fit.

Some comments are required on the relation between cooperativity, allosteric interaction and the classic phenomenon of thermodynamic linkage. The latter is most clearly illustrated by the following statement: if binding of ligand A affects the affinity for ligand B, then the binding of ligand B has the same effects on the affinity for ligand A. Cooperative and allosteric processes are usually associated with linkage, but the latter is certainly not restricted to the former.

Although whole books have been written on allosteric and cooperative effects on enzyme reactions, steady state kinetics has contributed little to our understanding of the mechanisms involved. Equilibrium binding studies are useful for diagnostic purposes and provide the information required for an understanding of the physiological phenomena. The rates of allosteric or cooperative transitions (see section 7.3) do not play an important role in metabolic control. The sort of kinetic data which give information about the mechanisms of transitions is obtained from the observation of various relaxation phenomena discussed in chapters 5 and 6. As an exploratory approach, however, binding experiments are useful for defining interesting phenomena for investigation by the techniques discussed in the next two chapters. If steady state kinetics is used for the determination of the dissociation constants of inhibitors they should be complemented with direct binding studies by equilibrium methods (see Edsall & Gutfreund, 1983). These approaches are valuable for the determination of the number of binding sites and give some information about their structure by correlation of ligand conformation to affinity. As always, discrepancies between results obtained by kinetic and thermodynamic techniques are a guide to inadequacies in the model.

Protons as inhibitors and activators

The study of the effects of hydrogen ion equilibria on enzyme reactions has a long and distinguished history. It made contributions to the reliability of kinetic information, to our knowledge of protein stability and to the identification of essential catalytic groups and their reactivities. Sorensen (1912) made major contributions to the study of reactions of proteins. They were based on the great tradition of the Copenhagen school of physical chemistry of electrolyte solutions (Brönsted, Bjerrum and Christianson). The realization by Sorensen that the rates of enzyme reactions are critically dependent on the hydrogen ion concentration made it possible for him to get more consistent results by controlling conditions. Furthermore his definition of the pH scale, and the proposal to use buffers to obtain a constant environment, opened up a new chapter in the study of the role of ionizing groups in reactions of proteins. As happened frequently, phenomena exposed by the behaviour of enzyme reactions led to their more general study in other types of proteins.

Protonic equilibria of isolated charged amino acids in aqueous solutions are established rapidly, compared to most steps of enzyme reactions, although the concentrations of proton donors and accepters have to be taken into account in these second order reactions (see below). However, if proton uptake or release is coupled to a conformation change of the protein, then the observed rate can be relatively slow, even rate limiting, for the overall reaction. In fact many conformation changes can be monitored by accompanying changes in pK values of the protein. Examples of such phenomena will be documented in section 5.1, where experiments are described on the observation of proton uptake and release during individual steps of enzyme reactions. An example of such a linkage is discussed in connection with the relaxation analysis of the binding of NAD^+ to alcohol dehydrogenase.

Before returning to the relation between rates of enzyme reactions and rates of establishing protonic equilibria we deal with the effects of ionic equilibria on the steady state kinetics of enzyme reactions. The effects of such equilibria on reactions of proteins in general will, however, be a recurring theme throughout this volume. The relation between rapid pre-equilibria and apparent rate constants for ligand binding is discussed in a number of sections. The study of the pH dependence of enzyme reactions has been a favourite pre-occupation of enzymologists in attempts to distinguish different enzymes and in the exploration of mechanisms. The discussion of the benefits and of the pitfalls of such investigations will concentrate on attempts to characterize and, sometimes, to identify the ionizing groups involved in substrate binding and catalysis. The simplest model adds a protonated form of each intermediate to the minimum mechanism:

The supposition that protonic equilibria are usually faster than binding, conformation changes or catalysis can be used to draw useful conclusions from the following argument. One can think in terms of the enzyme being in the protonated form for a fraction of the time, rather than that a fraction of the enzyme is in that form. The proton will go on and off many times during the lifetime of a particular enzyme intermediate. This enables one to draw the conclusion that a pH independent K_M and a pH dependent maximum velocity point to the fact that only steps prior to the rate limiting one contribute to K_M.

The effects on the kinetics of different steps of enzyme reactions, exerted by the state of ionization of amino acid residues, can be distinguished as follows: (a) charge attraction or repulsion during substrate binding; (b) the involvement of proton donors/acceptors in catalysis; and (c) linkage of changes in ionization to conformation changes affecting substrate binding or catalysis. As mentioned above, binding of substrates or any other ligands to a protein will usually perturb the pK of one or more ionizing groups through changes in the local environment, whether or not it is accompanied by a conformation change. Apart from effects of the local environment (the local dielectric constant and other nearby charges) there are two principal problems in the identification of a group at the active site of a protein through the pH dependence of steady state rates. First there are often two or three interacting polar amino acid residues involved in catalysis. Thus, even if the rate observed is that of a distinct step, the pH rate profile represents that of a system of ionizing groups. Second changes in the rate limiting steps with different pH dependence will give anomalous results. An example of this is the reaction of alkaline phosphatase (Trentham & Gutfreund, 1968). Studies of the individual steps, by transient kinetic techniques, showed that below pH 7.5 product release is rate limiting, while at more alkaline pH the formation of the enzyme–product complex determines the steady state rate. For the subsequent elucidation, by NMR studies, of the greater complexity of this mechanism the paper by Hull *et al.* (1976) should be consulted.

As already pointed out, it is very unlikely that the pKs of groups situated within the special environment of an active site of a protein will have the same values as those of the same residue in a small model compound. A classic, but by no means isolated, example for this comes from the studies on acetoacetate decarboxylase by Schmidt & Westheimer (1971). They found that $\varepsilon - NH_2$ of the lysine residue, which forms the imine with the substrate, has a pK of 5.9; this amounts to a perturbation of about 3 pK units. Sali, Bycroft & Fersht (1988) and Knowles (1991) discuss the interesting effects on pKs, which are due to the position of groups with respect to the direction and proximity of α-helices. Knowles reports a lowering of a pK for an imidazole residue in triose phosphate isomerase by more than two units, due to the positive end of a helix pointing at it. The effects of the local environment on ionizing groups and on reactivities in general, are part of an interesting but highly specialized subject. In the context of pH profiles of kinetic data it is best to illustrate a number of specific points with relatively simple examples in which it is clear that either substrate or chemical catalysis is affected by proton uptake or release.

Effects of the ionization of substrates are neglected in the present discussion, but one has to be aware of them in the interpretation of data. Protons can act over different pH ranges in different steps of the reaction as activators as well as inhibitors and as both. For the activation of the catalytic rate at constant K_M

$$\frac{V_{max}}{V_{maxH^+}} = K_1/c_{H^+} \qquad (3.4.4)$$

and for inhibition

$$\frac{V_{max}}{V_{maxH^+}} = 1 + c_{H^+}/K_2 \qquad (3.4.5)$$

where K_1 and K_2 (p$K_1 = -\log K_1$ and p$K_2 = -\log K_2$) are the two dissociation constants involved in activation and inhibition; V_{max} is the optimal velocity and V_{maxH^+} the velocity at the specified pH. Equation (3.4.5) corresponds to the expression

$$\frac{V_{max}}{V_{maxI}} = 1 + c_I/K_i$$

which can be derived for the models for non-competitive inhibition by the procedure shown on p. 88, where the inhibitor binds either to ES or to any of the subsequent states of the enzyme.

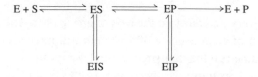

The pH dependence of steady state rates of enzyme reactions has been subjected to considerably more theoretical analysis and interpretation than is justified. Steady state equations for much more complicated schemes of ionizations of intermediates can readily be derived by the methods des-

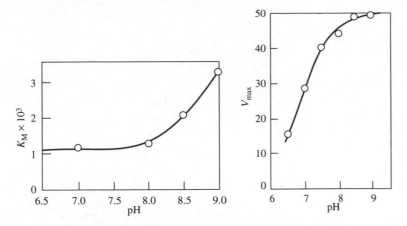

Figure 3.5   The pH dependence of k_M and V_{max} for the chymotrypsin catalysed hydrolysis of acetyl L-phenylalanine ethyl ester at 25 °C. The marked increase in K_M at alkaline pH made it difficult to determine at pH above 9. This led to erroneous interpretation of the maximum velocity at high pH (from Hammond & Gutfreund, 1955).

cribed in section 3.3. One can only repeat the warning that even direct analysis of individual steps by transient kinetic techniques only provides information about a sort of operational pK or pKs of the enzyme at a particular stage of the reaction. One early assignment of the pK for an essential group at the active site was that for the histidine of the proteolytic enzymes trypsin and chymotrypsin. In the case of trypsin the uptake of a single proton shifts the equilibrium from active to inactive enzyme–substrate complex (Gutfreund, 1955). The related enzyme, chymotrypsin, shows an additional feature. While the reactivity of the enzyme–substrate complex is also controlled by the dissociation constant for a single proton, the formation of the enzyme–substrate complex is inhibited when another proton dissociates at alkaline pH (Hammond & Gutfreund, 1955) resulting in an increase in K_M above pH 7.5 (see figure 3.5). This was later shown by Fersht (1985) to be due to a conformation change linked to the ionization of the terminal isoleucine. This amino group has to be protonated for the enzyme to go into a conformation which can form a complex with the substrate. The involvement of proton uptake and release for substrate binding and catalysis respectively can give misleading bellshaped pH profiles for enzyme activities if substrate saturation is not assured.

In many cases the complexity of the pH dependence of the overall reaction is increased by the involvement of two (or more) ionizing groups in catalysis and changes in rate limiting steps (see figure 3.6).

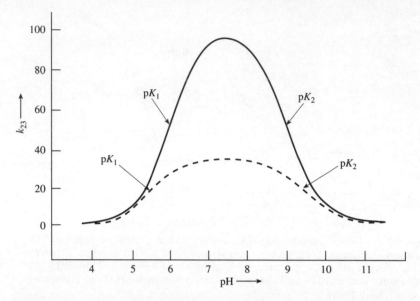

Figure 3.6   A simulated representation of the pH profile of the steady state rate of an enzyme reaction in which two steps have rate constants k_{23} and k_{31} (see mechanism p. 80) of comparable magnitude. The pH dependence of k_{23} is as indicated by the solid line (maximum 100 s^{-1}), while $k_{31} = 50$ s^{-1} is pH independent. The pK values for the pH dependence of k_{23} and the apparent pK values for the overall reaction are indicated on the diagram.

There are many more detailed descriptions of plausible interpretations of the pH profiles of enzyme reactions (Dixon & Webb, 1964 and Fersht, 1985). The usefulness of really complex equations, such as those of Templeton *et al.* (1990) is left to the reader to judge, but the advice of the sceptical but fair review by Knowles (1976) should be taken seriously.

4

Kinetic analysis of complex reactions

4.1 Introduction to the study of transients and reaction sequences

Definition of states

So far we have been largely concerned with two state reactions. The exceptions to this were steady states and equilibria of enzyme reactions and ligand binding to multiple sites. In this and the next chapter we are concerned with rates of transition between a larger number of states of complex systems. An important statement about multi-state systems, which will recur in different forms, is that:

A system of n distinct states linked by first order transitions will approach equilibrium with a time course described by the sum of $n-1$ exponential components.

The definition of a specific state, of a system undergoing a sequence of reactions, as being *kinetically distinguishable* always requires qualification with respect to time and signal resolution. However, even in a system in an apparently well-defined state, like haemoglobin molecules with three out of four oxygen binding binding sites occupied, the system will only be in that state *on average*. If one could obtain information about the site occupation of a small number of molecules and over short time frames one would identify different states of site occupation. Furthermore the protein molecule fluctuates between different conformations. This topic is discussed and documented in some detail in section 7.3.

The kinetic methods discussed in this volume only define the average time taken for transitions between average states. The observation of individual molecules on a short time scale has, until recently, only been achieved for ion channels. With the patch clamp technique it is possible to observe the time a channel molecule spends in the open or closed state (see Colquhoun & Hawkes, 1983). The signal for the transition *closed to open* is enormously amplified by the change in conductivity. However, even for such a system sufficient time resolution has not yet been achieved to measure the time taken for transitions between those two states. It must be emphasized that channels may have several open and closed states. During the last few years methods have been developed which allow the observation of individual molecules involved in a variety of motility mechanisms (for instance Finer, Simmons & Spudlich, 1994).

Elucidation of the number of steps

In this and the next chapter attempts at the definition of the number and lifetimes of intermediate states of biological processes will be demonstrated with examples from a wide range of systems. The transition of a system from one equilibrium state to another or one steady state to another, via a succession of intermediate states, must be studied by a variety of pertur- bation techniques, each of them using a number of different signals to characterize the nature of the events. Such combinations of many different experimental techniques for the initiation and observation of the sequence of events, together with appropriate methods for the analysis of the data (see section 2.3), are important for maximizing the information available. It is often said that kinetic studies can only disprove mechanisms and not prove them. Although this is literally true for most sorts of experiments, not only kinetics, it is also true that time resolved observations of an interme- diate, for example the direct record of the transient change in the spectrum, or of some other specific signal, can define the obligatory existence of an intermediate on the reaction path. Hence the emphasis on the use of a wide range of signals to monitor a reaction, as well as a range of methods for its initiation. It will be seen that different steps of a reaction are perturbed by changes in, for example, concentration (section 5.1) and temperature or pressure (chapter 6).

The detail in the information obtained about mechanisms will depend on what sort of measurements can be made. It is always helpful if changes of several states can be followed. This again is aided by the steady improve- ment and diversification of the signals used to follow the time course of reactions. Sensible use of kinetic analysis can help in the design of further experiments in the search for and identification of intermediate steps

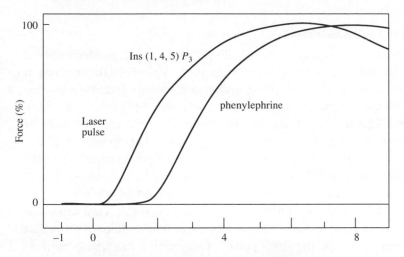

Figure 4.1   Illustration of lag phases in physiological responses: two force transients recorded after photolysis to liberate from a caged precursor (see section 8.4) inositol(1,4,5)-phosphate in a permeabilized muscle strip and phenylephrine in an intact muscle strip of guinea-pig portal vein. Receptor activation is followed by inositol phosphate liberation, which in turn results in Ca^{2+} liberation and force generation by the muscle. For detail see Somlyo *et al.* (1988).

between a physiological stimulus and the physiological response, for instance after phenylephrine stimulation as shown in figure 4.1. A particularly interesting example is the gradual exploration of the chain of events in the transmission of the signal consequent upon the absorption of photons by the visual pigment (rhodopsin) in the retina. Three interacting phenomena, excitation, adaptation and restoration, each involve a number of steps and form a complex kinetic scheme. Fuortes & Hodgkin (1964) recorded the time course of the electrical response of ommatidia from the limulus after a flash of light. They derived equations which could be used to fit the data to models with different numbers of steps as intermediate events. The results of such experiments could be interpreted in terms of a specific number of steps in the mechanism of transduction. Some kinetic studies of this system are discussed in more detail in section 4.2, where the sequence of steps involving protein–protein interactions and enzyme reactions will be illustrated. Although the mechanism of signal transmission and of light and dark adaptation is gradually becoming clear, it will be some time before a grand coupled kinetic scheme of events in the retina can be written.

Description of mechanisms

In chapter 6, where relaxations of equilibria are discussed, the derivation of the equations for time constants and amplitudes depends on the principle of 'small perturbations', which results in linear equations describing systems which would otherwise deviate from first order behaviour. The term 'transient kinetics', the topic of this chapter, is used for the study of transitions between steady states or equilibria. The treatment is not restricted to small changes in concentration, although linearization of the rate equations is usually achieved by maintaining high (buffered) ligand concentrations. The combinations and permutations of complex kinetic schemes which have potential applications are, of course, very large. The selection of those presented depended on the author's choice of those that have been useful in the description of interesting examples, both as physiological systems and for their likely further wide application.

Under first order or pseudo first order conditions (see section 3.2) the set of differential equations describing the time course of changes in concentrations of all species present can be solved for any mechanism. This is described with examples in the next two sections. The numerical methods involving matrix equations are illustrated there for consecutive reactions approaching equilibria and steady states. For non-linear equations describing reactions at low ligand concentrations, solutions can only be obtained by the numerical integration techniques referred to in section 2.2. Once a computer program has been written to describe specific reaction schemes, *numerical* predictions of the time constants, and amplitudes, of the exponential components are easily obtained. However, to obtain explicit algebraic (as distinct from numerical) solutions is often hard work, as illustrated for a number of enzyme catalysed reactions in section 5.1. It will rarely be feasible for systems with more than three states. Nevertheless it is nice to have algebraic solutions when they are feasible, because this allows generalizations to be made, and approximations to be found (for instance for low ligand concentrations and non-linear conditions) in a way that is impossible when relying entirely on numerical solutions. The merits of different approaches are discussed in sections 2.2, 4.2 and 5.1.

The geneticist Richard Levontine has been quoted as describing population genetics as 'the auto-mechanics of evolution' because most users do not like to be bothered with the nuts and bolts of it. The derivations are laborious and lead to inelegant solutions. The same can be said of the relationship between most biochemists and physiologists and the details of kinetic derivations. Such disenchantment with the field under discussion can be avoided if one restricts one's attention to the treatment of models which are related to and are testable on a system under experimental investigation. The important point to consider in proposing a reaction scheme is the formulation of a minimum mechanism which is thus economical in the necessary algebra (there is no point at all in postulating a mechanism which is so complicated as to be untestable). Such a mechanism contains the minimum number of intermediates which satisfy the kinetic data available at a particular point in an investigation.

Physiological problems

Perusal of the physicochemical chapters of textbooks on physiology and biochemistry, published up to the 1950s, reveals an overwhelming concern with the analysis of equilibria. This interest in the presentation of detailed and useful accounts of ionic processes and the energy balance of metabolic pathways left little space for attention to rate processes. Briggs & Haldane (1925) introduced the steady state treatment of simple enzyme reactions, as opposed to the earlier, unrealistic, equilibrium approach (see section 3.3). Since then, and especially from the 1950s onwards, there has been more appreciation of the fact that cellular processes are in a constant state of flux or are in a steady state. Individual reactions may be at or near equilibrium, but for the cell as a whole equilibrium is death.

In recent years more and more control phenomena have been discovered. Some of these cause gentle transitions from one steady state to another by feed-back, while others can result in sharp transitions (switches) through the influx of messengers. These regulatory functions apply both to systems of reactions and to steps within a single enzyme reaction. This has brought an interest in transients into biochemistry. Electrophysiologists and students of muscle contraction had been concerned with the methods for the analysis of transients for some time, for instance in connection with the study of action potential. In the author's experience physiologists are ready to use principles from enzyme kinetics side by side with their own approaches, to interpret related phenomena in neuro- and muscle physiology. Enzymologists have tended to develop their own algebra for the interpretation of the behaviour of systems and have often neglected or re-invented the solutions to problems available from other disciplines. It will be seen that there are many different routes and methods to derive the equations to describe formally identical systems.

Although there will be some references to reactions of nucleic acids it will become evident that most of the dynamic problems discussed in this volume involve the reactions of protein molecules. The study of enzymes, together with that of haemoglobin and myoglobin, lead the field of physicochemical investigations into the structure and reactivity of proteins. Abundant availability, like that of proteolytic and glycolytic enzymes, and the easy assays for monitoring their purification and stability, made them ideal paradigms for structure studies and the development of kinetic methods. The application of kinetic analysis to the elucidation of the temporal structure of such physiological functions as muscle contraction and signal transduction has a long history prior to considerations of molecular detail. In recent years more and more similarities with and references to enzymology can be found in investigations on these systems. Three physiological functions are used below to illustrate how the identification of the enzymes and other proteins involved in them, is leading to a combined effort to study the overall 'system kinetics' side by side with the 'molecular kinetics' of the reactions of isolated steps in solution. This forms a preface to more detailed discussions of these systems in subsequent sections.

The study of rates of muscle contraction and tension development under different metabolic conditions and work load are among the earliest kinetic investigations on physiological systems (see chapter 1). All the proteins thought to be required for the contraction of skeletal muscle fibres, after calcium release from the sarcoplasmic reticulum by excitation at the neuromuscular junction, are available in pure form. Again the comparison of the rate processes involved in the reactions of the isolated proteins and of the events in intact fibres form an important part of current research on muscle contraction. In this system the major question to be elucidated involves the resolution of the structure changes responsible for movement. Methods which permit the observation of the rates of structure changes are particularly valuable for correlation with rates of isolated steps. Time resolved X-ray scattering has been extensively applied to contracting muscle fibres (see figure 4.2) as well as to the interconversion of intermediates in enzyme reactions (section 5.1)

The investigation of the rates of signal transmission in nerve fibre has already been quoted as an example of the ingenious experiments carried out by Helmholtz (see chapter 1). Outstanding biophysical investigations on the rate processes and electrochemical events, as well as on the chemical and ionic excitation mechanisms, have been carried out in the intervening 150 years. Much has been achieved recently by molecular biological techniques in the identification and production of components of receptor and channel proteins. Several of the kinetic techniques described in this and subsequent

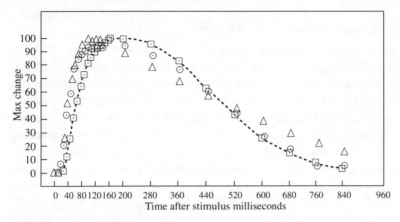

Figure 4.2   Time course of changes in intensities of the equatorial reflections and tension during single twitches of frog sartorius muscle. The [10] reflection (○) decreases in intensity, the [11] reflection (△) increases in intensity during the rise in tension (□). In this diagram all quantities are shown as percentage of their maximum value (from Huxley *et al.*, 1983).

chapters are being actively used to analyse, at the molecular level, the mechanisms involved in the function of receptors for acetylcholine and other transmitters. In such systems the signal that can be measured most conveniently is the current that flows through ion channels that are opened by the ligand. In the early 1970s useful information was obtained about such systems by measuring the fluctuations (*noise*) of such currents, but this method has now been largely superseded by single channel measurements (see Colquhoun & Hawkes, 1983).

Before discussing the types of kinetic problems encountered and their contribution to our understanding of the mechanisms of different phenomena in more detail, let us summarize the questions which we are asking. The first of these is an open ended one. As mentioned before, statements on the number of steps involved in an overall process depend both on the ever improving technology for signal detection and on the chemical and physical detail of interest to the investigator at that particular time. In the case of reactions between enzymes and substrates, when both are of known structure in atomic detail, kinetics can contribute to the resolution of questions asked by the physical organic chemist. On the other hand in the example of the transmission of signals from rhodopsin to the rod cell membrane, where several proteins of so far unknown structure are involved, the kinetic analysis is at present used to test a proposed mechanism at much lower chemical resolution.

Thermodynamic information from kinetics

In addition to the problems amenable to kinetic resolution which are mentioned above, kinetics can also provide thermodynamic information about individual steps which cannot be studied directly at equilibrium. Some specific examples of this will be given in two types of reactions discussed later in this chapter and in chapter 5. As a general rule ligand binding occurs in several steps. While equilibrium binding measurements give the overall dissociation constant and thus the total Gibbs energy, kinetic evaluation of all the individual forward and reverse rate constants allows the calculation of the equilibrium constants for each step. This has been called 'the art of titration' (Winkler-Oswatitsch & Eigen, 1979). Transient kinetic studies of enzyme catalysis provide information about the equilibrium between enzyme–substrate and enzyme–product complexes; the so-called 'on enzyme equilibrium' of the reaction between substrate and product was predicted by Haldane (1930) and first demonstrated in different systems by Gutfreund & Trentham (1975) (see sections 3.3 and 5.1). The same situation occurs when the binding of a ligand to an ion channel causes the channel to open: in this case the equilibrium constant for the opening and shutting reaction influences the total amount of binding, and it is only through kinetic studies that it has been possible to estimate the equilibrium constant for the initial binding of the ligand to the shut channel. The elucidation of the equilibrium constants, and hence the standard free energy changes, of individual steps of enzyme reactions is of particular interest when these are coupled to the conversion of chemical energy to mechanical work, ion transport, or information.

4.2 Consecutive reactions and physiological responses

Coupled and uncoupled sequences

This section is concerned with the analysis of signals obtained from physiological processes in terms of the number and kinetic significance of separate states. Problems of defining observation of a kinetically defined step in a complex sequence of reactions were discussed in the previous section. A step or state can be easily characterized if it fulfils the conditions for being uncoupled from the rest of the system by being at least both one order of magnitude slower than the preceding one and faster than the subsequent one. Some numerical examples of the effects of coupling are given in section 5.1. If an uncoupled step has a distinct signal, then it can be characterized unambiguously. In most cases all but the first step of a reaction sequence, which can often be made fast by high ligand concentration, require two or more coupled exponentials for their description.

Consecutive reactions (two irreversible first order steps)

A number of features of multiple exponential records of reactant concentrations and physiological responses will be discussed in the light of specific examples. The subject is best introduced with the simplest model for a process with two irreversible consecutive steps:

$$A \xrightarrow{k_{12}} B \xrightarrow{k_{23}} C \tag{4.2.1}$$

Kinetic equations for such a scheme were first derived by Bateman (1910) and Lowry & John (1910) and used more specifically by Rutherford, Chadwick & Ellis (1930) for the interpretation of their experiments on radioactive decay

$$radium \longrightarrow radon \longrightarrow polonium \tag{4.2.2}$$

These two sequential processes have half lives of 1622 years ($\tau = 2350$ years) and 5500 minutes ($\tau = 8000$ minutes) respectively. The rates of decay and formation of the three radioactive elements can be described by

$$dc_A/dt = -k_{12}c_A(t); \quad dc_B/dt = k_{12}c_A(t) - k_{23}c_B(t); \quad dc_C/dt = k_{23}c_B(t)$$

These equations can be integrated to give

$$c_A(t) = c_A(0)\exp(-k_{12}t) \tag{4.2.3}$$

and

$$c_B(t) = c_A(0)\frac{k_{12}}{k_{23} - k_{12}}\{\exp(-k_{12}t) - \exp(-k_{23}t)\} \tag{4.2.4}$$

Clearly equation (4.2.4) for the intermediate B cannot be applied to the special case when k_{12} is exactly equal to k_{23} since the denominator in the fraction $k_{12}/(k_{23} - k_{12})$ is then zero. In that case

$$c_B(t) = c_A(0)kt\exp(-kt)$$

The expression for $c_C(t)$ can be obtained from

$$c_C(t) = c_A(0) - c_A(t) - c_B(t)$$

and from (4.2.3) and (4.2.4)

$$c_C(t) = c_A(0) \left[1 - \frac{1}{k_{23} - k_{12}} \left[k_{23}\exp(-k_{12}t) - k_{12}\exp(-k_{23}t) \right] \right] \qquad (4.2.5)$$

or for the special case $k_{12} = k_{23}$ one obtains

$$c_C(t) = c_A(0)\{1 - \exp(-kt) - kt\exp(-kt)\} \qquad (4.2.6)$$

(see Boas, 1966, p. 330).

We shall later use specific examples to illustrate the practical consequences when expressions for the time course of changes in intermediate concentrations contain two or more exponentials. In reversible three state systems and other more complex reactions the expressions for the exponents (reciprocal time constants) and for the amplitudes contain several rate constants (see for instance equation (4.2.20)).

Experimental records of the changes in concentrations c_B or c_C can be fitted to equations (4.2.4) and (4.2.5), respectively, in terms of two exponents and two amplitudes (see section 1.4). This, however, does not allow one to assign the exponents to k_{12} and k_{23} without additional information. Equations (4.2.4) and (4.2.5) are symmetrical, they do not change at all if k_{12} and k_{23} are interchanged. For an assignment of the two rate constants to the first and second step respectively, one has to have either a separate record of the rate of decay of c_A, to obtain k_{12} independently, or the maximum concentration c_B relative to $c_A(0)$ must be determined. This is one of many reminders that amplitudes of reaction intermediates give important information in the interpretation of kinetic experiments. This point is demonstrated by the progress curves in figure 4.3, which show the results of evaluating the equations for pseudo first order – first order consecutive reactions with rate constants for the first and second steps interchanged.

The equations describing the time course for three state irreversible (see (4.2.1)) and for reversible reactions (see (4.2.15)) can be derived by a number of different methods.

Derivation of rate equations for consecutive reactions by matrix algebra

Let us first use the matrix method to derive the rate equations for the irreversible reaction described in (4.2.1).

The coupled differential equations for the time course of two out of three states, together with conservation of mass, are required for the solution describing the time course of the change in concentrations of all three states:

$$dc_A/dt = -k_{12}c_A(t) + 0$$

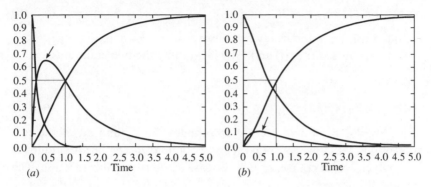

(a)   (b)

Figure 4.3   Simulation of two consecutive reactions to demonstrate the effect of interchanging the two rate constants from 1 and 5 s^{-1}, for the forward and reverse reactions, in (a) to 5 and 1 s^{-1} in (b). The time course of appearance of the final product is unchanged although the rate constant for the second step is increased fivefold. The most misleading feature of (b) is that the formation of the intermediate (see arrow) is apparently faster than in (a). The symmetry of the rate equation for consecutive reactions and the need to take amplitudes into account are discussed in the text.

$$dc_B/dt = k_{12}c_A(t) - k_{23}c_B(t) \tag{4.2.7}$$
$$c_C(t) = c_A(0) - c_B(t) - c_A(t)$$

In section 4.1 it was pointed out that a system with n distinct states linked by first order transitions will approach equilibrium (or a steady state) with a time course described by the sum of $n-1$ exponentials where the exponents are defined as λ (eigenvalues). The relation between λ, τ and k_{obs} are defined in section 2.1 and later in this section. We can write, for instance,

$$\frac{dc_A}{dt} = \lambda c_A \quad \text{and} \quad \frac{dc_B}{dt} = \lambda c_B,$$

where there are two solutions for λ. The time course of the concentrations are then described by

$$\lambda c_A(t) = -k_{12}c_A(t) + 0$$
$$\lambda c_B(t) = k_{12}c_A(t) - k_{23}c_B(t) \tag{4.2.8}$$

which can be expressed in the form of a matrix equation (see section 2.4), where the solutions for λ are the eigenvalues for the matrix of rate constants

$$\begin{bmatrix} -(k_{12}+\lambda) & 0 \\ k_{12} & -(k_{23}+\lambda) \end{bmatrix} \begin{bmatrix} c_A(t) \\ c_B(t) \end{bmatrix} = 0 \tag{4.2.9}$$

Since $c_A(t)$ and $c_B(t)$ are not zero, the determinant of the matrix of rate constants must be zero and the characteristic equation can be written as:

$$\lambda^2 + (k_{12} + k_{23})\lambda + k_{12}k_{23} = 0 \tag{4.2.10}$$

This quadratic equation has two solutions: $\lambda_1 = -k_{12}$ and $\lambda_2 = -k_{23}$.

The time course for $c_A(t)$ and $c_B(t)$ is given by the two equations:

$$c_A(t) = A_0 + A_1 \exp(\lambda_1 t) + A_2 \exp(\lambda_2 t) \tag{4.2.11}$$

$$c_B(t) = B_0 + B_1 \exp(\lambda_1 t) + B_2 \exp(\lambda_2 t) \tag{4.2.12}$$

A_0, A_1, A_2, B_0, B_1 and B_2 are the amplitudes, for c_A and c_B, of the two exponentials with time constants τ_1 and τ_2; A_0 and B_0 are the amplitudes at $t = \infty$, which are, of course, zero in the case of an irreversible reaction. The total amplitudes are

$$A_0 + A_1 + A_2 = c_A(0) \quad \text{and} \quad B_0 + B_1 + B_2 = c_B(0)$$

On substituting the eigenvalues in the rate equations (4.2.8) ratios of concentrations can be obtained:

$$c_A(t)/c_B(t) = [k_{23} - k_{12}]/k_{12} \tag{4.2.13a}$$

Setting the initial conditions (at $t = 0$) for c_A and c_B respectively $c_A(0)$ and 0, we can write

$$c_A(t) = c_A(0) \exp(-k_{12}t) \tag{4.2.13b}$$

and

$$c_B(t) = c_A(0) \left[\frac{k_{12}}{k_{23} - k_{12}} \right] \{\exp[-k_{12}t] - \exp[-k_{23}t]\} \tag{4.2.14}$$

and the expression for $c_C(t)$ is obtained from the conservation of mass as in (4.2.5).

The solution of coupled rate equations for reversible reactions (approach to equilibrium)

Two examples, one here and the other in the next section, will demonstrate how the complexity of the algebra increases when the matrix method is used to derive the rate equations for reversible reactions, even though we shall still restrict ourselves to three state systems.

Using the same rationale as for the three state, irreversible reaction (p. 113) we can set up the matrix equation for

$$A \underset{k_{21}}{\overset{k_{12}}{\rightleftharpoons}} B \underset{k_{32}}{\overset{k_{23}}{\rightleftharpoons}} C \tag{4.2.15}$$

we can write the three differential equations for the time course of the changes in concentrations of the three states:

$$dc_A/dt = -k_{12}c_A + k_{21}c_B$$
$$dc_B/dt = k_{12}c_A - (k_{21} + k_{23})c_B + k_{32}c_C$$
$$dc_C/dt = k_{23}c_B - k_{32}c_C$$

Setting $dc_i/dt = \lambda c_i$ for each state, we can write the three differential equations in the matrix form:

$$\begin{bmatrix} -k_{12} & k_{21} & 0 \\ k_{12} & -k_{21}-k_{23} & k_{32} \\ 0 & k_{23} & -k_{32} \end{bmatrix} \begin{bmatrix} c_A \\ c_B \\ c_C \end{bmatrix} = \begin{bmatrix} \lambda c_A \\ \lambda c_B \\ \lambda c_C \end{bmatrix} \tag{4.2.16}$$

Using the discussion of eigenvalues in section 2.4, equation (4.2.16) can be rearranged to give:

$$\begin{bmatrix} -k_{12}-\lambda & k_{21} & 0 \\ k_{12} & -k_{21}-k_{23}-\lambda & k_{32} \\ 0 & k_{23} & -k_{32}-\lambda \end{bmatrix} \begin{bmatrix} c_A \\ c_B \\ c_C \end{bmatrix} = 0 \tag{4.2.17}$$

Since the elements in the vector of concentrations are not zero, the determinant of the matrix of coefficients must be zero. The cubic equation resulting from the evaluation of this determinant (see section 2.4) has no zero order term and thus reduces, after division by λ, to a quadratic equation (4.2.18). It follows that $\lambda = 0$ is a solution of the cubic equation; its significance will be seen in the numerical example below.

$$\lambda^2 + \lambda(k_{12} + k_{21} + k_{23} + k_{32}) + k_{12}k_{23} + k_{12}k_{32} + k_{21}k_{32} = 0 \tag{4.2.18}$$

The two other solutions for λ obtained from equation (4.2.18) are the two exponentials characterizing the approach to equilibrium. The amplitudes for the two exponentials in the three equations

$$\left. \begin{aligned} c_A(t) &= A_0 + A_1\exp(\lambda_1 t) + A_2\exp(\lambda_2 t) \\ c_B(t) &= B_0 + B_1\exp(\lambda_1 t) + B_2\exp(\lambda_2 t) \\ c_C(t) &= C_0 + C_1\exp(\lambda_1 t) + C_2\exp(\lambda_2 t) \end{aligned} \right\} \tag{4.2.19}$$

are again obtained from the eigenvectors derived on substituting each eigenvalue and solving the three rate equations for scheme (4.2.15). The resulting three vectors give only the ratios of the amplitudes for each of the exponentials in the equations for the three states. The absolute values depend on the initial conditions (total amplitude).

This is illustrated below with a numerical example. Some idea of the complexity of the expressions obtained, even for this relatively simple reaction, is given by the equation for the final state, c_A and c_B are zero at $t = 0$:

$$\frac{c_C(t)}{c_A(0)} = k_{12}k_{23} \left[\frac{1}{\lambda_1\lambda_2} + \frac{1}{\lambda_1(\lambda_1 - \lambda_2)} \exp(\lambda_1 t) + \frac{1}{\lambda_2(\lambda_2 - \lambda_1)} \exp(\lambda_2 t) \right]$$

(4.2.20)

The terms

$$\frac{k_{12}k_{23}}{\lambda_1\lambda_2}, \quad \frac{k_{12}k_{23}}{\lambda_1(\lambda_1 - \lambda_2)} \quad \text{and} \quad \frac{k_{12}k_{23}}{\lambda_2(\lambda_2 - \lambda_1)}$$

are A_0, A_1 and A_2 respectively in equation (4.2.19).

For the approach to an equilibrium state, the amplitudes for each component, being the difference between the initial and final state of the reaction, can obviously also be obtained from the initial conditions and the equilibrium constants for each step.

A numerical example for the evaluation of time constants and amplitudes for the above model

If we now substitute numerical values for the rate constants in (4.2.15), taking the values 2, 0.1, 3 and 1 s^{-1} for k_{12}, k_{21}, k_{23} and k_{32} respectively, the following matrix is obtained

$$\mathbf{X} = \begin{bmatrix} -2 & 0.1 & 0 \\ 2 & -3.1 & 1 \\ 0 & 3 & -1 \end{bmatrix}$$

The eigenvalues for this matrix can be evaluated from the characteristic polynomial for λ, which is obtained from the determinant of $\mathbf{X} - \lambda\mathbf{I}$ (see section 2.4). The resulting cubic equation has no zero order term and one solution is $\lambda_1 = 0$; the other solutions are $\lambda_2 = -4.147$ and $\lambda_3 = -1.953$. The order is irrelevant, as long as the particular eigenvalues are kept together with the eigenvectors calculated from them. The eigenvectors corresponding to the three λs are obtained by solving, separately for each eigenvalue, the equations

$$-(2 + \lambda)c_A + 0.1c_B = 0$$
$$2c_A - (3.1 + \lambda)c_B + c_C = 0$$
$$3c_B - (1 + \lambda)c_C = 0$$

resulting in the following values for A, B and C:

Amplitudes	A	B	C
Eigenvalues			
$\lambda_0 = 0$	0.016	0.316	0.949
$\lambda_1 = -1.953$	0.545	0.254	-0.799
$\lambda_2 = -4.147$	0.034	-0.723	0.741

Each row, being the eigenvector corresponding to a particular eigenvalue, represents the ratios of the three amplitudes for the exponential phases of the reactants with that particular eigenvalue as the exponent. To obtain the absolute values for the three amplitudes for each component these numbers have yet to be adjusted using the boundary conditions at $t = 0$ and $t = \infty$. It is postulated for this example that at $t = 0$, $c_A = 1$ and the other reactant concentrations are zero. Using these conditions, the true amplitudes for each of the reactants are obtained as follows from the values in each of the columns, maintaining the ratios in each of the rows. The sum of the amplitudes for A is 1 and the sums for B and C are 0 (initial condition, $t = 0$). The row for $\lambda = 0$ must also have the sum $= 1$ (condition for $t = \infty$) and the sums in the rows for λ_1 and λ_2 are 0 (conservation of mass). We can now complete the general rate equation for each reactant by substituting the values for the exponents and amplitudes in (4.2.19)

$$c_A(t) = 0.012 + 0.928 \exp(-1.953t) + 0.0578 \exp(-4.147t)$$
$$c_B(t) = 0.247 + 0.432 \exp(-1.953t) - 0.679 \exp(-4.147t)$$
$$c_C(t) = 0.739 - 1.36 \exp(-1.953t) + 0.62 \exp(-4.147t)$$

These equations are illustrated in figure 4.4.

Two checks can be made for an equilibrium system. While the three amplitudes A_0 must add up to the initial concentration, each of the other sets of amplitudes, corresponding to the changes of the three states during a phase, must add up to zero (conservation of mass). Furthermore final equilibrium conditions are represented by the ratios of the three values of A_0. The two equilibrium constants $K_1 = 20$ and $K_2 = 3$ can be evaluated from the rate constants.

The convention used in the above derivation differs from that found in most texts on reaction kinetics. In equation (4.2.19) no assumption is made about the sign of the exponent. Numerical evaluation results in negative values for the two λs. In other texts λ is

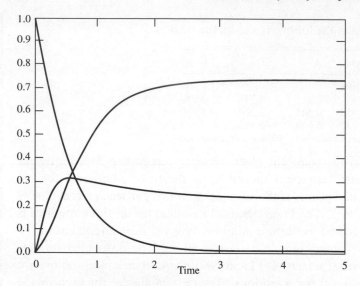

Figure 4.4   Representation of the time course of the approach to equilibrium
concentrations of the three states of a two step reversible reaction. The
parameters (amplitudes and exponential coefficients) for the three curves are
those calculated by the method described in the text.

usually introduced as a decay $(dc_i/dt = -\lambda c_i)$ and consequently
changes in the algebra result in positive values for the λs and the terms
in equation (4.2.19) are written in the form $\dots A_i \exp(-\lambda_i t) \dots$. The
latter convention moves from mathematical generality to the convenience of making λs equivalent to the much used term k_{obs} and hence
also to $1/\tau$, the reciprocal time constant. The present derivation
results in the λs having the opposite signs from k_{obs} and $1/\tau$ values.

There are now commercial computer programs available which provide
even symbolic solutions of eigenvalues and eigenvectors. Methods for their
numerical evaluation are readily available. These and the relative merits of
obtaining solutions for rate equations by the above method or by numerical
integration are discussed in section 2.3. It must be emphasized that the
algebraic method can only be used for linear systems. This is, of course, also
true for any hope of obtaining analytical solutions to rate equations. The
latter are often obtained by the approximation of linear conditions through
high ligand concentrations. Apart from linearization through small perturbations (see section 6.2) this is illustrated with many examples of the use of
pseudo first order rate constants. In the next section we shall compare the
procedure for analytical solutions of rate equations for the approach to the

steady state of enzyme catalysed reactions with the application of the matrix method illustrated above.

Several other methods are widely used to derive complex rate equations. Sets of first order differential equations with constant coefficients can be solved by a variety of methods listed in standard textbooks. The use of second order differential equations for this purpose, as well as of the method of Laplace transforms, is illustrated in section 5.3.

The sequence of events after calcium activation of aequorin

The following is an example of a consecutive reaction in which a rapid rise in signal is followed by a slow decay and yet k_{12} is smaller than k_{23}. Hastings *et al.* (1969) studied the rate of appearance and decay of bioluminescence of aequorin on addition of calcium. In the scheme (simplified from Hastings *et al.*, 1969), where A represents aequorin

$$A + Ca^{2+} \xrightarrow{\hspace{2cm}} A.Ca^{2+} \xrightarrow{k_{12}} X \xrightarrow{k_{23}} Y + hv \tag{4.2.21}$$

the rate of formation of $A.Ca^{2+}$ is sufficiently fast to be neglected in the kinetics of the subsequent events leading to the light emission. The time profile of light emission shows a rapid rise ($\tau = 0.01$ s) and a slow decay ($\tau = 1$ s). However, detailed studies showed that the rate of utilization of aequorin is characterized by $k_{12} = 1$ s^{-1}, while the light emitting step is characterized by $k_{23} = 100$ s^{-1}. The kinetic principle of sequestration of a reactant is discussed in section 5.3. The reason for the apparent rapid formation of the intermediate X is that its maximum concentration at the stationary point (see figure 4.5) is very small, that is it is reached rapidly because it does not have to go very far. In the calculation of the time profile the concentration of the intermediate is given as a proportion of the initial concentration $c_A(0)$. The maximum concentration can be related to the rate constants by the following derivation:

$$dc_X/dt = k_{12}c_{ACa^{2+}}(t) - k_{23}c_X(t) = k_{12}\{c_A(0) - c_X(t)\} - k_{23}c_X(t) \tag{4.2.22}$$

and since $dc_X/dt = 0$ at $c_{X_{max}}$, the stationary point, we obtain

$$c_{X_{max}} = \{k_{12}/[k_{12} + k_{23}]\}c_A(0)$$

The rate of approach to the inflection point (from (4.2.22)) gives an apparent rate constant

$$k_{obs} = k_{12} + k_{23}$$

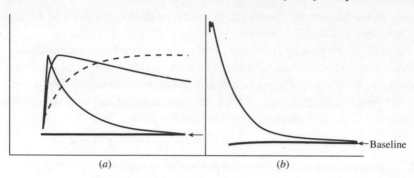

Figure 4.5   (*a*) Kinetics of the reaction of aequorin with maximally effective concentration of Ca^{2+} (0.02 M) at 20 °C in a stopped-flow apparatus. Aequorin 2.5 µg ml^{-1} in the reaction mixture. Three successive experiments were recorded with sweep speeds of 5, 50 and 500 ms per division. (*b*) A double stopped-flow experiment at 26 °C in which aequorin 0.7 µg ml^{-1} and calcium 0.1 M were first mixed and the resulting solution was mixed with 0.1 M EDTA; sweep speed 5 ms per division. (For further detail see Hastings *et al.*, 1969.)

This observed rate, like the rate of approach to equilibrium or steady state, is faster than any component step. Neglect of this piece of kinetic wisdom has caused frequent erroneous interpretations of reaction mechanisms (see for instance section 5.3): one must remember to distinguish rates from rate constants.

Long sequences of essentially irreversible reactions

Now let us return to sequences which can be interpreted in terms of essentially irreversible steps. A characteristic feature of the rate of formation of the final product of a sequential reaction is a lag phase, unless the last step is slower than any of the preceding ones. If all the steps are of similar rates the lag in the appearance of the end product of reactions with increasing number of intermediate steps can be used as a diagnostic (see figure 4.6). For a two step reaction the lag phase appears longest when the two rate constants are equal. Similarly, for longer sequences the lag phase for the final product is longest when all rate constants are equal. Attempts to fit the data to the equations for reaction sequences with different numbers of steps will indicate the minimum number required (see the discussion later in this section of protein folding intermediates and the reactions in the retina). In section 5.1 and in subsequent sections we discuss in some detail the analysis of the transient approach to the steady (or stationary) state. In that case an exponential approach to the steady state

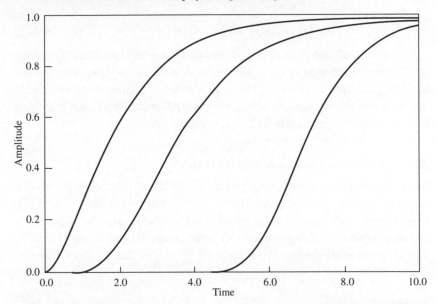

Figure 4.6   Illustration of increasing lag periods of product formation with increasing number of states of consecutive reactions. Simulation of the rate of appearance of the final product of (a) a two step (b) a three step and (c) a six step irreversible reaction. All rate constants are 1 s^{-1}.

rate occurs and a time constant can be defined for the lag phase. In the consecutive reactions discussed in this chapter only *stationary points* occur and in that case time constants of lag phases are not defined; an attempt to do so would correspond to defining a line by a single point.

The general solution for the rate equation with n irreversible steps is:

$$A_1 \xrightarrow{k_{12}} A_2 \xrightarrow{k_{23}} A_3 \longrightarrow \dots \longrightarrow A_n$$

(assume A_n is the final product and hence $k_{n(n+1)}=0$)

$$c_n(t) = c_1(0) \left[1 - \left(\frac{k_{23}k_{34}k_{45}\dots k_{(n-1)n} \exp(-k_{12}t)}{(k_{23}-k_{12})(k_{34}-k_{12})\dots(k_{(n-1)n}-k_{12})} \right) \right.$$

$$- \left(\frac{k_{12}k_{34}k_{45}\dots k_{(n-1)n} \exp(-k_{23}t)}{(k_{12}-k_{23})(k_{34}-k_{23})\dots(k_{(n-1)n}-k_{23})} \right)$$

$$\dots$$
$$\dots$$

$$\left. - \left(\frac{k_{12}k_{23}k_{34}\dots k_{(n-2)(n-1)} \exp(-k_{n-1}t)}{(k_{12}-k_{(n-1)n})(k_{23}-k_{(b-1)n})\dots(k_{(n-2)n}-k_{(n-1)n})} \right) \right]$$

$$= c_{A_1}(0) - \{c_{A_1}(t) + c_{A_2}(t) + \ldots + c_{A_{n-1}}(t)\} \tag{4.2.23}$$

For details of the derivation of this equation see Capellos & Bielski (1972). Any additional steps on the pathway must be fast compared to the two assigned rate constants. The solution for a sequence of reversible reactions can be obtained by one of the methods mentioned above, as long as all the steps approximate first order conditions.

The sequential reaction of a restriction enzyme

The reaction of the restriction enzyme *Eco*RI with circular DNA provides a simple example of a two step reaction (Halford, Johnson & Grinsted, 1979). This enzyme cleaves the two strands of DNA molecules and the appearance and disappearance of supercoiled (A), open circle (B) and linear DNA (C) provide good evidence that the open circle form is on the direct pathway of the reaction:

$$A \rightleftharpoons B \rightleftharpoons C \tag{4.2.24}$$
$$\quad S \qquad O \qquad L$$

(S is supercoiled, O is open circle and L is linear DNA).

Since the concentrations of all three reactants can be determined throughout the time course of the reaction, the individual rate constants can be calculated from

$$c_B(\max)/c_A(0) = k_{12}/\{k_{12} + k_{23}\} \quad \text{and} \quad c_A(t)/c_A(0) = \exp(-k_{12}t)$$

The time profile for the appearance and disappearance of the open circle form can be fitted to equation (4.2.4). There is therefore a double check on the open circle form being an obligatory intermediate.

The analysis of consecutive reactions involved in enzyme catalysis, single turnover and transient approach to the steady state, is the subject of section 5.1. In these reactions readily reversible steps are often encountered and the resulting rate equations will be discussed in that context. Some examples will now follow of sequential reactions involved in protein folding and unfolding and in physiological responses consequent on ligand binding or some other stimulus. They will be used to illustrate the range of information which can be obtained from kinetic analysis. It will also be shown how different, though sometimes equivalent, algebraic methods can be used to describe and analyse the observations. Simplification of the expressions through suitable choice of conditions can make formidable equations readily applicable.

Protein folding

There has been a considerable increase in interest and change in emphasis in the study of the kinetics of folding and unfolding of protein structures. Early reviews by Kauzmann (1959) and Tanford (1968) surveyed the problems involved in the elucidation of the thermodynamic and kinetic parameters of protein denaturation in what might be called the pre-structure period of protein science. Most of the phenomena they referred to involve the unfolding of the secondary and tertiary structures, which is sometimes preceded by the dissociation of oligomeric proteins into their component monomers. Denaturation is usually initiated by elevated temperatures and/or the addition of urea, guanidine hydrochloride, or other cosolvents. In the early days it used to be considered an irreversible process and, until recently, unfolding was treated as a two state system. The availability of protein structures at atomic resolution, the realization that they consist of distinguishable domains and the discovery of the reversibility of unfolding/folding presented new and interesting kinetic problems.

Some of the early kinetic work on denaturation was discussed by Johnson, Eyring & Stover (1974), who proposed the following scheme for the time course of protein unfolding. Let us assume that a protein can be stimulated into an active form, A, by some rapid change in conditions, like diluting out of urea. This form is in equilibrium with a protected form, P, and it decays into an inactive form, I,

$$I \xleftarrow{k_{13}} A \underset{k_{21}}{\overset{k_{12}}{\rightleftharpoons}} P$$

This scheme is applicable to the interpretation of many reactions involving two forms of a protein in equilibrium prior to an irreversible process. The rate equations for this mechanism can be derived by the method described at the beginning of this section. They can easily be extended to a more complex scheme like that of Goldberg, Rudolph & Jaenike (1991) for the competition between renaturation and aggregation during refolding of lysozyme. The present discussion is intended to illustrate the role kinetic investigations have played and should do so actively in the future, with new tools. Recent reviews of different aspects of protein folding have been presented by Kim & Baldwin (1982), Fischer & Schmid (1990), Matouschek *et al.* (1990) and Jaenike (1991). When it became apparent that proteins could be renatured under suitable reversal of denaturing conditions (Anfinsen, 1972) kinetic studies of folding of polypeptide chains into the native conformations of proteins became feasible. The early studies were carried out on monomeric proteins and at first only reactions independent of protein concentration were observed. The characteristic property of the change in fluorescence of tryptophan residues on transfer between environments of different polarity has proved to be an important signal for changes in protein conformation, especially for folding. In some proteins, which contained either single or distinguishable tryptophans, transfer into a non-polar environment during the folding process could be used as a monitor for local events. The now classic experiments of Epstein *et al.* (1971) on staphylococcal ribonuclease, which are described below, fall into this category. The recent introduction of the tools of site directed mutagenesis into the study of the kinetics of protein folding and of more local conformation changes makes it possible to introduce fluorescent markers into any required position in the domain under investigation (see for instance Smith *et al.*, 1991). The relation between these relatively large conformation changes and the general structural mobility of protein molecules is discussed in section 7.3.

As unfolding and folding studies were extended and a wider range of monitors used to follow the process in a particular protein, several intermediate states were found to be in reversible equilibrium with each other and with the native form under suitable conditions (see reviews quoted above). The conclusions of Badcoe *et al.* (1991), for instance, about the number of intermediates formed during folding serve as an example for the use of lag phases in the time course of the appearance of active enzyme from the unfolded protein. They studied the reversible steps in the folding and assembly of lactate dehydrogenase from *Bacillus stearthermophilus*. Numerical fitting of their data to rate equations for consecutive reactions with different numbers of steps provided good evidence for a sequence of two first order folding processes followed by second order assembly of the active dimer. Clearly only steps with rate constants of similar magnitude would be detected. The more detailed analysis of the folding process of hen lysozyme by Radford, Dobson & Evans (1992) characterized partially structured intermediates and multiple pathways. These authors combined the techniques of time resolved deuterium exchange, NMR identification of specific helices with protected amide groups and stopped flow CD measurements, to resolve folding processes with time constants ranging from $\tau <$ 1 ms to $\tau \simeq 500$ ms. The resolution of different phases in the time course of specific exchange and dichroism proved to be powerful in resolving this complex process.

Thermodynamic studies have given valuable information about the domain structures and stabilities of proteins (see reviews by Privalov, 1982, and Edsall & Gutfreund, 1983). The dogma that all the information about the three-dimensional structure of a protein is contained in the primary structure (amino acid sequence) depends on the argument that this structure is in a potential energy minimum. There are many possible energy troughs on the folding pathway and the thermodynamic argument alone is not sufficient. The conditions during folding have to be such that the pathway to the correct (native) trough is the fastest one.

The discovery of chaperon proteins, which regulate folding and assembly of oligomers, has added a new dimension to these problems (see Ellis & van der Vies, 1991, for a review of that subject). The mechanism of the involvement of the molecular chaperon GroE system in the kinetics of folding is discussed by Badcoe *et al.* (1991) and Buchner *et al.* (1991). The latter authors describe experiments on the refolding of citrate synthetase. They found that the yield of native protein during refolding experiments was inversely proportional to the concentration of unfolded protein. Buchner *et al.* found, in their study, that there was a competition between correct refolding and aggregation of unfolded protein. Increasing concentration favoured aggregation and the addition of GroE favoured refolding by protective binding to the unfolded protein. The folding process is first order (or the sum of first order reactions) while the aggregation will be a higher order reaction, and hence a concentration dependent process. Reducing the concentration of unfolded protein by complex formation will shift the balance of rates towards correct folding.

An additional complexity to the scheme comes from the involvement of an ATP binding/hydrolysis cycle in the dissociation of the complex between GroE and the unfolded protein. This presents a very interesting kinetic problem of the competitive protective action being controlled by the steps of the ATPase reaction. One is tempted to speculate that ATP binding to GroE is required to dissociate the unfolded protein and ATP hydrolysis is required to dissociate the nucleotide to make the chaperon molecule available for further protective action. This mechanism is similar to that suggested by Geeves, Goody & Gutfreund (1984) for the regulation of protein–protein interaction and by Gutfreund & Trentham (1975) for the interconversion of intermediates in some other systems involving ATP hydrolysis.

The experiments of Epstein *et al.* (1971), already referred to, on the reversible folding of staphylococcal nuclease on change of pH from 3.2 (unfolding conditions) to 6.7, provided a good example of a record with two time constants. They reported that the faster process (12 s^{-1} at 25 °C) was not significantly affected by temperature (13 s^{-1} at 38 °C) while the slower process ($2 s^{-1}$ at 25 °C) increased fourfold over this temperature range. This system was re-investigated with a sequential jump method, which could have a more general application. Unfolding was affected by pH jumps from neutrality to either acid or alkaline conditions. After different delay times a second pH jump returned the solution to neutrality. Chen, Martin & Tsong (1992) used this technique to follow refolding from different unfolded states D. They interpreted their results in terms of three unfolded and one folded (native) state N_0:

$$N_0 \rightleftharpoons D_1 \rightleftharpoons D_2 \rightleftharpoons D_3$$

The analysis of this reversible consecutive reaction relied on the fluorescence change during the first (N_0 to D_1) transition. Subsequent steps were inferred from the dependence of the kinetic profile of refolding on the delay prior to the second (neutralizing) jump and consequently on the progress of the unfolding prior to refolding. Studies of protein folding require the analysis of reactions in terms of several different models, even if only two distinct time constants are observed. The overall process can often be subdivided into events which can be identified by different techniques, such as NMR and optical dichroism, as well as by the selective introduction of fluorescent markers.

Some physiological examples

An ever increasing number of new functions are found to involve enzymes. At one time the only role of enzyme catalysis was thought to be the facilitation and regulation of intermediary metabolism. However, many of the enzyme reactions now being studied are involved in communication between cells and are part of a cascade regulating transduction of signals and energy. Details of consecutive reactions involving steps in the mechanisms of single enzymes, as well as of sequential multi-enzyme systems, are discussed in subsequent sections, but some points should be made here about the kinetics involving enzymes as part of a signalling system. The problems involved in receptor–messenger interactions and the consequent responses, somehow enter into the discussions of every system covered in this volume. For the specialized study of the connected problem of ligand binding and channel opening, which are treated by noise analysis and stochastic methods, the reader has to be referred to the volumes by Hille (1992) and Sakmann & Neher (1983).

We conclude this section with the description of some systems which present good examples of the potential for the kinetic analysis of steps involved in physiological responses. No detailed mathematical description is given for the models described, but the literature on attempts to do so is referred to. The present qualitative description is intended to illustrate studies in which the kineticist's approach has helped and should help further in the elucidation of the number and character of steps leading from stimulus to response in physiological systems.

The sequence of events in the retina

During the above discussion of consecutive reactions and in particular in connection with the example of protein folding, the point has been made that one can get information about the number of intermediates from the time course of the reaction. The rate of acceleration (or lag phase) of the final response can be analysed in terms of the number of events with approximately equal time constants. The form of the kinetic profile depends on whether the process consists of a sequence of steady states or a sequence of consecutive reactions. A classic example is the prediction by Fuortes & Hodgkin (1964) that there are a considerable number of steps in between the absorption of a photon by rhodopsin and the electrical response of limulus ommatidia. Hodgkin (1992) wrote 'We became interested in the long delay between a light flash and the appearance of an electrical response, which we thought might arise from the time taken for a signal to pass through a cascade of intermediate chemical reactions, possibly stages of amplification.' This approach to the kinetic analysis of photoresponse has been steadily improved during the last two decades. Figure 4.7 shows more recent observations on the response of turtle cones to weak flashes of light (Baylor, Hodgkin & Lamb, 1974). These authors fit their data to a model with six steps between light absorption and the effects on the sodium current across the cone membrane. Baylor *et al.* describe their procedure for deriving suitable kinetic equations in some detail. Lamb (1984) compares the six steps of turtle cones, derived from voltage recordings, to the four steps of toad rods, fitted to current records, with respective time constants of $\tau = 60$ ms and $\tau = 500$ ms. The difference between the conclusions from the two species can be due either to the actual number of events or to two faster steps in the cascade of toad rods, which do not have any influence on the time constant of the transient. It will be noticed that information on different aspects of visual response is often derived from experiments with material obtained from different species. This is not unusual in biochemical and biophysical research. Experimental investigation of some phenomena is often easier with material from one

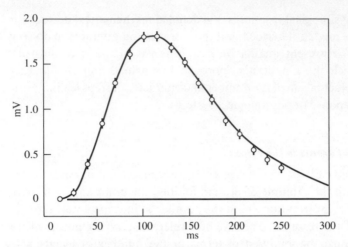

Figure 4.7   Response of red-sensitive turtle cones to weak flashes of white light of 11 ms duration. The abscissa is the time from the middle of the flash. The response was fitted to an equation for six steps (see text and Baylor *et al.*, 1974).

particular species. Once one knows what one is looking for, it is possible to investigate other tissues profitably and, as the field progresses, any possible anomalies due to such comparisons can usually be elucidated. The contribution of 'unusual animals' to biological research as tools and as special examples was emphasized by Hodgkin (1973).

During the last decade details in terms of protein–protein interaction, ligand exchange and enzyme reactions gave a meaning, in molecular terms, to the steps predicted from the results of the above investigations. Biochemical and biophysical studies in many laboratories (see the review of Stryer, 1991) established that phototransduction in vertebrate retinal rods involves the sequential activation of rhodopsin ($R \rightarrow R^*$), G-protein ($G \rightarrow G^*$) and phosphodiesterase ($PDE \rightarrow PDE^*$) on the disk membranes of rod outer segments and in the cytoplasm. This sequence of the reactions, which follows the absorption of photons by rhodopsin, is linked to the primary electrophysiological response at the cytoplasmic membrane of the rod by the rapid depletion (by PDE^* mediated hydrolysis) of the negative messenger cyclic GMP (cGMP). This scheme is illustrated in figure 4.8 and elaborated below in terms of examples for kinetic problems in the elucidation of the mechanism of visual response.

R^*, the pigment formed after photochemical and dark reactions with time constants $\tau < 1$ ms, is an integral membrane protein. It interacts via diffusion with G-protein, a peripheral membrane protein (these references are to disk membranes as distinct from the cytoplasmic membranes of the

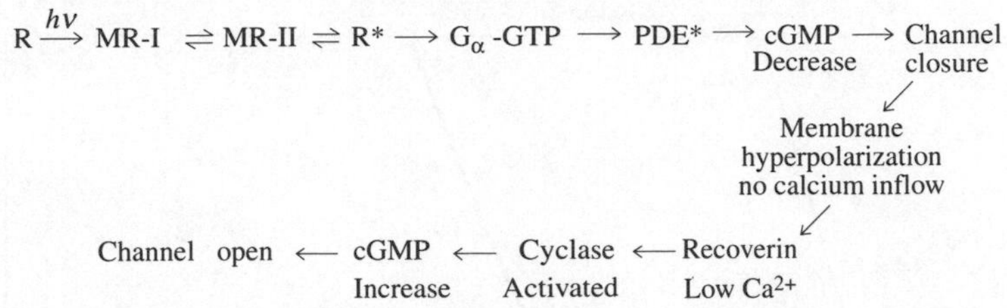

$$R \xrightarrow{h\nu} \text{MR-I} \rightleftharpoons \text{MR-II} \rightleftharpoons R^* \longrightarrow G_\alpha\text{-GTP} \longrightarrow \text{PDE*} \longrightarrow \underset{\text{Decrease}}{\text{cGMP}} \longrightarrow \underset{\text{closure}}{\text{Channel}}$$

$$\swarrow$$

Membrane
hyperpolarization
no calcium inflow

$$\swarrow$$

$$\underset{}{\text{Channel open}} \longleftarrow \underset{\text{Increase}}{\text{cGMP}} \longleftarrow \underset{\text{Activated}}{\text{Cyclase}} \longleftarrow \underset{\text{Low Ca}^{2+}}{\text{Recoverin}}$$

Figure 4.8 A partial scheme of the enzyme cascade involved in the transmission of the signal from rhodopsin bleaching to channel closure and the consequent membrane hyperpolarization. The reactions resulting in channel opening, as part of recovery, are also shown. The phosphorylation and dephosphorylation of rhodopsin, which are involved in the inactivation of R^*, are omitted.

rod). This results in the exchange of GDP for GTP at the binding site on the α-subunit of the hetero-trimeric G-protein, usually referred to as transducin in the literature on photoreceptors. As a result of the nucleotide exchange, the α-subunit (G^*) with bound GTP and the remaining dimeric transducin dissociate separately from their complex with rhodopsin. G^*-GTP proceeds to the activation of a phosphodiesterase (PDE). The activation step G-GDP to G_α-GTP (G^*), mediated by R^* is catalytic ($R^* + G \rightarrow G^* + R^*$); indeed, Bruckert, Chabre & Vuong (1992) analyse the kinetics of this reaction in terms of an expanded Michaelis–Menten model. Each molecule of R^* can activate approximately 5000 molecules per second to form G^*. In contrast, the subsequent step is a stoichiometric interaction between G^* and the γ-subunits of phosphodiesterase (PDE), the enzyme which hydrolyses cGMP to GMP. These three steps are each of a composite nature, involving additional states and diffusion processes. This can be illustrated by non-linear Arrhenius plots, which indicate changes in rate limiting step with increasing temperature (see section 7.2). However, under specified conditions, the above three steps can each be characterized by a single time constant and they can be used to illustrate a point before the rest of the cascade is discussed.

The steady state concentration of R^* is reached with a time constant τ_1, the activation G to G^* occurs with a time constant τ_2 and the steady state concentration of active PDE is reached with the time constant τ_3. These time constants are additive; remember that $\tau = 1/k$ for a first (or pseudo first) order reaction, and in an essentially irreversible steady state sequence (see section 3.3)

$$1/k_{obs} = 1/k_{12} + 1/k_{23} + 1/k_{34}\ldots$$

and therefore

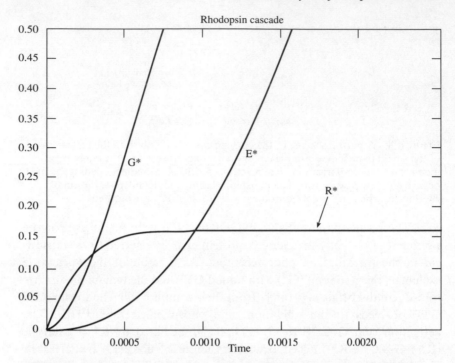

Figure 4.9   Simulation of a cascade corresponding qualitatively to the sequence
of steps: (1) rhodopsin activation by light $R \rightarrow R^*$, (2) G-protein activation
$R^* + G \rightarrow R^* + G^*$, (3) activation of phosphodiesterase $G^* + E \rightarrow G^* + E^*$.

$$\tau_{obs} = \tau_1 + \tau_2 + \tau_3 \dots$$

This is illustrated in figure 4.9.

The PDE molecule has two active sites, each controlled by an inhibitory
γ-unit. The stoichiometric interaction of a molecule of G^* with the γ-units
is required for activation of the hydrolysis of cGMP. This nucleotide is a
negative messenger for the photoreceptor ion channel, a high concentration
of cGMP keeps the channels open (note: this is specific for vertebrates; the
concentration dependence may be inverse for invertebrates). If fully
activated, the PDE can halve the concentration of cGMP in 0.6 ms and the
nucleotide controls the resting sodium current in a highly cooperative
interaction (Hill constant $n = 3$) with the ion channels. Closing of the
sodium channels results in hyperpolarization. The formulation of a complete kinetic mechanism has to await the results of still more quantitative
investigations into several steps of this cascade: enzyme concentrations,
rates of activation and catalytic turnover, etc. These activation processes
are balanced by dark and light adaptation and recovery.

Recent studies are beginning to add explanations for signal sensitivity on dark adaptation and increased time resolution on light adaptation. A mathematical model of the adaptation processes compatible with the experimental evidence available has been presented by Forti *et al.* (1989). Lamb & Pugh (1992) have modelled the cascade under the limiting condition of neglecting the inactivation reactions. With such linear systems they obtained analytical solutions and compared them to responses of the photocurrent.

The feedback effects of the ion movements on the two enzymes, diesterase and cyclase, which are respectively responsible for the hydrolysis during the excitation and the synthesis of cGMP during the recovery, are of particular interest. The cyclase reaction is inhibited by calcium and cGMP synthesis is accelerated when the Ca^{2+} concentration is decreased by closure of the ion channels. The calcium cycle and relaxation is another interesting story.

The primary events of GDP/GTP exchange, catalysed by R*, are thought to be very similar for invertebrate photoreceptors. However, the cascade initiated by the liberation from transducin of the subunit with bound GTP differs from that observed for vertebrates. The α-unit activates the integral membrane enzyme phospholipase C to hydrolyse phosphotidy-linositol; the product of this reaction, in turn, activates calcium release. In invertebrates the sodium channels on plasma membranes are closed in the resting state and open on excitation, with the consequent depolarization of the cell, in contrast to the hyperpolarization of vertebrate photoreceptors. It is postulated that the increase in current (as distinct from a decrease in vertebrates) results from Ca^{2+} release. The direct effects of Ca^{2+} on the channels and indirect effects on the enzymes controlling cGMP concentration, have yet to be clarified. The specific properties of invertebrate photo-receptors and their variations have been described by Nagy (1991).

The sequence of reactions in the retina provides a number of interesting examples of special kinetic problems. The steps from the primary photochemical event to the formation of R*, the state of rhodopsin responsible for transducin nucleotide exchange, are discussed above. The rate limiting reaction of the formation of metarhodopsin-II in homoisothermic animals has a very large temperature dependence (see section 7.2), the major effect being on the forward rate. Below 10 °C the equilibrium between MR-I and MR-II can be followed spectrophotometrically by the change in absorbency at 380 nm (MR-II formation) or 480 nm (MR-I disappearance). The formation of R*, prior to interaction with and activation of transducin, can be studied by light scattering (Kohl & Hofmann, 1987) or spectroscopi-

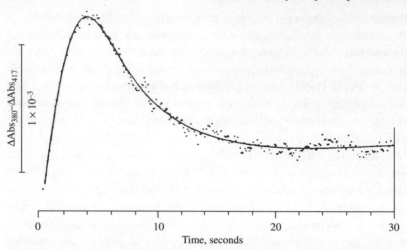

Figure 4.10   Kinetic analysis of the consequences of the following sequential processes: a light flash results in the rapid equilibration MR-I⇌MR-II + MR-II(bound G-protein). Absorbence at 380 nm measures the concentrations of both MR-II and its complex with G-protein. In the presence of a GTP analogue the G-protein becomes activated and dissociates from the complex; as G-protein dissociates equilibrium results in a decrease in total MR-II and the rate of dissociation can be followed in terms of absorbence at 380 nm (from Kohl & Hofmann, 1987).

cally. Figure 4.10 illustrates one such experiment in the form of a record of the formation and disappearance of MR-II. The absorbency at 380 nm indicates the equilibration of MR-I with MR-II and the MR*-transducin complex. As the transducin dissociates, after the GDP/GTP exchange, the absorbence due to the extra (complexed) MR-II and R* disappears due to partial reversal to MR-I. The data can be interpreted in terms of a sum of exponentials. It is not clear whether the degeneracy of rate constants will prevent a more detailed interpretation of this part of the sequence of events during phototransduction. However, the use of another signal, the spectral change of a pH indicator, has enabled Arnis & Hofmann (1993) to detect the additional R* state between MR-II formation and activation of G-protein. Under suitable experimental conditions of lipid environment and temperature, proton uptake during the formation of R* can be kinetically distinguished from MR-II formation. The former has a much larger temperature dependence than the latter.

The eventual synthesis of a dynamic picture of the sequence leading to excitation requires the modelling of the protein–protein interactions between R*, G-protein and PDE leading to the removal of the negative messenger (cGMP) and the closing of ion channels. This has to be coupled to the resulting decrease in calcium concentration and the two consequent events necessary for recovery, the inactivation of R* by phosphorylation and by the synthesis of cGMP. This involves the enzymes phosphokinase and cyclase as well as the control proteins arrestin and recoverin. As all the functional components of this system (enzymes, control proteins and ion channels) are identified and their specific activities in situ are determined, the problem is changing from being a purely electrophysiological one to one that includes the kinetics of protein–protein interaction, of enzyme reactions and of their regulation.

The above system is one of many in which reactions of nucleoside phosphate esters, coupled to affinity changes at protein binding sites, are involved in energy transduction (through ATP hydrolysis) and information transfer (through GTP binding proteins). The development of novel techniques for the study of time resolved light scattering changes in solutions and membranes has helped to elucidate the important role of protein–protein interaction in signalling (Schleicher & Hofmann, 1987) and should contribute to the kinetic study of related systems. The techniques for the study of nucleoside phosphate hydrolysis in terms of enzyme kinetics are discussed in the next section. The further application of these approaches to the detailed analysis of the individual states of the reaction of GTP with α-G and of the function of the intermediates, will be illuminating (see Ting & Ho, 1991, for some such investigations).

Molecular kinetics and muscle contraction

The chain of reactions involved in muscle contraction, from the signal at the neuro-muscular junction and the release of calcium by the sarcoplasmic reticulum, to the force development by the contractile apparatus, is currently studied by a wide variety of techniques. These involve experiments with preparations at every level of organization from pure proteins in solution to intact muscle. Some of the important new developments of methods, which can be used for the study of the same reaction step in solution and within cells, have used the contractile system as a paradigm. The application of pressure relaxation to the study of actin–myosin interaction in solution as well as of tension changes in fibres is discussed in section 6.4. The initiation of reactions by the photochemical release of ATP from inactive precursors, discussed in section 8.4, has provided data for the rates of steps of myosin ATPase in contracting muscle fibres. Many different signals can be used for the kinetic resolution of steps involved in the transduction of chemical energy (from the hydrolysis of ATP) to mechanical energy believed to be derived from the dynamic interaction between myosin heads (see figure 4.11) and actin filaments. The coupled processes of the hydrolysis of ATP and the interactions between proteins and ligands in solution are studied by optical techniques, are discussed in section 5.1. Changes in the mechanical properties of muscle fibres, recorded by time resolved changes in stiffness and tension during the contractile processes, should be correlated with the kinetics of the molecular events. The study of processes at these different levels of organization must, eventually, be correlated. The kinetics of different phases of muscle contraction vary considerably between muscle fibres isolated from different tissues. However, comparisons will only be made when they illustrate some point of interest to kinetic analysis. In the present section we regard the hydrolysis of ATP as a single step. In the next section the intermediates in this reaction are discussed in terms of enzyme kinetics.

As emphasized above, the wide interest in the elucidation of the mechanism of muscle contraction at the molecular level has stimulated the development of many of the techniques now available for the recording of physical and chemical events; among these is time resolved X-ray diffraction. Hugh Huxley added to his great contributions to the elucidation of the sliding filament model (see Huxley, 1969) of muscle contraction with further X-ray investigations on intact muscle fibres. Huxley and his colleagues have been able to plot simultaneously the time course of the movement of the relatively large mass of the cross bridges from myosin filaments to actin filaments and of the force development and relaxation. The distinct time course of cross bridge (myosin heads) movement, recorded by changes in the equatorial diffraction pattern, prior to tension development is illustrated in figure 4.2. A corresponding lag phase in tension response is also seen. This phase should correspond to the time constants of the steps preceding tension. When these experiments were performed the time resolution of data collection was restricted to photon counting in 1 ms bins. This limitation also affected deduction from the reflection of the low angle X-ray diffraction pattern at the meridian, which is assumed to be a monitor for the movements or change of shape of myosin heads attached to actin filaments. Huxley *et al.* (1983) showed that the intensity of this reflection changed considerably during the time of rapid shortening of active muscle. On current theories of the mechanism of muscle contraction, force development is due to a change in shape and/or orientation of the attached head. It is, therefore, of particular interest to establish whether the reflections arising from the axial repeat of the myosin

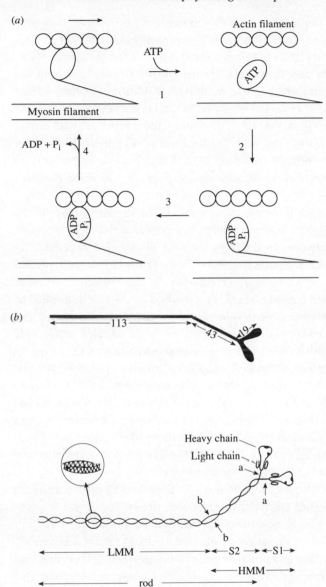

Figure 4.11 (*a*) Diagrammatic representation of the myosin filament and its interaction with the actin filament during the ATP hydrolysis cycle. More detail of the steps involved is given in section 5.1. (*b*) Outline of the composition of the myosin molecule indicating the S1 head of each chain, which contains both the active site for the ATPase reaction and the actin binding site. The proteolytic cleavage sites, a and b, are involved in the production S1 and the double headed heavy meromyosin (HMM) respectively.

heads change truly in phase with the rise in tension. Improvements in time resolution to photon collection in 0.2 ms bins have enabled Irving, Lombardi & Piazzesi (1992) to provide more evidence for the synchronous movement of heads and force development. More detailed information often introduces more problems to be solved. Additional evidence from these structure studies, as well as from mechanical investigations (repetitive rapid length changes), showed that the stroke can be repeated with a time constant of about 10 to 15 ms, while the turnover of ATP hydrolysis, under their conditions, is thought to be about 100 ms. This is, at present, an unresolved problem: *can more than one step occur per ATP molecule hydrolysed?*

During the last decade a new development, the in vitro assay of the movement of biological motor molecules, has increased the tools available for kinetic investigations. In different variants of detailed experimental arrangements Sheetz & Spudich (1983), Finer *et al.* (1994) and several others (see Huxley, 1990) have observed optically the movement of actin filaments on a myosin coated surface. This method could also be applied to the study of the movement due to other motor proteins such as dynein and kinesin, which also utilize the energy derived from coupled nucleoside triphosphate hydrolysis. With tracking arrangements linked to computer analysis both movement and force fluctuations (Ishijima *et al.*, 1991) can be subjected to stochastic analysis to obtain information about the lifetimes of different states of the actomyosin complexes. Analysis of the step length has also added to the problem of interpreting the quantitative relation between the interaction cycle and the number of ATP molecules hydrolysed. These novel kinetic techniques promise to stimulate more interest in the interpretation of the kinetic behaviour of 'small systems'.

From the point of view of kinetic analysis these are all problems in the resolution of systems of consecutive reactions, similar to several encountered above. The specific lessons to be learned from the way different biological systems have been successfully investigated are that it is important to observe each phenomenon in terms of a range of parameters and that each of them illustrates additional monitors which help to overcome the, apparently insurmountable, complexities of the problem. While much emphasis is put on the elucidation of the number of identifiable steps on a reaction pathway, sometimes it is of importance to demonstrate that a pair of events are tightly coupled (concomitant), as in the example of force development and the change in configuration of myosin heads.

The solution of another problem in physiology, which depends on the correct design and analysis of kinetic experiments, is the correlation of calcium release and transmitter release. The time course of the sequential events of changes in voltage potential, calcium concentration and neuro-transmitter release by exocytosis is a subject of some controversy (Parnas, Parnas & Dudel, 1986; Yamada & Zucker, 1992). The question whether calcium is necessary but not sufficient for determining the time course of transmitter release is being studied by a number of time resolved analytical techniques and by computer modelling. The model presented by Yamada & Zucker (1992) is an example of a problem in which the kinetic behaviour depends on diffusion and very local concentration changes.

5

Transient kinetics of enzyme reactions

5.1 Kinetic identification of reaction intermediates

Single and multiple turnover of enzyme reactions

For the purpose of the present discussion the term 'transient kinetics' is applied to the time course of a reaction from the moment when enzyme and substrate are mixed, $t = 0$, until either a steady state or equilibrium is established. The difference between the kinetic problems discussed in section 3.3 and in the present section is, respectively, the presence of catalytic as distinct from catalytic concentrations of enzyme. Here we are concerned with the stoichiometry of enzyme states. Transient kinetic experiments with enzymes can be divided into two types. The first of these (multiple turnover) is carried out under the condition that the initial concentrations of substrate and enzyme are $c_S(0) \gg c_E(0)$ and $c_S(t)$ can, therefore, be regarded as constant throughout the course of the reaction until a steady state is attained. Alternatively, in a single turnover reaction, when $c_S(0) \leq c_E(0)$, the transient formation and decay of reaction intermediates is observed until the overall process is essentially complete. These two possibilities will be illustrated with specific examples. In connection with a discussion of the approach to the steady state, in section 3.3 it was emphasized that, at $t = 0$, the concentrations of the intermediates, enzyme–substrate and enzyme–product complexes, are zero and, therefore, the rate of product formation is also zero. Under the experimental conditions used for steady state rate measurements and for enzyme assays, the first few seconds after the initiation of a reaction are ignored. However, when the experimental techniques and interpretation discussed below are used, events during the first few milliseconds of a reaction can be analysed and provide important information. With suitable monitors it is possible to follow the formation and decay of enzyme complexes with substrates and

138

products until they reach their steady state concentrations. Alternatively, under certain conditions, the intermediates during a single turnover of the reaction between stoichiometric amounts of substrate and enzyme can be observed.

Jencks's (1989) comments in his minireview on 'How does a calcium pump pump calcium' are worth quoting extensively as a testimony to the transient kinetic study of enzymes:

The investigations of individual steps of the reaction catalysed by the calcium ATPase that have been carried out in many laboratories have clarified what this enzyme does and have provided a good deal of information on how it does it. In contrast, relatively little has been learned about the enzyme from investigations of steady state kinetics. . . . In the meantime it is important to characterize the steps of these reactions in order that information about the structure of the protein can be interpreted when it becomes available.

These comments apply to all enzyme mechanisms to varying degrees and particularly also to a problem closely related to the calcium pump, that of the myosin ATPase coupled to muscle contraction. The methods used to study the latter enzyme are discussed in some detail below and are also the ones which helped to elucidate the steps involved in the coupled process of ATP hydrolysis and calcium translocation.

The choice of conditions for simplifying rate equations

Analytical solutions of the rate equations for the whole course of such reactions can readily be obtained under two conditions:

(1) when $c_S(0) \gg c_E(0)$ and the substrate concentration remains essentially constant, so that the binding process is pseudo first order until the steady state is reached.
(2) when $c_S(0) \leq c_E(0)$ and substrates are in rapid equilibrium with the enzyme complex prior to the remaining steps of a single turnover reaction and the substrate dissociation constant is smaller than concentration of binding sites. Linearity of the differential equations is thus preserved.

A number of authors have described general solutions to the equations for transients. As with the sale of cheap airline tickets 'certain restrictions may apply' for analytical solutions. If this fails, numerical solutions can be used to model the transients (see section 2.2). With the general availability of computers and suitable programs for the numerical solution of systems of differential equations, it is sometimes useful to use such methods to compare the results with those obtained from analytical solutions, which require that the experimental conditions are put into the straitjacket of linear equations. It is always important to ascertain whether the linearity approximation is justified. A simple example of this precaution has already been described for the case of pseudo first order reactions (see section 3.2).

Rapid reaction techniques for the study of enzymes

Eigen (1968) wrote an article on enzyme kinetics under the title 'It is all over with the overall'. This was intended to demonstrate the power of temperature jump and other chemical relaxation techniques (see chapter 6). At that time one interesting, but specialized, problem of enzyme kinetics had been solved by temperature jump experiments on glyceraldehyde-3-phosphate dehydrogenase (Kirschner *et al.*, 1966). In the event it turned out that relaxation techniques are very powerful tools for the study of ligand binding and conformation changes of proteins and nucleic acids. They have, however, had limited application to the study of enzyme catalysis. For the study of these essentially irreversible reactions rapid flow techniques have been the methods of choice. In the present section examples of enzyme systems are selected for their suitability to illustrate different kinetic approaches. The study of reactions by flow techniques has, perhaps, given more information through the direct observation of the transient appearance of intermediates than from the determination of rate constants. The period from 1955 to 1975 saw a large increase in the application of these techniques to the study of reactions of enzymes and other proteins available in moderately large quantities. After a relatively static period the advent of molecular genetics is responsible for a resurgence. The expression and overproduction of previously scarce proteins with interesting biological functions has widened the potentialities of the approaches described below. The mapping of specific replacements of amino acid residues, introduced by site directed mutagenesis, has also been successfully combined with modern kinetic techniques. An outstanding example of this, outside enzymology, is the investigations on both natural and site directed mutants of haemoglobin (Gibson, 1990; Perutz, 1992).

Flow techniques (see Gutfreund, 1965, 1972) used for the study of transient phenomena have a time resolution of approximately 1 ms. More rapid mixing has been achieved, but this involved the use of larger volumes of reactant solutions than those normally acceptable for biological materials. The time resolution of some relaxation techniques can be more than three orders of magnitude better than this. However, the limitations of these methods will become apparent in the discussion of specific examples in chapter 6. The direct observation of the rates of formation and interconversion of transient intermediates provides unique and unambiguous evidence for their position in the reaction sequence. In favourable cases they can be characterized by specific time resolved signals. Such information cannot usually be obtained from steady state kinetic investigations; even if the methods of analysis used are much more sophisticated than those described in section 3.3. However, when transient approaches to the steady state are observed, steady state kinetic data are obtained from the asymptote. The equations describing the postulated mechanism must provide agreement between the individual rate constants, obtained from transients, and the steady state parameters K_M and k_{cat}. Another important feature of rapid mixing techniques, which distinguishes them from observations of equilibrium perturbations and fluctuations, is the fact that binding sites make their first contact with substrate on mixing. The initiation of reactions in a stopped-flow apparatus avoids possible hysteresis effects, which may be observed in solutions of ligand and receptor at equilibrium. This point has been emphasized by Hess, Udgaonkar & Olbricht (1987) in connection with the time dependent desensitization of acetylcholine receptor in the presence of agonists.

The chemical reaction mechanisms of the systems discussed in this section are understood in greater detail than those used as examples for kinetic problems elsewhere in this volume. It is difficult to decide how much chemistry to present since there are other monographs more specifically concerned with that aspect of enzymology (Jencks, 1969; Walsh, 1979; Fersht, 1985). However, a description of the structure of the intermediates adds life to a kinetic analysis and demonstrates its powers. As one goes from numbers (defined intermediates and rate constants) to reaction mechanisms the statements become more ephemeral. Hopefully the kinetic data obtained, under well-defined conditions of composition and intensive parameters, remain of permanent value for the design of models to be tested. Kinetic data, which should be facts, can be interpreted in terms of several plausible chemical models. Each of the models should make predictions to be tested by further kinetic experiments. The purpose of the presentation of transient kinetics is to describe the tools and their application to the study of enzyme mechanisms. No attempt is made to present the most recent studies if earlier examples give a good understanding of the development of the field.

The study of enzyme mechanisms on a time scale of milliseconds became possible when the rapid flow techniques of Hartridge & Roughton

(1923a, b) were modified for economy of reactants. The original continuous flow technique was developed for the study of the reactions of heamoglobin with oxygen and other ligands; large volumes of solutions were easily available. The article by Roughton & Chance (1963) should be consulted for a technical and historical survey of rapid mixing techniques. The turnover numbers of enzymes, defined in terms of an observed rate constant (k_{cat}) with units of s^{-1} (see p. 84), range from about $0.1\ s^{-1}$ to $10^8\ s^{-1}$ (see section 7.3). The range is continuously extended at both ends by new observations, but most turnover numbers are within an order of magnitude of $100\ s^{-1}$. For the resolution of the events during a single turnover or the approach to the steady state it is, therefore, necessary to observe reactions from 1 ms after mixing. As pointed out above, attempts at faster mixing have so far not been successful for material of biological origin. Other techniques with better time resolution can be applied to the study of selected reactions. These will be discussed in chapters 6 and 8. Some tricks can be used to slow down reaction steps found to be too fast for rapid mixing techniques under conventional conditions. Kinetic studies at low temperatures have solved a number of problems (Douzou, 1977; Barman & Travers, 1985; Millar & Geeves, 1988). When using this approach, or the introduction of any other suboptimal conditions, such as the study of modified substrates, changes of pH or solvent composition, one has to bear in mind that alternative reaction pathways could be favoured under the altered conditions.

The classic paper of Chance (1943) describes how his stopped-flow modification of the Hartridge & Roughton continuous flow method could be used to study the formation and decomposition of enzyme–substrate complexes on a millisecond time scale. It had been shown by Keilin & Mann (1937) that the haemenzyme horse-radish peroxidase formed a complex with hydrogen peroxide. The spectral changes during the formation of the complex and of its decomposition in the presence of an oxygen acceptor, were resolved by Chance's apparatus. He was thus able to show that the individual rate constants were compatible with parameters obtained from steady state kinetics. This was the first direct evidence for the Michaelis–Briggs–Haldane mechanism and set the scene for the kinetic analysis of the formation and interconversion of spectrally distinct intermediates during enzyme reactions. Another 'first' of Chance's (1943) paper was the use of a differential analyser, built in the School of Electrical Engineering of the University of Pennsylvania, to simulate the non-linear differential equations describing the transient formation of reaction intermediates. Various points of this historical diversion will be taken up in connection with the descriptions of similar investigations of enzyme reactions. Since that time developments in electronics have not only contributed to refinements in the detection and recording of spectral changes, they have also made it possible to monitor reactions by recording changes in fluorescence, conductivity and temperature. Examples presented below will demonstrate that a multiplicity of monitors adds considerably to the elucidation of mechanisms and emphasize the point made about the potential of transient kinetics in proving the existence of intermediates. For instance, the mechanism of the reaction of lactate dehydrogenase with NAD^+ and lactate will be explained by 'kinetics without numbers' when the sequence of events is displayed as a sequence of signals for the different intermediates formed on the pathway to the steady state.

Optical monitors have been used to obtain transient kinetic data under three principal conditions. First, some enzymes have a 'built in' monitor for substrate binding and product dissociation. This can be either a chromophoric prosthetic group such as haem or riboflavin, or a fluorescent side chain near the active site like those of myosin, lysozyme and alcohol dehydrogenase. Second, transients can often be monitored when substrates absorb light or fluoresce in the visible or near ultraviolet wave band. Spectral properties usually change characteristically during binding and catalytic turnover. Sometimes small modifications of the substrate can provide such signals. A third approach to the analysis of transients uses the time profile of the appearance of products during the approach to the steady state. The successful use of this method will be illustrated by the elucidation of the mechanism of a number of hydrolytic enzymes.

Discussions of relaxation kinetics (see section 6.2) and of transient kinetics, often contain the following general statements. In principle the relaxation spectrum of a reaction contains the necessary information to evaluate all the rate constants of the elementary steps of the reaction. Similarly one can state that in principle the time profile, and its concentration dependence, of the appearance of products during the transient approach to the steady state, contains all the information for the evaluation of the individual rate constants of the formation and interconversion of intermediates. However, in both cases there are important limitations. The theoretical limitations are that the degeneracy of the sequential time constants and the position of the rate limiting step within the sequence of events can reduce the information contents, even if the record of the reaction has an unlimited signal to noise ratio. In real life, noise and restricted time resolution further reduce the number of steps which can be resolved in any particular experiment. The time resolution of different

kinetic techniques from slow test tube experiments to the fastest photochemical techniques, is best defined as the accuracy with which zero time can be
determined. Recording the subsequent course of events does not normally
present a limitation, except that care has to be taken with the time constants
introduced by the electronic components of the measuring and recording
circuits. Study of any text on electronic monitoring devices shows that there
is a trade off between response time and signal to noise ratio resulting from
the design of instruments.

Derivation of equations for the transient approach to steady state using the matrix method

The resolution of the initial acceleration of reactions (lag phases) presents a
special experimental problem since it is difficult to define the starting point
of a reaction which, by definition, has zero rate at zero time. However,
studies of lag phases, to be discussed below, have proved important for the
elucidation of enzyme catalysis, conformational transitions and the behaviour of coupled enzyme systems.

   A number of authors have provided general methods for the solution of
rate equations describing the formation and decay of n intermediates
during the transient state of the reaction

$$E + S \rightleftharpoons ES \rightleftharpoons X_1 \rightleftharpoons X_2 \rightleftharpoons \ldots \rightleftharpoons EP \rightleftharpoons E + P \qquad (5.1.1)$$

(see for instance Pettersson, 1976; Roberts, 1977; Woledge, Curtin &
Homsher, 1985). Here several examples are treated individually to provide
the reader with a variety of approaches to these problems. The important
lessons to be learned from these examples are the choice of experimental
conditions justifying assumptions of linearity and the use of boundary
conditions to obtain integration constants which determine individual
amplitudes. Three different algebraic techniques are in common use for the
solution of equations describing complex kinetic mechanisms. The choice
between matrix, Laplace and analytical solutions for a set of differential
equations is largely one of personal preference, although one or other
method may be more suited to a particular problem.

> I shall first illustrate how the exponents and amplitudes for the
> approach to the steady state can be derived from eigenvalues and
> eigenvectors by the method outlined in the previous chapter. The
> problem is to obtain the rate equations for the approach to the steady
> state concentrations of enzyme and its complexes in the simple model:

$$\begin{array}{c} \overset{k_{12}^*}{S + E \; \rightleftharpoons \; ES} \\ k_{21} \end{array}$$

The conditions are (a) that the substrate concentration is sufficiently high to be regarded constant and $k_{12}^* = k_{12}c_S(0)$ and (b) the product concentration remains sufficiently small for the step characterized by k_{31} to be neglected. We are interested in the evaluation of the rate constants for the formation of the intermediates from the exponential terms of the experimental records for the time course of one of the states. From the matrix set up for this scheme in the same manner as that for the three step linear reaction (previous section p. 114)

$$\begin{bmatrix} -(k_{12}+\lambda) & k_{21} & k_{31} \\ k_{12} & -k_{23}-k_{21}-\lambda & k_{32} \\ 0 & k_{23} & -(k_{32}+k_{31}+\lambda) \end{bmatrix} \tag{5.1.2}$$

we obtain the characteristic equation by evaluating its determinant. We again obtain a cubic equation without a zero order term and can divide by $-\lambda$ (one solution is $\lambda = 0$), which results in the quadratic equation:

$$\lambda^2 + \lambda(k_{12}+k_{23}+k_{21}+k_{32}+k_{31}) + k_{12}(k_{23}+k_{31}+k_{32})$$
$$+ k_{31}(k_{21}+k_{23}) + k_{21}k_{32} = 0 \tag{5.1.3}$$

As for the equilibrium example in section 4.2, the two solutions for λ (the eigenvalues) are the exponential terms, which would result from a given set of rate constants. The eigenvectors obtained by solving the equations after substitution of the eigenvalues give the ratios of the amplitudes for the three states.

A numerical example

As a numerical example the rate constants $k_{12}^*, k_{21}, k_{23}, k_{32}, k_{31}$ are set at $1000 \text{ s}^{-1}, 0.01 \text{ s}^{-1}, 100 \text{ s}^{-1}, 20 \text{ s}^{-1}$ and 50 s^{-1} respectively. The matrix

$$\begin{bmatrix} -(1000+\lambda) & 0.01 & 50 \\ 1000 & -(100.01+\lambda) & 20 \\ 0 & 100 & -(70+\lambda) \end{bmatrix}$$

has three eigenvalues (λ_i, $i = 0$ to 2) and corresponding eigenvectors as listed in the following table

Amplitudes	E	ES	EP
Eigenvalues			
$\lambda_0 = 0$	0.041	0.573	0.819
$\lambda_1 = -176.1$	0.042	-0.727	0.685
$\lambda_2 = -993.9$	-0.663	0.744	-0.081

As in the previous example the amplitudes have to be obtained from the eigenvectors for each λ. Because the values in each vector only give ratios, each row has to be multiplied by a factor which is deduced from initial and final conditions. The values in the row corresponding to λ_0 are normalized to the final condition that the sum is the total concentration of the three components ($c_E + c_{ES} + c_{EP} = 1$) and the final amplitudes for the three concentration profiles corresponding to the reactants are 0.028, 0.4 and 0.57 respectively. The sums of each of the other two rows are 0. Furthermore the sum of the pre-exponential terms have to be 1 for c_A and 0 for c_B and c_C. Some rows have to be multiplied by a negative factor to obtain the above conditions at constant ratios. From these operations we obtain the following equations:

$$c_A(t) = 0.029 + 1.021 \exp(-994t) - 0.05 \ \exp(-176t)$$
$$c_B(t) = 0.4 \ \ \ \ - 1.137 \exp(-994t) + 0.737 \exp(-176t)$$
$$c_C(t) = 0.57 \ \ \ + 0.123 \exp(-994t) - 0.694 \exp(-176t)$$

While the complexity of reactions with larger numbers of steps makes the derivation of amplitudes prohibitively difficult, much can be gained from obtaining numerical or symbolic expressions for eigenvalues. In the case of pseudo first order steps, such as the one characterized by k^* above, useful information can be obtained from the concentration dependence of time constants. This will be illustrated with examples of experiments in the literature. The time course of the approach to the steady state is illustrated in figure 5.1 for this numerical example.

Derivation of transient kinetic equations via second order differential equations

The analytical solutions for lag phases, or acceleration of product formation, give a more direct physical description of the reactions.

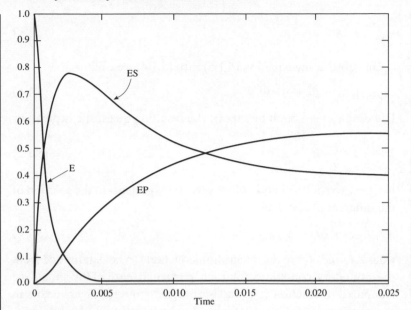

Figure 5.1   Representation of the time course of the concentrations of the three states of the enzyme during the approach to the steady state. The parameters (amplitudes and exponential coefficients) for the three curves are those calculated by the method described in the text.

This requires some familiarity with second order differential equations. If the rate of product formation is dc_P/dt then the acceleration of product formation is d^2c_P/dt^2. The general equation, which will be modified below for particular mechanisms, can be written as:

$$\alpha \frac{d^2c_P}{dt^2} + \beta \frac{dc_P}{dt} + \gamma c_P = 0 \tag{5.1.4}$$

where α, β and γ are constant coefficients (i.e. rate constants and constant concentrations). Mathematical problems are also often solved by experiment and Euler (1707–1783) made the intuitive assumption that the solution has the form

$$c_P = \exp(\lambda t) \tag{5.1.5a}$$

and hence

$$\frac{dc_P}{dt} = \lambda \exp(\lambda t) \tag{5.1.5b}$$

$$\frac{d^2 c_P}{dt^2} = \lambda^2 \exp(\lambda t) \tag{5.1.5c}$$

On substituting equation (5.1.5) into (5.1.4) we obtain

$$\exp(\lambda t)\{\alpha\lambda^2 + \beta\lambda + \gamma\} = 0 \tag{5.1.6}$$

Dividing this equation by $\exp(\lambda t)$ the resulting quadratic expression

$$\lambda^2 + \frac{\beta}{\alpha}\lambda + \frac{\gamma}{\alpha} = 0 \tag{5.1.7}$$

has the two roots λ_1 and λ_2 which give the exponents in the solution of the differential equation

$$A = A_1 \exp(\lambda_1 t) + A_2 \exp(\lambda_2 t) \tag{5.1.8}$$

where A_1 and A_2 are the integration constants to be determined from the boundary conditions for each particular case. This should be compared to reaction amplitudes obtained from the eigenvectors in the matrix method. When $[(\beta/\alpha)^2 - 4\gamma/\alpha] > 0$ the two real solutions λ_1 and λ_2 correspond to the eigenvalues obtained from the matrix method described above. When $[(\beta/\alpha)^2 - 4\gamma/a] < 0$ the imaginary solutions result in trigonometric terms and hence in oscillations. Maynard Smith (1968) presents interesting insight into the mode of oscillations of solutions of second order differential equations under a range of conditions.

The equation

$$\alpha \frac{d^2 c_P}{dt^2} + \beta \frac{d c_P}{dt} + c_P = 0 \tag{5.1.9}$$

which is used for the solution of two problems below, reduces to a first order differential equation by change of variable (see p. 152).

The application of Laplace operators to the solution of kinetic equations

Another method for solving differential equations, which is favoured by some investigators, deserves attention. This involves the use of Laplace transforms. Physiologists, who often think of rate processes in terms of electrical circuits (see section 2.1), tend to introduce the transform method into their derivations, since this is the frequent choice of electrical engineers.

The Laplace transform $L(f)$ of a function $f(t)$ is defined by the equation

$$L(f) = \int_0^\infty f(t)e^{-st}dt$$

where s is the Laplace operator. The elementary principles for the practical application to many problems can be explained quite easily in the form of a 'black box method'. However, before applying the method it is advisable to become conversant with the theoretical background by consulting a specialized text on the subject (Churchill, 1958; Boas, 1966), or to get help from an experienced user. There is a wide variation in the notation and in the tabulation of transforms. The examples given below are only intended as illustrations and not as instructions. The application of Laplace transforms to the solution of a wider range of kinetic equations can be found in Capellos & Bielski (1972), Roberts (1977) and Woledge, Curtin & Homsher (1985). All equations which can be solved by the Laplace transform method can also be solved by other techniques. However, the Laplace transform provides the integration constants, and with that reaction amplitudes, directly.

The essential practical feature of the method is the use of the operator 's' in place of d/dt (the derivative with respect to time). This transforms a differential equation into an algebraic expression, to be solved in the usual way. The solution of the algebraic equation has to be subjected to inverse transformation to obtain the integrated rate equation. The latter operation is simplified by the fact that tables of transforms for different expressions are available (see table 5.1) and the difficult mathematical problem of obtaining the transform is avoided. The Laplace method can also be used to solve second order differential equations by replacing the operator d^2/dt^2 by s^2 and solving the resulting algebraic equation. An attempt to do so will quickly show that a quadratic equation in s gives the same roots as those obtained as the characteristic equation (for instance (5.1.3)) for eigenvalues.

Two examples will illustrate the operations which have been performed when it is stated 'the equation has been solved by the Laplace transform method'. First we solve the rate equation for a case when the function equals zero at $t = 0$. We re-investigate the behav-

Table 5.1. *Some selected Laplace transforms*

Laplace form	Original form
$1/s$	1
$1/s^2$	t
$b/(s+a)$	$b\exp(-at)$
$s/(s+a)$	$a\exp(-at)$
$(s+b)/(s+a)$	$(b/a)-[(a-b)/a]\exp(-at)$
$1/[(s-b)(s-a)]$	$[1/(a-b)][\exp(-at)-\exp(-bt)]$
$s/[(s-a)(s-b)]$	$[1/(a-b)][a\exp(-at)-b\exp(-bt)]$
$b/[s(s+a)]$	$[b/a][1-\exp(-at)]$
$c/[s(s+a)(s+b)]$	$c[1/ab-\exp(-at)/(ab-a^2)-\exp(-bt)/(ab-b^2)]$
$1/(s^2+a^2)$	$(1/a)\sin(at)$
$s/(s^2+a^2)$	$\cos(at)$

iour of the approach to equilibrium of the ligand binding process (see section 3.2)

$$A+R\underset{k_{21}}{\overset{k_{12}}{\rightleftharpoons}}AR$$

under the condition $c_A(0)\gg c_R(0)$.

The conservation equation, $c_R(t)=c_R(0)-c_{AR}(t)$, is used to eliminate free binding sites from the equations. The rate of formation of AR is given by

$$dc_{AR}/dt=k_{12}c_A(0)\{c_R(0)-c_{AR}(t)\}-k_{21}c_{AR}(t)$$

On rearranging this equation and introducing the operator $s=d/dt$, we obtain

$$sc_{AR}=k_{12}c_A(0)c_R(0)-c_{AR}(t)\{k_{12}c_A(0)+k_{21}\}$$

and

$$c_{AR}(t)=k_{12}c_A(0)c_R(0)/[s+\{k_{12}c_A(0)+k_{21}\}]$$

From table 5.1 we see that this transforms to the original function

$$c_{AR}(t)=[k_{12}c_A(0)c_R(0)/\{k_{12}c_A(0)+k_{21}\}][1-\exp(-\{k_{12}c_A(0)+k_{21}\}t)]$$

If we wish to solve a function which is not zero at $t=0$ '$sc_x(0)$' has to be added to the algebraic equation for the initial concentration of the dependent variable x. Writing the equation for the time course of the equilibration of R in the above reaction,

$$dc_R/dt = -c_R(t)\{k_{12}c_A(0)+k_{21}\}+k_{21}c_R(0)$$
$$sc_R(t)-sc_R(0) = -c_R(t)\{k_{12}c_A(0)+k_{21}\}+k_{21}c_R(0)$$
$$c_R(t)\{s+k_{12}c_A(0)+k_{21}\} = c_R(0)\{k_{21}+s\}$$
$$c_R(t)/c_R(0) = \{k_{21}+s\}/\{k_{12}c_A(0)+k_{21}+s\}$$

From the fifth line in table 5.1 we obtain the transformation to

$$c_R(t)/c_R(0) = \{1/[k_{12}c_A(0)+k_{21}]\}\{k_{21}+k_{12}c_A(0)\exp(-[k_{12}c_A(0)+k_{21}]t)\}$$

Note the introduction of $sc_R(0)$ for the initial condition $c_R(0)\neq0$ at $t=0$. To check whether the correct solution is obtained one can substitute for $t=0$, when $c_R(t)/c_R(0)=1$ and for $t=\infty$, when $c_R(t)/c_R(0)=k_{21}/k_{12}c_A$.

When the complexity of the mechanism is increased to a two step reaction, then the solutions of the Laplace forms of the equations involve the two roots of a quadratic as indicated above for second order differential equations. Recently Zhang, Strand & White (1989) have suggested how a general matrix solution of rate equations in the Laplace form can be used to model kinetic mechanisms. Zhang *et al.* (1989) suggest this method as an alternative to numerical integration, but its use is, of course, restricted to linear equations like that of the more elegant matrix method described in section 4.2.

The kinetics of transients in product formation

Chance (1943) and Theorell & Chance (1951) observed the formation and decomposition of complexes of enzymes with substrates and products by following changes in light absorption. In these pioneering studies the theories and techniques of pre-steady-state kinetics were only applied to reactions in which the complexes had distinct absorption spectra. It became apparent that transients of a much wider range of enzyme reactions could be studied when the initial rate of product formation is analysed (Gutfreund, 1955). Observations with a time resolution of milliseconds showed that there are often three distinct phases in product formation. These are determined in turn by the rate of formation of the enzyme–substrate complex, the enzyme–product complex and of free product. Of course, as we shall see, the most fruitful investigations into enzyme mechanisms resulted when it was possible to combine the observation of transients of product formation with those of spectral changes of complexes.

The solution of the differential equations given below was developed initially by Gutfreund (1955) and Gutfreund & Sturtevant (1956) for the study of the pre-steady-state kinetics of proteolytic enzymes like chymotrypsin, trypsin and ficin. It was possible to determine the position of the rate limiting step of the sequence of interconversions of enzyme complexes. When enzyme and substrates are mixed in a stopped flow apparatus the following conditions prevail: at $t=0$ the concentrations of enzyme and substrate are $c_E(0)$ and $c_S(0)$, the concentrations of products and of all intermediates are zero and $dc_p/dt=0$. The rate of product formation will increase as the concentrations of the intermediate enzyme complexes rise to their steady state concentration. The principle of this approach can be illustrated with what we have called the minimum mechanism, with only the first step reversible at negligible product concentration:

$$E+S \underset{k_{21}}{\overset{k_{12}}{\rightleftharpoons}} ES \xrightarrow{k_{23}} EP \xrightarrow{k_{34}} E+P$$

In the steady state $dc_p/dt = k_{23}c_{ES} = k_{34}c_{EP}$. During the approach to the steady state, at constant substrate concentration $[c_E(0) \ll c_S(t) = c_S(0)]$

$$dc_{ES}/dt = k_{12}c_S(0)\{c_E(0) - c_{ES}(t)\} - \{k_{21}+k_{23}\}c_{ES}(t) \qquad (5.1.10)$$

The acceleration in product formation can be derived from equation (5.1.10) by substituting as follows for dc_{ES}/dt and $c_{ES}(t)$:

$$dc_{ES}/dt = \frac{d^2c_p/dt^2}{k_{23}} \quad \text{and} \quad c_{ES}(t) = \frac{dc_p/dt}{k_{23}}$$

and thus we obtain

$$d^2c_p/dt^2 + dc_p/dt\{k_{12}c_S(0) + k_{21} + k_{23}\} = k_{12}k_{23}c_S(0)c_E(0) \qquad (5.1.11)$$

This second order differential equation with constant coefficients is in the form already referred to above (5.1.9.)

$$\frac{d^2c_P}{dt^2} + \alpha\frac{dc_P}{dt} = \beta \quad \text{(with constant coefficients } \alpha \text{ and } \beta\text{)}$$

This equation can be integrated to give

$$c_P = \frac{\beta}{\alpha}t + A_1\exp(-\alpha t) + A_2 \qquad (5.1.12)$$

The integration constants A_1 and A_2 can be evaluated from the boundary condition $c_P(0)=0$ and $dc_P/dt=0$ at $t=0$

$$\frac{\beta}{\alpha}-\alpha A_1=0 \quad \text{and} \quad A_2=c_P(0)-A_1=\frac{\beta}{\alpha^2}$$

and therefore $c_P(t)=\dfrac{\beta}{\alpha}\,t+\dfrac{\beta}{\alpha^2}\left[\exp(-\alpha t)-1\right]$ \hfill (5.1.13)

Substitution for α and β from equation (5.1.11) into (5.1.13) results in

$$c_P(t)=\frac{k_{12}k_{23}c_S(0)c_E(0)}{k_{12}c_S(0)+k_{21}+k_{23}}\,t+$$

$$\left\{\frac{k_{12}k_{23}c_S(0)c_E(0)}{[k_{12}c_S(0)+k_{21}+k_{23}]^2}\right\}\{\exp(-[k_{12}c_S(0)+k_{21}+k_{23}]t)-1\} \quad (5.1.14)$$

The physical meaning of the three terms of this equation is illustrated in figure 5.2. If k_{23} characterizes the rate limiting step, then equation (5.1.14) can be written (see section 3.3)

$$c_P(t)=\frac{k_{12}c_E(0)c_S(0)}{K_M+c_S(0)}\,t$$

$$+\left\{\frac{k_{12}k_{23}c_S(0)c_E(0)}{[k_{12}c_S(0)+k_{21}+k_{23}]^2}\right\}\{\exp(-[k_{12}c_S(0)+k_{21}+k_{23}]t)-1\} \quad (5.1.15)$$

The time constant, τ, of the exponential characterizing the acceleration contains a substrate concentration dependent term

$$\tau^{-1}=k_{12}c_S(0)+k_{21}+k_{23} \hfill (5.1.16)$$

(this term is illustrated in figure 5.2).

In the case of this simple model $K_M=(k_{21}+k_{23})/k_{12}$ and the intercept on the time axis of figure 5.2 is $[k_{12}(c_S(0)+K_M)]^{-1}$. In principle it is thus possible to evaluate k_{12} from the observation of the initial acceleration of product formation, the so-called lag phase. If experiments are carried out over a range of substrate concentrations and the exponent of the lag phase

$$k_{12}c_S(0)+k_{21}+k_{23}$$

is plotted against $c_S(0)$, the slope is k_{12} and the intercept at the ordinate is $k_{21}+k_{23}$.

With the techniques available at present it would be difficult to evaluate, even approximately, a transient time constant of less than 3 ms. The combined limitations due to the magnitude of K_M, the need for the fulfilling of the condition $c_S(0)\gg c_E(0)$ and the signal to noise ratio have resulted in obtaining only lower limits for the values of k_{12} from most of such

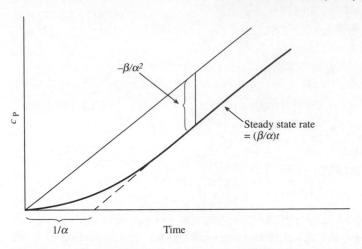

Figure 5.2   Simulation of the time course of the transient approach to the steady state rate of product formation under the condition when ES → EP is rate limiting. The terms in equations (5.1.13) and (5.1.14) define α and β.

experiments (Gutfreund, 1955). The small temperature dependence of this diffusion controlled step (see section 7.4) makes it difficult to resolve it from subsequent events. At the time when the method was developed, the opportunity to set a lower limit of $k_{12} > 10^7\ \mathrm{M}^{-1}\mathrm{s}^{-1}$, for the chymotrypsin catalysed hydrolysis of amino acid esters, was of importance for current thinking on enzyme mechanisms. Much more information about the rates of initial steps in substrate binding is available now (see chapter 7.4).

If the initial Michaelis assumption had been a universal truth and the rate limiting step of every enzyme reaction followed immediately after initial complex formation, this step would have been the only one of significance for transient kinetic studies. Under those conditions free enzyme and the initial complex ES would be the only species to occur at any significant concentrations during transients or in the steady state. However, one of the major achievements of transient kinetics was the demonstration that several complexes of enzyme with substrate or product accumulate and can be characterized during different stages of the reaction. In the examples discussed below it will be demonstrated that rate limit can be, and often is, due to any one of many steps in the reaction sequence. Fortunately for the practitioners of transient kinetics a late step, a conformation change controlling product release, is frequently the slowest step. This permits the observation of the maximum number of intermediates (see for instance the reaction of lactate dehydrogenase described below). Frequently two or more steps have similar rates and observed rate constants have to be distinguished from true rate constants (see the discussion in section 3.2). Changes in conditions can alter the relative magnitudes of the rate constants and thus separate them.

When the transient acceleration of the formation of free product is observed at high substrate concentrations $[c_S(0) \gg K_M]$ and the lag phase is independent of $c_S(0)$, one can consider the following cycle

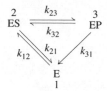

k_{13} is ignored since $c_P = 0$. This is the same model as that treated by the matrix method on p. 145 above.

On mixing enzyme and substrate the formation of ES is essentially instantaneous and the acceleration of the rate of formation of free product is determined by the rate of approach to the steady state concentration of EP. This will be determined by an observed rate constant $k_{obs} = k_{23} + k_{32} + k_{31}$. As more transitions between intermediate states are considered, the equation for k_{obs} becomes more complex. Theoretically it should be composed of several exponential terms. These complexities are best discussed side by side with practical examples. The successful use of such 'lag phase' experiments for the determination of rate constants for the interconversion of intermediates is described below, pp. 161 and 164.

The identification of reaction intermediates

Some of the most informative experiments on transient product formation have turned out to be those in which it was possible to record the transient formation of both enzyme bound and free products. The record shown in figure 5.3 illustrates a hypothetical rate profile of product bound to an enzyme complex as well as the steady state appearance of free product. The first observation of rate limiting product release was made by Theorell & Chance (1951) using Chance's rapid spectrophotometric method for the study of the reaction catalysed by horse liver alcohol dehydrogenase

$$\text{ethanol} + \text{NAD}^+ \rightleftharpoons \text{acetaldehyde} + \text{NADH} + \text{H}^+$$

Their finding that the rate of dissociation of NADH was rate limiting for the steady state of the oxidation of alcohol was extended considerably 20

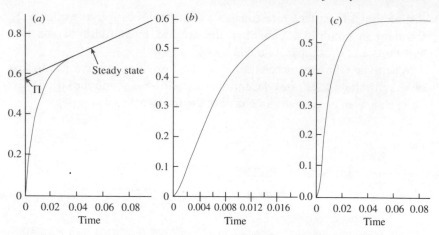

Figure 5.3   Simulation of the time course of product formation (enzyme bound and free) under the condition when the dissociation of product from the enzyme complex is rate limiting. Diagrams (*a*) and (*b*) show the approach to the steady state on different time scales and (*c*) shows the formation of the enzyme product complex. The intercept Π in diagram (*a*) can be used to calculate the concentration of active sites according to equation (5.1.23).

years later (Shore & Gutfreund, 1970), when both the interconversion of the ternary complexes

$$E^{NAD^+}_{ethanol} \rightleftharpoons E^{NADH}_{aldehyde} + H^+$$

and product dissociation was recorded. The following specific examples of enzyme reactions will serve to illustrate a range of experimental conditions and methods of analysis, which have provided information about kinetic mechanisms:

(1) In the study of NAD^+ linked dehydrogenases spectral and fluorescence changes of both protein and nuceotides made it possible to record the time course of many steps in the reactions. The variety of signals recorded during the transient approach to the steady state of the reaction catalysed by lactate dehydrogenase is illustrated in figure 5.4.

(2) In the reactions of hydrolytic enzymes, like chymotrypsin or alkaline phosphatase, rapid sampling and optical observations of the fate of chromophoric substrates, gave complementary information. The recording of the time course of the formation of an enzyme–product complex, as well as of free product helped to map out the reaction mechanism.

(3) Myosin ATPase, which was studied extensively in relation to steps in muscle contraction, had all possible monitors applied to the elucidation

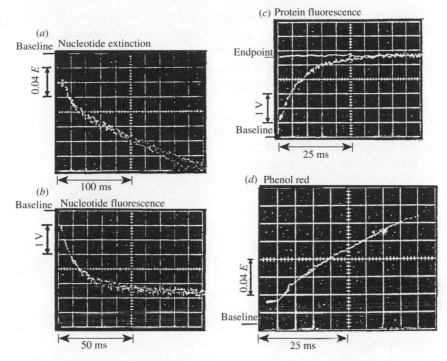

Figure 5.4   The reaction of pig heart lactate dehydrogenase with lactate and NAD$^+$. The four panels show (a) two phases of NADH formation (enzyme bound and free), (b) NADH fluorescence, which is predominated by the enzyme–NADH complex after pyruvate dissociation, (c) protein fluorescence quenching monitoring concentration of all enzyme–NADH complexes and (d) phenol red absorbence monitoring two phases of proton liberation. (For detail see text and Whitaker *et al.*, 1974.)

of the intermediates of the reaction. The relation between experimental conditions of different experiments and the resulting information obtained on the mechanism of this reaction will be used as a paradigm.

(4) Creatine phosphokinase will be described as an example where experiments at low temperature made it possible to separate transients of the reaction, which could not be distinguished at 25 °C.

(5) In the study of haem proteins and flavoproteins, which provide a series of distinct spectral signals during the reaction pathway, the detection and characterization of the intermediates is more of a problem in spectroscopy than in kinetics.

A derivation which has found wide application in the interpretation of hydrolytic reactions and of dehydrogenases, is the so-called 'product burst analysis'. Let us consider the condition of high substrate concentration, so that the reaction $E + S \rightarrow ES$ is sufficiently rapid to be considered instantaneous on the time scale of the experiment and goes to completion prior to any subsequent reaction. If we then observe, either spectrophotometrically or by rapid sampling, the sum of all states of a product, enzyme bound and free, we can proceed from the scheme

$$E + S \xrightarrow{k_{12}} ES \xrightarrow{k_{23}} EP \xrightarrow{k_{31}} E + P$$

where all steps are considered essentially irreversible.

Since the total concentration is

$$c_E(0) = c_{ES}(t) + c_{EP}(t)$$
$$dc_{EP}/dt = k_{23}c_{ES}(t) - k_{31}c_{EP}(t) = k_{23}c_E(0) - \{k_{23} + k_{31}\}c_{EP}(t) \tag{5.1.17}$$

It follows that

$$\frac{dc_{EP}}{\left[\dfrac{k_{23}c_E(0)}{k_{23} + k_{31}}\right] - c_{EP}dt} = (k_{23} + k_{31})dt \tag{5.1.18}$$

Under the boundary conditions that at $t = 0$, $c_{EP}(0) = 0$, integration of (5.1.18) gives

$$c_{EP}(t) = \frac{k_{23}c_E(0)}{k_{23} + k_{31}} \left[1 - \exp[-(k_{23} + k_{31})t]\right] \tag{5.1.19}$$

and the total product formation $c_P(t) + c_{EP}(t) = c_P(\text{total})$ is given by

$$\frac{dc_P(\text{total})}{dt} = c_E(0) \frac{k_{23}}{k_{23} + k_{31}} \left[k_{31} + k_{23}\exp[-(k_{23} + k_{31})t]\right] \tag{5.1.20}$$

which is integrated to give

$$c_P(\text{total}) = c_E(0) \left[\frac{k_{23}k_{32}}{k_{23} + k_{31}} t - \left\{\frac{k_{23}}{k_{23} + k_{31}}\right\}^2 \exp[-(k_{23} + k_{31})t]\right] + \text{constant} \tag{5.1.21}$$

When $c_S(0)$ is large enough the intercept of the steady state with the ordinate at $t = 0$ is

$$\Pi = \{k_{23}/(k_{23} + k_{31})\}^2 \tag{5.1.22}$$

This condition is useful for the titration of active site concentrations. If $c_S(0)$ is not large enough for the condition $c_E(0) = c_{ES}(t) + c_{EP}(t)$ to hold, then the intercept is

$$\Pi = c_E(0) \left[\frac{k_{23}c_S/(k_{23}+k_{31})}{c_S + (k_{23}+k_{21})/k_{12}} \right]^2 \tag{5.1.23}$$

The constant of integration at $t = 0$ is obtained from $c_P(0) = 0$:

$$\text{constant} = [k_{23}/(k_{23}+k_{31})]^2$$

and then we obtain

$$\frac{c_P(t)}{c_E(0)} = \left[\frac{k_{23}k_{31}}{k_{23}+k_{31}} t + \left\{ \frac{k_{23}}{k_{23}+k_{31}} \right\}^2 \{1 - \exp[-(k_{23}+k_{31})t]\} \right] \tag{5.1.24}$$

Simple extension of these equations by introduction of k_{32} is required if the interconversion of enzyme–substrate and enzyme–product complexes is significantly reversible. This arises in some of the examples discussed below.

'Product burst kinetics' was originally developed for the study of hydrolytic enzymes using substrates with chromophoric leaving groups (Gutfreund, 1955). Figure 5.3 shows a simulation of the type of result obtained in such experiments. The hydrolysis of nitrophenyl esters by chymotrypsin (Gutfreund & Sturtevant, 1956) and of nitrophenyl phosphate by alkaline phosphatase (Trentham & Gutfreund, 1968) results, under certain conditions, in a rapid release (burst) of nitrophenol, during acylation or phosphorylation of the enzymes. This is followed by a steady state due to the rate limiting de-acylation or de-phosphorylation. This basic principle was subsequently widely used for the investigation of the kinetic mechanism of many enzymes.

The extension of such product burst experiments to other groups of enzymes can be illustrated with the results of the study of Shore & Gutfreund (1970) on liver alcohol dehydrogenase (LADH) already referred to above. When a solution of LADH was mixed in a stopped flow apparatus, with one containing high concentrations of NAD^+ and ethanol (substrate saturation conditions) the pre-steady-state appearance of (nearly) one mole of NADH per mole of enzyme sites occurred at a rate of $130\ s^{-1}$ at 20 °C. The steady state rate of $4\ s^{-1}$ corresponds to the rate of product (NADH) dissociation. When deutero-ethanol was used as a substrate, the transient rate of formation of enzyme bound NADH was reduced to $23\ s^{-1}$. This large deuterium isotope effect indicates that the transient describes the chemical step of hydride transfer. With this enzyme, as with lactate dehydrogenase and myosin ATPase (see below), the multiplicity of signals for monitoring reaction steps helped to elucidate the mechanism. When the rate of appearance of protons was monitored, in the presence of pH indicators, it was found that this slightly preceded the formation of the enzyme–NADH complex and this rate was identical for ethanol and deutero-ethanol as the substrate. These results, together with those reported earlier in this section, allowed the characterization of the following steps:

$$
\text{EH}^+ \underset{\substack{\pm\text{EtOH}}}{\overset{\substack{\pm\text{NAD}^+}}{\rightleftharpoons}} \text{E} \overset{\text{NAD}^+}{\underset{\text{EtOH}}{\rightleftharpoons}} \text{EH}^+ \underset{\pm\text{Ald}}{\overset{\text{NADH}}{\rightleftharpoons}} \text{EH}^+ \overset{\text{NADH}}{}
$$

As indicated above, this kind of experiment is also often used for the determination of the concentration of enzyme active sites by application of equation (5.1.22). While this has been useful in many cases, it has also led to one major confusion in enzymology in the 1970s. This was the so-called 'half of sites' reactivity hypothesis. It was based on the fact that with some enzymes reactions with certain substrates resulted in burst amplitudes Π corresponding to approximately half the number of enzyme sites in the solution. It was postulated that in many oligomeric enzymes, in any pair of sites, the states of the two partners alternated, one active and one inactive form. In many cases this has turned out to be due to trivial artefacts, which were summarized by Gutfreund (1975). In some special cases, notably in membrane enzymes, it may be correct. In a group of interesting cases the low burst yield is due to the phenomenon of 'on enzyme equilibration':

$$
\underset{\text{I}}{\text{E} + \text{S}} \rightleftharpoons \underset{\text{II}}{\text{ES}} \rightleftharpoons \underset{\text{III}}{\text{EP}} \rightleftharpoons \text{E} + \text{P}
$$

In several NAD^+ linked dehydrogenases, ATPases and phosphokinases the equilibrium constant for step II, the interconversion of the enzyme–substrate to the enzyme–product complex is near unity (see Gutfreund & Trentham, 1975; Edsall & Gutfreund, 1983). The thermodynamic rationale for this phenomenon is also given in section 3.3 and it will be documented further when the reaction pathways of myosin ATPase and of lactate dehydrogenase are described below.

There are some examples of reactions in which both a lag phase and a burst phase have been observed. Among them are the myosin ATPase reaction (discussed later in this section) and the interconversion of creatine (Cr) and creatine phosphate (Cr-P), catalysed by creatine kinase:

$$
\text{creatine} + \text{ATP} \rightleftharpoons \text{creatine phosphate} + \text{ADP}
$$

Travers, Barman & Bertrand (1979) extended the experiments of Engelborghs, Marsh & Gutfreund (1975) by working over a wide temperature range, 4 °C to 35 °C. There is a fourfold increase in the temperature of activation of the overall reaction below 25 °C. Using a development of the rapid quench flow sampling technique of Barman & Gutfreund (1964), Travers *et al.* (1979) showed that at 4 °C a record of the reaction over 40 ms could be divided into three phases: a lag phase, a burst phase and a steady state phase (figure 5.3 illustrates this type of record). Another good example of such a complete three phase time course is seen with myosin ATPase, which is discussed below. For creatine kinase the proposed mechanism includes a complex X, which does not contain creatine phosphate:

$$
E \underset{}{\overset{+ATP}{\rightleftharpoons}} E \underset{\pm Cr}{\overset{ATP}{\rightleftharpoons}} E \underset{Cr}{\overset{ATP}{\rightleftharpoons}} EX \rightleftharpoons E \underset{Cr-P}{\overset{ADP}{\rightleftharpoons}} \overset{+ADP}{\underset{\pm Cr-P}{\longrightarrow}} E
$$

The ternary complex is in rapid equilibrium with its free components. Since product concentrations remain negligible throughout the experiment, the last step is assumed to be irreversible. The integrated rate equation can be derived by the method given on p. 145

$$c_{CrP}(t)/C_E(0) = A_1 + k_{cat} + A_2\exp(\lambda_1 t) - A_3\exp(\lambda_2 t) \tag{5.1.25}$$

where c_{CrP} is the sum of the concentrations of bound and free creatine phosphate. The integration constants A_1, A_2 and A_3 give the amplitudes of the steady state intercept and of the two exponentials with exponents λ_1 and λ_2. The original paper of Travers *et al.* (1979) should be consulted for details of the derivation and of the remarkable achievement of extracting the five rate constants from the data.

Intermediates in the reaction of myosin ATPase

Without any doubt, one of the most extensively studied enzyme reaction by transient kinetics is the ATPase activity of the subfragment-1 (S1) of myosin. This proteolytic product constitutes the heads of the myosin molecule, which has, in addition, a hinge and a tail component (see figure 4.11). S1 contains the fully competent enzymatic site and also the site for attachment to actin, the thin filament of the myofibril. The interaction between S1 and actin, discussed in section 6.4, is the key process in muscle contraction, and is controlled by the steps in ATP hydrolysis (Geeves *et al.*, 1984; Geeves, 1991). The reciprocal relation between the modulation of the affinity of myosin heads for attachment to actin by the progress of ATP hydrolysis and the resulting conformation changes, are the seminal problems of the molecular basis of muscle contraction. This interest, and the relative ease with which large quantities of muscle proteins can be prepared, made the ATPase system a paradigm for the development and application of just about every modern kinetic technique. There are many reviews which describe the reaction itself, its relation to the interaction of myosin with actin in solution and the coupling to mechanical events in muscle fibres (Hibberd & Trentham, 1986). We shall concern ourselves here with the application of a range of methods for the transient kinetic analysis of the elementary steps of the reaction of S1 with ATP in solution. The presence of actin modifies the kinetics by accelerating product dissociation, but it is subject to the same methods of investigation. The procedures developed have had a considerable influence on studies of ATPases and GTPases involved in energy and signal transduction as well as cation transport in many physiological processes (see Eccleston *et al.*, 1992).

It is probably best to present first the sequence of major events as we now see them and then to describe how the information about each of the steps has been deduced from different methods used. New monitors and better resolution from old ones, will no doubt provide even greater detail of this reaction in the future. Firm evidence has been provided for the following minimum mechanism:

$$
\begin{array}{cccc}
\text{M + ATP} \xrightleftharpoons{K_1} \text{M·ATP} & \underset{k_{32}}{\overset{k_{23}}{\rightleftharpoons}} & \overset{*}{\text{M·ATP}} & \underset{k_{43}}{\overset{k_{34}}{\rightleftharpoons}} \text{M·ADP·P}_i \\
& & & \Big\updownarrow\, {k_{54}}\ {k_{45}} \\
\text{M + ADP} \xrightleftharpoons{K_7} \text{M·ADP} & \underset{k_{76}}{\overset{k_{67}}{\rightleftharpoons}} & \overset{*}{\text{M·ADP}} & \underset{k_{65}}{\overset{k_{56}}{\rightleftharpoons}} \overset{*}{\text{M·ADP· P}}_i
\end{array}
$$

Three types of reactions were identified. These are: rapid equilibrium binding (1,5 and 7), chemical catalysis (3) and isomerizations (2,4 and 6). There is evidence that steps 4 and 5 are preceded by an isomerization of the complex. M designates myosin and the stars indicate increase in fluorescence of the protein.

Transient kinetic studies on the hydrolysis of ATP by S1 were initiated by Taylor (Finlayson & Taylor, 1969; Lymn & Taylor, 1970) who established the temporal correlation between ATP binding and hydrolysis and the cycle of dissociation–association of the acto–myosin complex. Bagshaw & Trentham (1973) extended these studies with particular attention to the detail of the hydrolysis mechanism of S1 alone. The rapid sampling technique of Barman & Gutfreund (1964) used by both Lymn & Taylor (1970) and Bagshaw & Trentham (1973) demonstrated that a phosphate burst occurred prior to steady state product release. Bagshaw & Trentham also carried out single turnover experiments, mixing radioactive (^{32}P) ATP with excess S1. Such experiments were also carried out by Tesi *et al.* (1991) at low temperature to characterize the first three steps of the reaction (see figure 5.5). Since the ATP concentration can be more reliably determined than the active site concentration, this allows accurate interpretation of the stoichiometry of the reaction intermediates. Step 4 is about two orders of magnitude slower than any of the preceding steps and the rates and equilibrium constant of the 'on enzyme equilibrium' (see p. 91) between the complexes of myosin with ATP and its hydrolysis products can be deduced. The equilibrium constant for step 3 is approximately 10 at 25 °C, falling to near 1 at 4 °C. As discussed above, intermediate equilibrium constants can only be determined by kinetic analysis, but values appreciably different from unity are difficult to determine accurately. The slow nucleotide (ATP off) dissociation, step 2 (Mannherz, Schenck & Goody, 1974), is similar to the events with lactate dehydrogenase and NADH, and results in very tight binding of ATP ($K_1 \gg 10^{11}$ M).

The observed rate constant ($k_{obs} = k_{34} + k_{43} + k_{45}$) for step 3, determined from such rapid sampling experiments, is approximately 160 s^{-1} at 20 °C, and can also be obtained from observation of the increase in fluorescence during during the transition from $\overset{*}{M}$.ATP to $\overset{*}{M}$.ADP.P$_i$. The constants for step 2 have been determined by the cold chase technique. This technique adds considerable power to the rapid quench sampling technique in the elucidation of the kinetic constants for the interconversion of enzyme intermediates. When rapidly mixed enzyme–radioactive substrate solutions are, after various time intervals, in turn mixed with cold substrate, the analysis for radioactive products leads directly to the kinetic constants describing the formation of a productive enzyme–substrate complex (step 2). The success of the method depends on the enzyme turnover (k_{cat}) being at least as great as the rate of substrate dissociation (k_{32}). This technique was first used by Bagshaw & Trentham (1973) (Mannherz *et al.*, 1974, determined $k_{32} < 10^{-6}$ s^{-1}) for the study of the ATPase reaction and this system serves as a good example for the principles involved. Barman & Travers (1985) have developed the method further and this is illustrated with an experimental record in figure 5.6. This is a way of titrating the active site concentration. In this example S1 and ATP(^{32}P) were mixed in the flow device and, after suitable short time intervals, this solution goes through a second mixer when 500fold to 1000fold excess cold ATP is added. This

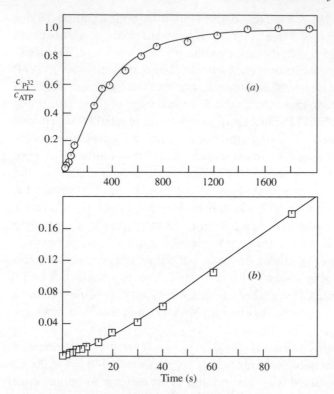

Figure 5.5   Time course of the burst of P_i formation during a single turnover of the myosin(S1) ATPase reaction in 40% ethylene glycol at $-15\,°C$. The complete reaction shown in (a) and the initial acceleration is shown in (b). The time constants for the acceleration and the burst reaction are $\tau_1 = 34.5$ s and $\tau_2 = 333$ s respectively (from Tesi *et al.*, 1991).

reaction mixture is incubated for some minutes, for all the radioactive ATP to be converted to product. The amount of radioactive phosphate found in samples quenched with cold ATP at different time intervals is equal to the amount of $ATP(^{32}P)$ bound as *ATP during that time. The observed rate constant for ^{32}P formation is

$$k_{obs} = k_{23}K_1 c_{ATP}(t)/\{1 + K_1 c_{ATP}(t)\} \tag{5.1.26}$$
$$\text{labelled}$$

This is the familiar situation (see p. 66) of a step preceded by a rapid pre-equilibrium. If such experiments are carried out over a range of $ATP(^{32}P)$ concentrations, k_{obs} can be extrapolated to the value for k_{23} at infinite concentration and K_1 is given by the slope.

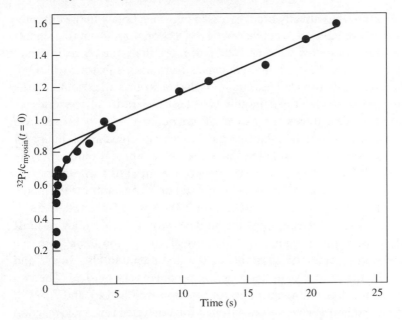

Figure 5.6   Illustration of a cold chase experiment: samples of a reaction mixture of myosin S1 (0.8 µM) and [γ^{32}P]ATP in 5 mM KCl and 40% ethylene glycol at pH 8 and 15 °C were quenched in 5 mM unlabelled ATP at the time points in the graph. The samples were then incubated for 2 min, the reaction quenched with 5.4% trichloroacetic acid and assayed for ^{32}P$_i$. The first order approach to the steady state ($k - 5.2$ s^{-1}) corresponds to the rate of formation of $\overset{*}{M}$ATP in the scheme shown on p. 162 (from Barman *et al.*, 1983).

Coupling and uncoupling of reactions

The derivations presented on p. 145 show that the interpretation of records of reactions involving more than two reversible steps requires the use of excessively cumbersome equations with three or more roots. One can often adjust the conditions of an experiment so that two of the steps of a complex reaction sequence are isolated from each other and/or the rest. Such uncoupling of part of a reaction occurs when all prior steps are much faster and the subsequent step is much slower. A numerical example can be used to demonstrate the reaction profile of different intermediates. The time course of a three state reversible reaction (see figure 4.4) is equivalent to three states of an enzyme, and this is illustrated with different effects of the two time constants on the three progress curves. The elucidation of the mechanism and evaluation of the individual rate constants depends on a number of factors. The problem of assigning the two time constants to the two transitions has already been discussed in connection with the interpretation of irreversible consecutive reactions (section 4.2). If suitable signals can be found to follow the time course of more than one of the intermediates, then experiments at different concentrations of substrate will often permit such evaluations. This is amply demonstrated with some of the enzyme reactions described in this chapter. The ratio of the rates of equilibration determines the extent of uncoupling of two consecutive reversible steps. From the values assigned to the rate constants in figure 4.4 the two steps would, in isolation, have the two reciprocal time constants 2.1 s^{-1} and 4 s^{-1} (the sums of their respective forward and reverse rate constants; p. 56). Evaluation of the reciprocal time constants for the model by one of the above methods (see section 2.3) gives 1.95 s^{-1} and 4.15 s^{-1}; which can be considered almost uncoupled, within the accuracy of most experiments. If one assigns rate constants of 10 s^{-1} and 0.5 s^{-1} for the forward and reverse rate constants of the first step and also 10 s^{-1} and 0.5 s^{-1} for the second step, the coupling becomes very marked. The two calculated reciprocal time constants for this case are 12.74 s^{-1} and 8.26 s^{-1}. Using the method of eigenvalues outlined for a two step reversible reaction in section 4.2, other examples can be easily evaluated.

Examples of other enzyme mechanisms

Some more examples are given below of the range of signals obtained from either cofactors (nucleotide substrates or prosthetic groups), or from protein fluorescence changes during enzyme reactions. NAD^+ linked dehydrogenases have led to the initiation and extension of several aspects of transient kinetics. As already mentioned liver alcohol dehydrogenase was the subject of several pioneering experiments in this field. The study of the transient kinetic behaviour of lactate dehydrogenase (LDH) is of special interest and serves as a paradigm for the elucidation of three phenomena: (1) it was the first enzyme for which the 'on enzyme equilibrium' between the enzyme–substrate and the enzyme–product complex was demonstrated; (2) the method for kinetic equivalence of the four sites of this oligomeric protein was developed and (3) the multiplicity of monitors permitted the direct identification of all the important intermediates. These three topics will be briefly discussed with reference to the records shown in figure 5.4.

The reaction catalysed by lactate dehydrogenase

$$\text{lactate} + NAD^+ \rightleftharpoons \text{pyruvate} + NADH + H^+$$

has been followed under transient kinetic conditions using absorbance of NADH at 340 nm, the fluorescence of NADH, the quenching of protein fluorescence by energy transfer to bound NADH and the uptake and release of protons by absorbance changes of an indicator (Whitaker *et al.*, 1974). Since there are some differences in the kinetic behaviour of LDH from different mammalian tissue it must be noted that the scheme outlined here is for the pig heart isoenzyme. This serves as an example to demonstrate the power of multiplicity of signals for the analysis of mechanisms. An interesting point about this particular study is that the mechanism can be elucidated from the qualitative information from the sequential signals, without having recourse to the numerical values of rate constants. This is an example of transient kinetics proving the mechanism in terms of intermediates.

When the wavelength for absorption measurements is chosen so as to monitor the rate of appearance of all forms of NADH, enzyme bound and free, then the record (see figure 5.4) shows three phases. The essentially instant formation of NADH, corresponding to approximately 10% of active site concentration at pH 8 and 30% at pH 6, is followed by a further, slower, transient and finally the steady state appearance of NADH. A record of the appearance of free protons follows only the second transient and the steady state. From this information a reaction sequence could be proposed, starting with the rapid formation of the ternary complex.

$$
E \underset{\pm Lact}{\overset{\pm NAD^+}{\rightleftharpoons}} E \overset{NAD^+}{\underset{Lact}{\rightleftharpoons}} F \overset{NADH}{\underset{Pyr}{\rightleftharpoons}} E \overset{NADH}{\rightleftharpoons} E
$$

$$
\text{I} \qquad \text{II} \qquad \text{III}
$$

This proposal could be tested by following the nucleotide fluorescence of NADH. The relative fluorescence of free NADH and its binary complex with enzyme and ternary complex with enzyme and pyruvate is 1:4:0.2. Since the large fluorescence increase must be due to the formation of the enzyme–NADH complex III it is evident that the second, slower, transient is due to the dissociation of pyruvate from the product complex. Furthermore the record of proton liberation, figure 5.4, shows that this also occurs during the slower transient. The instantaneous transient corresponds to the 'on enzyme' equilibration. The small amplitude change of this transient with a 100fold change in proton concentration (pH 6 to 8), as well as the subsequent liberation of protons, demonstrates that the equilibration between enzyme–substrate and enzyme–product complex is not directly linked to free protons in solution. Following the time course of protein fluorescence, which is quenched by bound NADH (see section 8.2), provides a record of the sum of the concentrations of intermediates II and III.

The study of this enzyme with pyruvate and NADH as substrate could be carried out under single turnover conditions, the reduced nucleotide binds very tightly ($K_D = 1$ μM). The reaction

$$E^{NADH} + \text{pyruvate} + H^+ \longrightarrow E + \text{lactate} + NAD^+$$

was carried out by mixing E^{NADH} [condition: $c_E(0) > c_{NADH}$] with a large excess of pyruvate. A single transient was observed for both NADH oxidation and proton uptake. The rate constant obtained from this exponential was not affected when deuterated NADD was used as a substrate and it was identical to k_{cat} for the steady state reaction in this direction. Single turnover experiments (binding sites in excess of substrate) were of importance in the study of many kinetic mechanisms. In enzymes or other biological systems one does not always know the concentration of sites as accurately as that of ligand.

The conclusion from the above experiments was that the rate limiting step of the reaction is a conformation change of the ternary complex prior to the chemical reaction of hydride transfer. Additional experiments (see Holbrook & Gutfreund, 1973 and Clarke *et al.*, 1985) have shown that the protonation of a histidine, linked to pyruvate binding, is coupled to this conformation change. Whitaker *et al.* (1974) also obtained some evidence that a conformation change occurred during lactate oxidation between pyruvate dissociation and NADH dissociation. This is likely to be the reverse of the rate limiting step of pyruvate reduction.

Single turnover experiments can also be used to test whether the sites of oligomeric enzymes are kinetically identical. For instance, lactate dehydrogenase has four active sites per molecule. The following parallel experiments were carried out

(1) Enzyme 5 μM in sites, NADH 4 μM, pyruvate 100 μM
(2) Enzyme 20 μM in sites, NADH 4 μM, pyruvate 100 μM

In the first case three to four sites of each tetramer were occupied by NADH; in the second case one site per tetramer was occupied; remember that the macroscopic binding constant for the first site is 16 times greater than that for the last site (see p. 68). The results of these parallel experiments with several isoenzymes of lactate dehydrogenase showed identical single exponentials in each pair. Similar experiments were also carried out with liver alcohol dehydrogenase and, again, it was shown that there is no difference in the behaviour of a single site on its own, from that of all sites reacting. The results of such tests have helped to set a standard for testing whether 'half of sites reactivity' can be proposed as the correct mechanism for a particular enzyme (Gutfreund, 1975).

So far, examples to illustrate experimental methods for following the time course of the approach to steady states and of their kinetic interpretation have been restricted to enzymes which do not have a natural chromophore attached to the protein; although reference has been made to the classic studies of Chance with peroxidase (see p. 142). Clearly the application of these techniques to the study of enzymes with built in chromophores, such as the prosthetic groups riboflavine, pyridoxal phosphate or haem, contributed considerably to the elucidation of reaction mechanisms. However, the progress in the identification of the number and character of intermediates depended more on the improvements of spectral resolution of stopped-flow equipment than on any kinetic principles additional to those enunciated above. This is illustrated, for instance, by the progress made between the first transient kinetic study of the flavoprotein xanthine oxidase by Gutfreund & Sturtevant (1959) and the much more detailed spectral analysis of intermediates by Olson *et al.* (1974) and Porras, Olson & Palmer (1981).

Some aspects of several other enzyme reactions are discussed in the next chapter in connection with investigations by relaxation techniques.

5.2 Sequential enzyme reactions in metabolism and analysis

Applications of the study of sequential reactions

The development of kinetic schemes for sequences of enzyme reactions contributes to the resolution of two problems. The first of these, the more complex one, concerns the study of the control of metabolic pathways and has been of major interest to biochemists for some time. Two related approaches to the problem have been developed for the interpretation of the behaviour of large assemblies of coupled enzyme reactions. Models can be made which contain the differential equations for the progress of the reactions for all the enzymes of a system. The numerical solutions of this set of equations can be compared with the experimental data for the concentrations of intermediates and their rates of change. Iterative improvements of the model can then be made. Alternatively, if data are only available for a small segment of a pathway, control theory can be applied to the analysis of the consequences of local perturbations. This specialized topic was the subject of a recent symposium volume (Cornish-Bowden & Luz Cardenas, 1990). Alas, the large amount of theory does not compensate for the paucity of experimental data and we shall not concern ourselves with this specialized topic, except for the elementary case of two sequential reactions.

A simpler, but equally important application of the study of sequential enzyme reactions is their use simply as an analytical tool, which depends critically on correct kinetic interpretation. This latter application is concerned with two related problems. The first of these concerns the linked assays of the rates of reactions by subsequent ones, which provide a suitable signal for monitoring the time course. Second, sequential enzyme reactions have been used for the estimations of the rates of dissociation of products from one of the enzymes, or of a ligand from its specific binding site. The problem can be defined as the determination of the rate of appearance of some free intermediate in a consecutive reaction. A topical problem requiring a careful application of such analyses is the recent proposal of the hypothesis of substrate channelling discussed in section 5.3. The proponents of this phenomenon (Srivastava & Bernhard, 1986) have rightly drawn attention to the fact that enzymes of the glycolytic pathway occur at very high (approaching millimolar) concentrations and consequently a large fraction of many reaction intermediates is enzyme bound. Several authors have not been sufficiently critical in the interpretation of the kinetic equations and of the required accuracy of kinetic constants necessary to distinguish between different models. This problem is discussed below in some detail.

Rates of product dissociation

Different methods for investigating the rates of dissociation of ligands, or the products of enzyme reactions, have applications under different circumstances and conditions. Displacement reactions, discussed in section 5.3, result in rates of dissociation from host molecules under conditions when all binding sites are occupied. For the study of rates of dissociation from oligomeric proteins under conditions of fractional occupation of binding sites one has to use one of the methods for sequestering free ligand. These are also discussed in the next section and do include the use of enzyme reactions which remove free ligand. Relaxation techniques (see chapter 6) can only be used on systems with equilibrium positions which can be significantly perturbed. Neither of these techniques allows the investigation of product dissociation rates during transients or steady states of enzyme reactions. For these problems the continuous assay of the rate of appearance of free product by its reaction with a suitable enzyme system has been used with success.

A few examples will illustrate the potentialities of the use of one enzyme for monitoring the rate of product dissociation from another one. The resolution of the mechanism of the reaction sequence during the hydrolysis of ATP by myosin, or its enzymatically active fragments, was advanced by this method. Some aspects of this reaction have already been discussed in section 5.1 and the use of enzymes for the determination of product dissociation from myosin has been reviewed in detail by Trentham, Eccleston & Bagshaw (1976). More recent developments (Webb, 1992 and Brune *et al.*, 1994) are discussed below. The transient kinetic experiments, described in the previous section, showed that the initial formation of products (ADP and phosphate enzyme bound) is more rapid than the steady state rate. When single turnover experiments (myosin site concentration $>$ ATP) are carried out, by rapid mixing of myosin and ATP and quenching of the reaction with acid at short time intervals, data for the total amount of phosphate and ADP formed (free and myosin bound) are obtained. The bound products are released by the acid quench. The distinct time course of formation of free and bound products can be determined when, in addition to the flow quench experiments, optical stopped-flow experiments are carried out to monitor the appearance of free phosphate (P_i) and ADP with linked enzyme systems. The following sequential reactions were used to monitor ADP (1) and P_i (2) respectively:

(1)
$$\text{phosphoenolpyruvate} + \text{ADP} \xrightarrow{\text{Catalysed by pyruvate kinase}} \text{pyruvate} + \text{ATP}$$
$$\text{pyruvate} + \text{NAD} \xrightarrow{\text{Catalysed by lactate dehydrogenase}} \text{lactate} + \text{NADH}$$

(2)
$$\text{glyceraldehyde-3-phosphate} \xrightarrow[\ +P_i, \text{NAD}^+\ ]{\text{Catalysed by glyceraldehyde-phosphate dehydrogenase}} \text{1,3-diphosphoglycerate} + \text{NADH}$$
$$\text{diphosphoglycerate} + \text{ADP} \xrightarrow[\text{phosphoglyerate kinase}]{\text{Catalysed by}} \text{3-phosphoglycerate}$$

The last enzyme serves to remove diphosphoglycerate from an unfavourable equilibrium.

Both the above reaction sequences can be monitored by the change in absorbence or fluorescence during the interconversion of NAD^+ and NADH. More recently Webb (1992) developed another continuous spectrophotometric assay for measuring the kinetics of phosphate release using purine nucleoside phosphorylase. On phosphorolysis the substrate 2-amino-6-mercapto-7-methylpurine ribonucleoside gives a large increase in absorbance at 360 nm. From the same laboratory comes a procedure for measuring rapid changes in free phosphate concentration using a phosphate binding protein with a fluorescent label (Brune *et al.*, 1994). Another example of the use of linked enzymes (see Whitaker *et al.*, 1974) is the coupling of two dehydrogenases, which can serve for the determination of the rate of dissociation of NADH. This is discussed further in section 5.3. In all these cases the second enzyme, the one intended to monitor the rate of product dissociation from the first one, must work fast; how fast it has to work can be calculated from the derivations given below.

Other methods for the determination of product dissociation come within the topics of the next section, where the importance of using several approaches to each problem is emphasized.

Linked assays of enzyme reactions

A typical example of the use of coupled enzymes for assays of a metabolite concentration is the determination of glucose by the use of the couple hexokinase and glucose-6-phosphate dehydrogenase:

$$\text{glucose} + \text{ATP} \xrightarrow{\text{catalysed by hexokinase}} \text{glucose-6-phosphate} + \text{ADP}$$

$$\text{glucose-6-phosphate} + \text{NADP}^+ \xrightarrow{\substack{\text{catalysed by glucose-6-} \\ \text{phosphate dehydrogenase}}} \text{gluconate-6-phosphate} \\ + \text{NADPH}$$

Many kinases are assayed by coupling to a dehydrogenase (for instance pyruvate kinase to lactate dehydrogenase, phosphoglycerate kinase to glyceraldehyde phosphate dehydrogenase). Reactions involving the oxidation or reduction of one of the two pyridine nucleotides are in fact favourites for coupled assays of those enzymes which cannot be followed directly by a continuous recording technique. Continuous recording is not only labour saving, as compared to sampling techniques, it is also more reliable for the estimation of initial velocities, provided the transients due to coupling (see below) are taken into account.

Assaying the steady state rate of one enzyme reaction by another can be taken as an example for deriving the equation for sequential reactions. The potential rate of the second one, $k_{cat}c_{E2}(\text{total})$, has to be appreciably greater than that of the first one. After the reaction is initiated by addition of the first enzyme, E1, to a reaction mixture containing initial substrate and the second enzyme, E2, with its cofactors, there will be a lag phase, which represents the transient approach to the steady state concentration of the product of E1 and substrate for E2. During this approach to the steady state the concentration of the intermediate reaches the level at which E2 turns over as fast as E1. The condition that E2 must have the potential to turn over much faster than the E1, is essential for a steady state to be reached and assures that it works under first order conditions, $c_S \ll K_M$, for E2. Let us consider the model:

$$E1 + S1 \xrightarrow{\quad v \quad} E1 + S2$$

$$E2 + S2 \xrightarrow{\quad k \quad} E2 + P$$

For these two consecutive reactions, we define the steady state rate for the first enzyme as $v = dc_{S2}/dt$ and the rate for the second enzyme as $dc_P/dt = kc_{S2}$, where $k = k_{cat}c_{E2}(\text{total})/K_M$ (which has the dimension of a first order constant, section 3.3). Consequently we can write for the coupled reaction

$$dc_{S2}/dt = v - kc_{S2} \quad \text{and} \quad dc_P/dt = kc_{S2} \tag{5.2.1}$$

The initial conditions at $t = 0$ are $c_{S2} = 0$ and $dc_P/dt = 0$. As the reaction proceeds c_{S2} accumulates, while the second reaction accelerates until $dc_P/dt = v$.

$$\frac{d^2 c_P}{dt^2} = k \frac{dc_{S2}}{dt} = kv - k^2 c_{S2} = kv - k \frac{dc_P}{dt} \tag{5.2.2}$$

This acceleration can be described by the same simple second order equation which was used for the approach to the steady state of a single enzyme reaction (see p. 152):

$$\frac{d^2 c_P}{dt^2} + \frac{dc_P}{dt} k = vk \tag{5.2.3}$$

which can be solved for $c_P(t)$ to give

$$c_P(t) = vt + C \exp(-kt) + D \tag{5.2.4}$$

where C and D are the integration constants, which can be solved using the boundary conditions $c_P(0) = 0$ and $c_P(\infty) = vt - v/k$, hence

$$c_P(t) = v\{t - k^{-1}[1 - \exp(-kt)]\} \tag{5.2.5}$$

Figure 5.7 shows the progress curve for the appearance of P for a linked

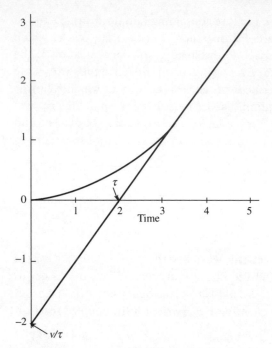

Figure 5.7   Simulation of the acceleration of final product (P) formation during a linked assay: $E1 + S1 \rightarrow E1 + S2$; $E2 + S2 \rightarrow E2 + P$. If a steady state is reached, then the rate of formation of P corresponds to the rate of formation of S2, that is the rate of the first enzyme reaction. The time axis intercept corresponds to $1/k$ and the ordinate intercept to $-v/k$.

reaction under the conditions specified in the legend and demonstrates the significance of the two intercepts. The rate at which the steady state is reached depends on k, as defined above.

As long as the conditions for reaching a steady state (k is large enough) are maintained, its level in terms of intermediate concentration can be varied. This is done by increasing $c_{E2}(0)$ above the lower limit to change the steady state flux. This makes it possible to assay E1 under conditions of different constant product concentrations and thus obtain information about product inhibition and equilibria. A comprehensive review of equations describing sequential enzyme reactions has been presented by Brooks & Suelter (1989).

Analysis of interconverting forms of substrates

Kvassman, Pettersson & Ryde-Pettersson (1988) showed how another interesting phenomenon affects the interpretation of the kinetic constants of sequential enzyme reactions. Many metabolic intermediates occur in spontaneously interconvertible forms. If the equilibration between these forms is slow compared to the time scale of the measurement, the intermediate in a coupled reaction will be essentially all in the form produced by the first enzyme. A quasi steady state will be reached before significant equilibration of the two forms of the intermediate occurs. If, however, the equilibration is on the time scale of the measurement or faster, then it will affect the steady state rate. This is best illustrated with a numerical example.

If we consider the coupled reaction

$$\text{fructose-1,6-bisphosphate} \xrightarrow{\text{catalysed by aldolase}} \text{glyceraldehyde-3-phosphate}$$
$$+ \text{dihydroxyacetone phosphate (DHAP)}$$

$$\text{dihydroxyacetone phosphate} \xrightarrow[\text{phosphate dehydrogenase (E2)}]{\text{catalysed by glycerol-}} \text{glycerol-3-phosphate}$$
$$+ \text{NADH} \qquad\qquad\qquad\qquad\qquad + \text{NAD}^+$$

DHAP, which is the product of the first reaction, forms a diol in aqueous solution and this is not a substrate for E2:

The equilibrium constant of the diol formation is $K = 1.22$ and the observed rate constant for equilibration is $k_{\text{obs}} = k_{12} + k_{21} = 0.8 \text{ s}^{-1}$. If the coupled reaction is carried out at a very high enzyme concentration and the steady state rate is observed on a very short time scale in a stopped-flow apparatus (say in about 100 ms), then the diol formation will have no effect on the time constant of the transient, which is $k_{\text{cat}} c_{\text{E2}}(0)/K_M$. At lower concentrations of E2 ($k \ll 0.8 \text{ s}^{-1}$), a more likely condition for such experiments, the rate constant obtained from the transient approach to the steady state will be smaller than the theoretical $k_{\text{cat}} c_{\text{E2}}/K_M$. The reason for this is that the apparent K_M for E2 reacting with equilibrium mixture of the keto/diol forms is

$$K_M(\text{apparent}) = K_M(1 + K) \tag{5.2.6}$$

where K is the equilibrium constant for the diol formation. If very slow equilibration occurs between active and inactive forms of a substrate, then a different K_M will be obtained by analysis through equation (5.2.5) (coupled reaction) than if an aged substrate solution is used for determining the K_M for E2 independently. This phenomenon can have interesting consequences for the modelling of metabolic pathways and it has to be taken into account when substrate channelling through enzyme–enzyme interactions is postulated. Kvassman *et al.* (1988) showed that this invalidates some of the evidence for the suggestion that some of the enzymes of the glycolytic pathway interact to pass on the product of one as the substrate to the other directly, instead of by free diffusion (see for instance Srivastava & Bernhard, 1986). If substrate channelling, as distinct from free diffusion of intermediates, in glycolysis were a reality, the linked assays widely used in clinical and in general biochemical laboratories, would not have a firm foundation. Pettersson (1989) provided a test for the validity of linked enzyme assays. He showed that the rate of production of dihydroxyacetone phosphate, on addition of aldolase to fructose bisphosphate is the same before and after the addition of glycerol phosphate dehydrogenase. Reference to methods for distinguishing between channelling and free diffusion will be found in several sections of this chapter. The papers of Kvassman & Pettersson (1989a, b) also contribute significantly to this problem.

The use of enzymes for the analysis of products can involve the coupling to chemical rather than enzyme catalysed reactions. An example of substrate isomerization, similar to that encountered above in connection with intermediate isomerization, occurs in aqueous solutions of glyceraldehyde 3-phosphate

$$
\begin{array}{ccc}
& & \text{H} \\
\text{HCO} & & \text{HOCOH} \\
| & \pm\text{H}_2\text{O} & | \\
& 2.5\ \text{s}^{-1} & \\
\text{HCOH} & \rightleftharpoons & \text{HCOH} \\
& 0.087\ \text{s}^{-1} & \\
| & & | \\
\text{H}_2\text{COPO}_3\text{H}^- & & \text{H}_2\text{COPO}_3\text{H}^-
\end{array}
$$

where again the diol is not a substrate for the dehydrogenase. Reynolds, Yates & Pogson (1971) and Trentham *et al.* (1969) used a kinetic approach for the determination of the equilibrium and rate constants for diol formation of dihydroxyacetone phosphate and glyceraldehyde 3-phosphate respectively. The procedure, which is illustrated in figure 5.8 for the dihydroxyacetone phosphate/glycerol-3-phosphate dehydrogenase system, is as follows. An aqueous solution of the substrate is mixed in a stopped-flow apparatus with a solution containing NADH and a high concentration of glycerol-3-phosphate dehydrogenase. When the resulting rate of NADH oxidation is monitored spectrophotometrically, a biphasic record is observed. The amplitude of the fast phase provides information about the

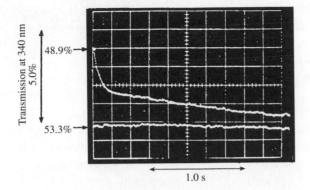

Figure 5.8   Stopped-flow record of NADH oxidation at 20 °C in a reaction mixture containing dihydroxyacetone phosphate (DHAP) (5 μM), α-glycerol-phosphate dehydrogenase (30 μN), NADH (50 μM), in Tris–HCl buffer (0.1 M) and EDTA (2 mM) adjusted to pH 7.5 with NaOH. One syringe contained DHAP and the other enzyme and NADH. The baseline shown is the end of the reaction. The ratio of the fast step to the slow step shows as the ratio of available substrate to precursor which is slowly converted to substrate (for detail see text and Reynolds *et al.*, 1971).

amount of the keto form in the solution and thus gives the equilibrium constant. The slow phase should correspond to the rate of conversion of the diol to the keto form and, together with the equilibrium constant, permits evaluation of the rate constant for the diol formation. The requirement for correct determination of the constants, by the above procedure, is sufficient enzyme concentration to assure that the reduction of the keto form of the substrate is very fast compared to the diol → keto isomerization.

Glaser & Brown (1955) studied the reaction

$$\text{glucono-6-phosphate} + \text{NADP}^+ \xrightarrow[\text{phosphate dehydrogenase}]{\text{catalysed by glucose-6-}} \text{gluconolactone-6-phosphate} + \text{NADPH}$$

They showed how varying the rate of equilibration of a reaction, by increasing the concentration of the relevant enzyme, allows the determination of the equilibrium constant independently from the subsequent hydrolysis of the lactone to gluconic acid. They found that the latter reaction had a time constant of about 25 min at pH 6.4, but at pH above neutrality the increased rate required the use of stopped-flow techniques. Many other metabolic intermediates occur as different isomers, anomers and conformers (see for instance Rose, 1975; Middleford, Gupta & Rose, 1976; Benkovic, 1979). In many cases kinetic techniques have to be used to determine which are the correct substrates and products before thermodynamic interpretations of metabolic sequences are possible. NMR analysis can either complement the information obtained from transient kinetics, or may be the method of choice for the determination of equilibria and rates of the interconversions of different forms of metabolites.

Carbonic anhydrase as a tool for product identification

Another application of product analysis by coupled enzyme reactions arises when CO_2 is involved as one of the substrates or products. In such cases it is of interest to establish whether the spontaneous reaction

$$CO_2 + H_2O \rightleftharpoons HCO_3^- + H^+$$

occurs before or after the enzymatic one. Before we can discuss the rationale for the use of the enzyme carbonic anhydrase for this purpose, a review of the above reaction is necessary. It has already been mentioned that the occurrence of an enzyme for the catalytic interconversion of CO_2 and carbonate had been predicted, prior to its isolation from kinetic analysis of respiratory processes. It is often a matter of surprise to chemists that the hydration of CO_2 and superoxide dismutation ($2O_2^-$ $+ 2H^+ \rightarrow H_2O_2$), the simplest of all chemical reactions which occur in biological systems, are controlled by specific enzymes and in fact in both cases are among the fastest ones known.

The atmosphere contains 0.03% (by volume) CO_2. The partial pressure of the gas under normal aerobic conditions is 3×10^{-4} atm. The solubility of CO_2 at 25 °C in pure water (it is decreased by the presence of salts) is 0.0344 M under 1 atm of the gas and 10.35 μM under aerobic conditions (if these can still be regarded as standard). The total carbon dioxide content of a solution in equilibrium with a given pressure of gas is given by the sum of the concentrations of CO_2, H_2CO_3, and HCO_3^-. The distribution between these species is pH dependent. The second acid dissociation constant

$$HCO_3^- \rightleftharpoons CO_3^{2-} + H^+; \quad pK = 10.33$$

is omitted from consideration in the present discussion. The apparent first $pK = 6.35$, when calculated from

$$K_A = c_H + c_{HCO_3^-}/[c_{CO_2} + c_{H_2CO_3}] = 4.455 \times 10^{-7} \text{ M}$$

However, for the reaction

$$H_2CO_3 \rightarrow HCO_3^- + H^+$$

the true $pK = 3.60$ can be calculated from the apparent equilibrium constant for the hydration of CO_2 at 25 °C, with pure water being at unit concentration. The above information, which is available in an extended form in a symposium volume (Forster *et al.*, 1968), is of general practical value for the interpretation of rates of biochemical reactions. The uncatalysed reaction is slow compared with the accompanying changes in ionization. This means that the kinetics of hydration and dehydration can be followed by monitoring pH changes. When a slightly acid solution of CO_2 is adjusted to pH 8, the rate limiting step for the formation of $HCO_3^- + H^+$ is the hydration process ($k = 0.038 \ s^{-1}$). The adjustment of a bicarbonate solution from pH 8 to 6.35 results in a negligible proton uptake by HCO_3^- ($pK = 3.6$) prior to the rate limiting dehydration

$$H_2CO_3 \rightarrow CO_2 + H_2O; \quad k = 30 \ s^{-1}$$

(see also Gibbons & Edsall, 1963 and Ho & Sturtevant, 1963).

Apart from its physiological interest in controlling the rates of hydration and dehydration of CO_2 in the red blood cell, carbonic anhydrase has also been used as an analytical tool. There are many enzyme catalysed carboxylation and decarboxylation reactions in metabolic pathways. From the point of view of their reaction mechanisms it is of interest to know whether CO_2 or HCO_3^- are the primary substrates or products in different cases. The first use of carbonic anhydrase for this purpose was made by Krebs & Roughton (1948), who showed that the primary products of the reaction of urease ($NH_2 \cdot CO \cdot NH_2 + H_2O$) are $CO_2 + 2NH_3$. This was demonstrated by the fact that the decrease in pH during the reaction was accelerated by the addition of carbonic anhydrase. Hall, Vennesland & Kezdy (1969) summarize such studies on a number of similar reactions, one of which is illustrated in figure 5.9. While CO_2 appears to be the substrate/product for most enzymatic carboxylation/decarboxylation reactions, those catalysed by enzymes with biotin prosthetic groups are reported to utilize carbonate (Walsh, 1979).

In connection with the last topic of this section, attention should also be drawn to some experiments of Roughton & Rossi-Bernardi (1968) who showed how kinetic techniques can be used to determine pK values for transient intermediates which react only in one of two possible states of ionization.

5.3 Kinetics of ligand dissociation and exchange between binding sites

Criteria for specificity

It is very tempting to associate some kinetic parameter of the overall binding equilibrium with the specificity of a ligand for its host site. It is,

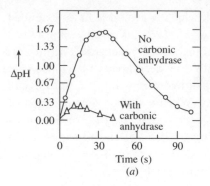
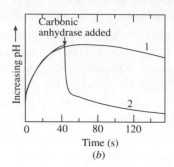

Figure 5.9   The use of carbonic anhydrase for the analysis of the reaction product of an enzyme reaction. Hall, Vennesland & Kezdy (1969) followed the reaction catalysed by glyoxylate carboligase: $2CHOCOO^- + H^+ \rightarrow CHO \cdot CHOCOO^- + CO_2$; $CO_2 + H_2O \rightarrow HCO_3^- + H^+$. Graph (a) shows experiments in which proton uptake is followed spectrophotometrically in the presence of cresol red and graph (b) shows proton uptake recorded potentiometrically. Both sets of experiments show that free CO_2 is liberated by carboxyligase. In the absence of carbonic anhydrase the second reaction is slow.

however, usually misleading to use equilibrium binding constants, or the individual rate constants by which they are determined, even for a comparison of the specificities or affinities of a set of ligands for their target. Specificity does not only involve attachment to a site (recognition) but also the initiation of a chain of molecular events for physiological transduction or catalysis. Thus biological specificity includes the efficiency of triggering a function. This is a recurring theme in this volume whenever the interpretation of kinetic data relates to mechanisms of such processes as enzyme action or the transmission of signals or energy.

In a related problem, when the specificity of different sites for the same ligand on a particular protein are compared, another criterion may have to be used. For instance troponin, the regulatory protein for muscle contraction, has four binding sites for calcium. The sites which are specific for calcium, that is they are not as readily occupied by the abundant magnesium present, have a considerably lower affinity for calcium than the ones which do not select between the two divalent ions. The role of calcium as a trigger (messenger) to initiate a process or transmit information has been mentioned in connection with a number of reaction sequences. The rate at which calcium can become available or dissociate from storage sites or from points of action is, therefore, of great importance and calcium dissociation kinetics is used as an example for one of the methods described at the end of this section and also in section 6.3.

Problems investigated by sequential reactions

Sequential reactions which can be used to determine the rates of dissociation of ligands from their binding sites fall into two classes: (1) displacement reactions by exchange with another ligand and (2) rapid removal of free ligand from the solution by some scavenging reaction. There is some overlap between the methods discussed here and those developed in the previous section. Examples of two types of displacement methods are the exchange of oxygen from oxyhaemoglobin by carbon monoxide and the rapid removal of free oxygen from an oxyhaemoglobin solution by mixing it with a solution of dithionite. The rapid removal of free ligand can be achieved either by such an (irreversible) chemical method or the binding of the ligand to another site, preferably one of high affinity. These methods involve a number of pitfalls and the design of such experiments and the analysis of results obtained are therefore discussed in some detail below with different examples. Other approaches to the study of rate constants for association–dissociation reactions are discussed in several sections throughout this text. Only with the use of more than one of them and the observation of different signals, can one elucidate the complexities of what are usually multi-step processes and sometimes cooperative interactions between sites. Section 6.3 deals with the application of relaxation techniques to the evaluation of the spectrum of time constants which characterize such equilibria. Deviations from oversimplified models of a reaction can often be detected by discrepancies between results obtained by different methods.

The role of product dissociation in enzyme mechanisms

In the case of enzyme reactions the results of the determination of rates of substrate and product association and dissociation have led to an interesting revision of previous generalizations. We have already discussed in a different context, that the early work of Theorell & Chance (1951), as well as later studies on a wider range of systems (Gutfreund, 1975), indicated that, contrary to the originally accepted Michaelis–Menten assumption, the step (or steps) involving chemical catalysis is frequently not rate limiting for the overall reaction. The rates of formation and decomposition of the reactive complex, which is formed after initial binding, are often slower than the steps involved in the covalent chemistry of the enzyme catalysis. This highlighted a problem in the interpretation of rates of substrate dissociation. Just as substrate induced conformation changes occur prior to the chemical interconversion of enzyme–substrate to enzyme–product complexes, so the reverse conformation change occurs before the product can dissociate from the latter complex. The enzyme must, of course, return to its original structure at the end of the turnover. In some cases this obligatory isomerization, before the product can dissociate, can only be observed during the overall reaction of the enzyme. This can best be explained using as an example an enzyme mechanism already discussed in section 5.1, namely the steps after lactate oxidation by lactate dehydrogenase:

$$^{+}HE^* \begin{array}{c} \text{Pyruvate} \\ \diagup \\ \diagdown \\ \text{NADH} \end{array} \underset{\pm H^+}{\overset{}{\rightleftharpoons}} E^* \cdot NADH \rightleftharpoons E \cdot NADH \underset{\pm NADH}{\overset{}{\rightleftharpoons}} E$$

(the asterisk marks a different conformation of the enzyme).

Pyruvate and NADH can be regarded as substrates or products for the reversible reaction

$$lactate + NAD^+ \rightleftharpoons pyruvate + NADH + H^+$$

This enzyme requires ordered addition of its substrates – the nucleotide has to bind prior to lactate or pyruvate. Similarly the nucleotide dissociates after the other substrate. The conformation change to form the reactive complex occurs when both NADH and pyruvate are bound. In the reverse direction, when the concentration of free pyruvate is negligible, an isomerization step ($E^* \rightarrow E$) has to occur after pyruvate has dissociated, but before NADH can dissociate. Therefore an additional step is involved in NADH dissociation after catalytic turnover. Some evidence for two step binding of NADH to lactate dehydrogenase has been found by Wu *et al.* (1991) even in the absence of pyruvate. Similar phenomena are observed during the dissociation of the products after ATP hydrolysis by myosin (see section 5.1). Some of these events may still be subject to revision, but it is clear that product dissociation from enzymes requires quite detailed analysis. Some of the approaches to this problem have been outlined in section 5.2.

Haemoglobin: displacement of O_2 by CO (Roughton & Gibson experiments)

Haemoglobin and myoglobin, are readily available in large quantities and their spectra are sensitive to changes in liganding. They have become paradigms for the development of many biophysical techniques. To quote

Perutz (1987) 'studies on haemoglobin have covered the whole range of subjects from the quantum chemistry of iron to the molecular basis of diseases'. The interesting features and complexities of the reactions of this protein are still one of the favourite areas of kinetic investigations and are discussed in several sections of this volume. The dissociation of oxygen from oxyhaemoglobin by competition (displacement) with carbon monoxide was first studied by Gibson & Roughton (1955), who developed the necessary equations for the analysis of such reactions. For simplicity we shall consider this process in terms of a scheme involving one of the four binding sites alone:

$$HbO_2 \rightleftharpoons Hb + O_2$$
$$Hb + CO \rightleftharpoons HbCO$$

The displacement method is usually carried out under conditions which provide only information about the rate of dissociation from fully liganded protein; that is, all sites are occupied either by the replaced or the replacing ligand. It will be shown that this is necessary for a quasi steady state solution of the rate equations. We have, therefore, neglected the cooperativity of the binding process in the case of haemoglobin. Ligand dissociation resulting from the sequestration of free ligand and some exchange techniques can be used to determine rates of dissociation from partially liganded oligomeric proteins.

Changes in the spectrum in the region of the Soret bands (410 − 430 nm) are used to monitor the concentrations of the different haemoglobin species, liganded and free, as a function of time. Other techniques for the observation of such reactions will be discussed with different examples; these involve changes in the optical properties of the ligand or of the binding site as a result of complex formation.

Displacement reactions can be visualized as re-equilibration after a concentration jump. This consideration will help us to see the similarity between the exchange of two ligands on one type of binding site and the exchange of one ligand between two different binding sites, which will be discussed later in this section. In both cases one can only get analytical solutions to the kinetic equations under limiting steady state conditions. In the general case graphical or numerical solutions have to be extracted. For very small concentration jumps relaxation equations (see section 6.2) can be applied. Wu, Gutfreund & Chock (1992) have calculated the deviations of the real time course of relaxations from the analytical solution, as the amplitude of the perturbation increases.

Gibson & Roughton's (1955) experiments were carried out by mixing solutions of oxyhaemoglobin and CO in a stopped-flow apparatus. Due to the limited solubility of CO at 1 atm of pressure, the range of concentrations of the displacing reagent was restricted. It will be seen that in other cases the concentration range is limited by the optical properties of the ligand. The following derivations will illustrate how these restrictions affect the forms of the equations describing the time course of the equilibration of the two coupled reactions of haemoglobin. The system

$$R + X \underset{k_{21}}{\overset{k_{12}}{\rightleftharpoons}} RX; \qquad R + Y \underset{k_{31}}{\overset{k_{13}}{\rightleftharpoons}} RY$$

where R represents empty binding sites on the protein and X and Y are the two alternate ligands, is described by the following rate equations:

$$\frac{dc_{RX}}{dt} = k_{12}c_R(t)c_X(t) - k_{21}c_{RX}(t) \tag{5.3.1}$$

and

$$\frac{dc_{RY}}{dt} = k_{13}c_R(t)c_Y(t) - k_{31}c_{RY}(t) \tag{5.3.2}$$

The aim of the experiments under discussion is to determine k_{21} from the observed rate of displacement of X from RX by Y. An analytical solution for the single exponential rate of disappearance of RX or appearance of RY can only be obtained when a set of conditions holds which results in linear rate equations, such as the steady state solutions given below. For the particular case of the reaction of haemoglobin with the two ligands, the following conditions were sought by Gibson & Roughton (1957):

(1) The concentrations of free X and Y are high enough to be considered constant throughout the time course of the observation.

(2) The concentration of free binding sites is negligible throughout the reaction and it can be assumed that they are in a steady state and that the total site concentration

$$c_R(0) = c_{RX}(t) + c_{RY}(t)$$

and hence that

$$dc_{RY}/dt = -dc_{RX}/dt$$

A number of related conditions will be used later to linearize the equations as examples with different properties are discussed. Under the above conditions

$$k_{12}c_R(t)c_X(0) - k_{21}\{c_R(0) - c_{RY}(t)\} = -k_{13}c_R(t)c_Y(0) + k_{31}c_{RY}(t) \qquad (5.3.3)$$

From this equation we get the expression for $c_R(t)$

$$c_R(t) = [k_{21}\{c_{RY}(t) - c_R(0)\} - k_{31}c_{RY}(t)]/[k_{12}c_X(0) + k_{13}c_Y(0)] \qquad (5.3.4)$$

which can be used to substitute for $c_R(t)$ in equation (5.3.2) to give

$$\frac{dc_{RY}}{dt} = c_R(0)\left[\frac{k_{21}}{1 + [k_{12}c_X(0)/k_{13}c_Y(0)]}\right]$$

$$- c_{RY}(t)\left[\frac{k_{13}}{1 + [k_{13}c_Y(0)/k_{12}c_X(0)]} + \frac{k_{21}}{1 + [k_{12}c_X(0)/k_{13}c_Y(0)]}\right] \qquad (5.3.5)$$

When $k_{13}c_Y(0) \gg k_{12}c_X(0)$, with large excess of displacing ligand, equation (5.3.5) reduces to

$$dc_{RY}/dt = k_{21}[c_R(0) - c_{RY}(t)] \qquad (5.3.6)$$

Using the boundary conditions that $c_{RY}(0) = 0$ and $c_{RX}(0) = c_R(0)$ (total binding sites) used above, or that at $t = \infty$, $c_{RY}(t) =$ total binding sites, equation (5.3.6) is integrated to give

$$c_{RY}(t) = c_P(0)[1 - \exp\{-k_{21}t\}]$$
$$c_{RX}(t) = c_R(0)\exp\{-k_{21}t\}$$

which describes the rate of dissociation of X from the complex RX by a single exponential.

This derivation shows that, if the solubility of both X and Y are high enough to fulfil the conditions stated above (p. 184), then the first order rate constant k_{21} for the dissociation is given by the exponential decay of the signal for c_{RX} or the rise of the signal for c_{RY}. If conditions (1) and (2) are fulfilled, but it is not possible to raise $c_Y(0)$ sufficiently for $k_{34}c_Y(0) > > k_{12}c_X(0)$, then equation (5.3.5) can be solved in terms of an observed rate constant $k_{obs} = k'_{21} + k'_{31}$,

where

$$k'_{21} = k_{21}/[1 + [k_{12}c_X/k_{13}c_Y]]$$

and $\qquad (5.3.7)$

$$k'_{31} = k_{31}/[1 + [k_{13}c_Y/k_{12}c_X]]$$

Limitation of ligand concentrations can be due to insufficient solubility, as in the case of Gibson & Roughton's experiments, or it can be due to the optical properties of the ligands; a problem encountered in the studies reported below. The use of condition (1), above, is only practical when the reaction is monitored through absorbence or fluorescence changes due to a signal provided by the binding site. If the signal is due to changes in the ligand, high concentrations result in small differences between two large quantities; consequently it is difficult to obtain accurate data.

Insufficient solubility of O_2 and CO, at 1 atm, led Gibson & Roughton to a graphical solution for k_{21} by using the equation

$$\frac{1}{k_{obs}} = \frac{1}{k_{21}}\left[1 + \frac{k_{12}c_{O_2}}{k_{13}c_{CO}}\right]$$

They plotted $1/k_{obs}$ for the dissociation of oxygen from oxyhaemoglobin (0.095 mM) by displacement with CO at its maximum concentration (0.51 mM) against a range of O_2 concentrations. From the intercept and the slope of this plot both k_{21} and k_{12}/k_{13} could be evaluated. We shall see, in the discussion of another displacement reaction, that the sign of the slope of this plot depends on the relative affinity of the two ligands for the binding site. This will be demonstrated below with the, now more appropriate, method of fitting the data directly to a hyperbola.

It is of historical interest that the application of the stopped-flow method to the above experiments with haemoglobin was a major stimulus for Gibson to develop an apparatus (Gibson & Milnes, 1964) specifically for investigations of the complex series of events during the dissociation of ligands from the four interacting binding sites of haemoglobin; a problem still under intensive investigation. The use of Gibson & Milnes's apparatus, and many modifications, has revolutionized the study of enzyme mechanisms (see section 5.1) as well as investigations on ligand binding in general.

Two step ligand dissociation

During all the discussions of ligand binding, throughout this volume, it has been emphasized that these processes involve at least two steps: (1) the formation of a collision complex and (2) a subsequent rearrangement to form the specific complex. As a corollary one should often find two steps in ligand dissociation. Such a situation can be analysed by a simplified model of the two step reversible reaction discussed in section 5.2. If the equilibrium

$$A + R \underset{k_{21}}{\overset{k_{12}}{\rightleftharpoons}} AR \underset{k_{32}}{\overset{k_{23}}{\rightleftharpoons}} \langle AR \rangle$$

is perturbed by the addition of a high concentration of X, which takes the place of R at the binding site, the decay of AR and $\langle AR \rangle$ will be represented by the sum of two exponentials. The expression derived for the behaviour of such a system (equation (5.2.18)) will be simplified by the assumption that $k_{12} = 0$ when $c_X \to \infty$. As an exercise the reader should substitute this condition in the matrix (4.2.17) and compare its determinant with the simplified equation (5.2.18). The method used to calculate the amplitudes (pre-exponential factors) will depend on the information available about the equilibrium and on the initial conditions. An investigation of ligand dissociation in which such a two step model is analysed has recently been described by Török & Trentham (1994).

Fluorescence signals and ligand binding

There are a number of other ways in which spectral or fluorescence changes can be used to monitor the time course of ligand binding and dissociation reactions. Each of these has its own problems of experimental design and data analysis. A frequent procedure for the study of the interactions of ligands with proteins is the attachment of a reporter group. An artificial chromophore can be attached and used as a monitor in the absence of an endogenous one near the site of ligand interaction. Attention must be drawn to a non-linearity of endogenous or artificial fluorescence signals from oligomeric proteins. This concerns the use of fluorescence quenching by non-radiative energy transfer. This is not a linear phenomenon and can give a false impression of the mechanism of sequential binding or dissociation (see Holbrook, 1972; and section 8.2).

Many studies are carried out by monitoring the effects of the environment (charge, dielectric constant, etc.) of the binding site on the spectrum of the chromophore. For such experiments the range of ligand concentrations that can be used is often limited. The concentration of free chromophoric ligand has to be kept low to maximize the differential signal. Consequently, in experiments like the exchange of nucleotides on the binding sites of enzymes (see below), the usual conditions are

$$c_R(\text{total}) \approx c_X(0) \ll c_Y(0)$$

It will be seen that the kinetic equations take different forms, depending on conditions imposed on the experiments by the relative affinities for the binding site of the displaced and the displacing ligand. The descriptions of some reactions with their analysis will serve as examples for the type of assumptions which have to be made in different cases. When conditions do not permit an analytical solution of the rate equation, extrapolations should be done with caution.

Displacement of NADH by NAD⁺

A specific example will illustrate some important points about the rate of displacement of a ligand with high affinity by another with relatively lower affinity. The rate of dissociation of NADH from its complex with the enzyme glycerol phosphate dehydrogenase (E) has been investigated by the displacement of NADH by NAD^+. At $10\,°C$ the two reactions can be described by the following equilibrium constants:

$$E + NADH \underset{k_{21}}{\overset{k_{12}}{\rightleftharpoons}} E \cdot NADH; \quad K_1 = k_{21}/k_{12} = 1.07\ \mu M$$

$$E + NAD^+ \underset{k_{31}}{\overset{k_{13}}{\rightleftharpoons}} E \cdot NAD^+; \quad K_2 = k_{31}/k_{13} = 90\ \mu M$$

Under the initial conditions that $c_{NADH}(total) < c_E(total)$, the rate of disappearance of $E \cdot NADH$ can be described by

$$dc_{ENADH}/dt = k_{12}c_E(t)c_{NADH}(t) - k_{21}c_{ENADH}(t)$$

When a series of displacement experiments are carried out by mixing a solution of enzyme and NADH (as defined above) with one of increasing concentrations of NAD^+, the concentration of free enzyme will decrease, after a very short initial period and at the limit, when $c_{NAD^+}(0) \to \infty$, $k_{12}c_E$ is negligible and

$$dc_{ENADH}/dt = -k_{21}c_{ENADH}(t) \quad \text{and} \quad c_{NADH} = const \cdot \exp(-k_{21}t)$$

It can be demonstrated by simulation of the reactions that the above steady state assumption holds adequately in spite of the twofold change in the concentration of free enzyme concentration during the displacement reaction. A set of values for k_{obs} is obtained from the exponential records of experiments at different concentrations of NAD^+. A plot of k_{obs} against NAD^+ results in a hyperbola with the asymptote k_{21} as $c_{NAD^+} \to \infty$. This is shown in figure 5.10.

The signal used for displacement experiments involving NAD^+ and NADH relies on the large fluorescence changes during binding of the reduced nucleotide to most dehydrogenases. The converse of the above experiment, replacing bound NAD^+ with NADH, cannot be followed using nucleotide fluorescence. At excess NADH concentrations the differential signal becomes too small. However, it would be possible to use quenching of protein fluorescence as a signal for $E \cdot NADH$ formation (see

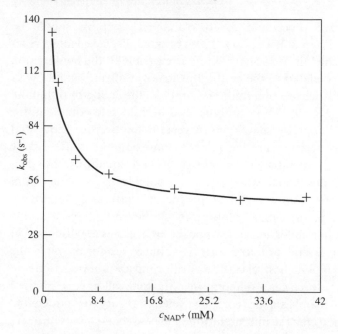

Figure 5.10 The observed rate constant for the displacement of NADH by NAD$^+$ from the complex with α-glycerol-phosphate dehydrogenase as a function of the concentration of NAD$^+$. (Extracted from the data of Chock & Gutfreund, 1988.)

section 8.2). This complements the experiments with haemoglobin described above. The conditions would be like those in the displacement of O_2 by CO at the haem groups: a weaker binding ligand being displaced by a stronger binding one. If both ligands are at sufficiently high concentrations to be considered constant and we assume (justifiably) that k_{12} is approximately equal to k_{34}, we can rewrite the Gibson & Roughton equation (5.3.7)

$$k_{obs} = k_{21}/\{1 + [k_{12}c_{NADH}(t)/k_{13}c_{NAD} + (t)]\}$$

In this case, as in Gibson & Roughton's example, k_{obs} would increase hyperbolically to the true value for k_{21}, as the concentration of the displacing reagent is increased. This is the inverse of the effect on k_{obs} compared to displacing a tight binding reagent with a weak one. This feature, which is also observed with transfer of ligands between sites of different relative affinities (see p. 193), has been noted by Fantania, Matthews & Dalziel (1982). It can be verified by numerical integration of the differential equations describing the rates of exchange of ligands with different relative affinities.

Some specialized procedures for displacement reactions should be mentioned. Kvassman & Pettersson (1979) coupled displacement to reactions which attached the displacing ligand irreversibly to the binding site. This procedure is related to the method of removal of free ligand by an irreversible sequential reaction (mopping up) like the action of dithionite on free oxygen ($k_{21} = 0$). These authors also derived rate equations for displacement reactions by simplifying a general solution for sequential reactions. Their conclusions are equivalent to those obtained above. It may be intellectually more satisfying to derive general equations and then introduce conditions under which they are soluble in a useful form. However, it is often more useful in practice to introduce simplifying conditions, which permit simple solutions for the interpretation of specific experiments. Similar philosophically opposite approaches are discussed in the treatment of several problems in this volume; see for example the solution of transient kinetics, relaxation kinetics and absolute reaction rate theory. In each of these cases different investigators will have a different balance between emphasis on complex general theory or pragmatic, simplified equations for the interpretation of experiments.

The exchange of ligands between binding sites

A problem very similar to the exchange of two different ligands at a binding site (displacement of one ligand by another) is the exchange of a single ligand between two different binding sites. This latter process is of interest not only for the evaluation of association–dissociation rate constants, but also for the interpretation of the behaviour of sequential enzyme reactions and signalling between different ligand donor and receptor sites. The interpretation of sequential enzyme reactions, including intermediate transfer, was discussed in section 5.2. Here we shall concern ourselves only with respective dissociation and association phenomena. The problem is of interest in connection with the possibility that products of one enzyme reaction can be transferred directly through enzyme–enzyme interaction, instead of going via free diffusion in solution. In structurally defined multi-enzyme complexes this phenomenon is well established. In the oxidative decarboxylation of pyruvate and 2-oxoglutarate the mobility of the lipoyl domain can shuttle intermediates between different active sites of the complex (see for instance Perham, Duckworth & Roberts, 1981). Indole, the intermediate in the two step reaction of tryptophan synthesis from indole-3-glycerol phosphate and serine, can diffuse in a channel from one active site to another within the protein complex. Both the crystal structure (Hyde *et al.*, 1988) and kinetic investigations (Anderson *et al.*, 1990; Dunn *et al.*, 1991) provide evidence for intermediate channelling in this reaction. For some other systems like fatty acid synthetase such phenomena are also well documented. A recent report that a nucleotide diphospate kinase can use G-protein bound GDP as a substrate (Randazzo, Kahn & Northup, 1991) had to be withdrawn (Randazzo *et al.*, 1992). The concentration of free nucleotide diphosphate in solution was underestimated and was sufficient to account for the amount of GTP produced.

A case for which good experimental data are available serves as an example of a test for the mechanism of ligand transfer. The rate of dissociation of NADH from the enzyme glycerol-3-phosphate dehydrogenase, E1, has been determined ($k = 35\,\mathrm{s}^{-1}$) as described above. Srivastava & Bernhard (1986) studied the apparent rate of transfer of NADH from its complex with E1 to the binding site on lactate dehydrogenase, E2, by addition of a high concentration of the latter enzyme. The considerable difference in the fluorescence signal between the two NADH complexes allows accurate kinetic observations on the transfer reaction. At concentrations of E2 in threefold excess over E1 the rate of NADH transfer was found to be very much faster than the rate of NADH dissociation from E1. This led Srivastava & Bernhard to the conclusion that the nucleotide was transferred within a complex, E1—NADH—E2, rather than via free solution. However, using established rate constants for the two reversible reactions:

$$E1 + NADH \underset{k_{21}}{\overset{k_{12}}{\rightleftharpoons}} E1 \cdot NADH$$

$$E2 + NADH \underset{k_{43}}{\overset{k_{34}}{\rightleftharpoons}} E2 \cdot NADH$$

the progress of the reaction can be simulated by numerical integration of the relevant differential equations. The exponential decay of the simulated profile for the decrease in the concentration of E1·NADH and the concentration changes of the other reactants are shown in figure 5.11. The rate of equilibration of the two complexes, at constant initial concentration of E1·NADH is dependent on the concentration of E2 added to initiate the reaction. Increasing the concentration of E2 results in a decrease of the apparent rate constant and, in the limit of high concentration of E2, it approaches the true rate constant for the dissociation of NADH from its

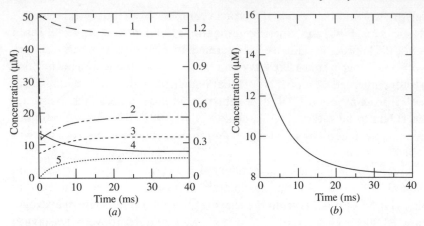

Figure 5.11   Computer simulation of the time course of transfer of NADH from
α-glycerol-phosphate dehydrogenase (GPDH) to lactate dehydrogenase (LDH)
on addition of LDH to a solution containing GPDH and NADH. (*a*) Curves 1 to
5 represent the concentrations of LDH, GPDH, GPDH–NADH and LDH–
NADH respectively. In (*b*) the time course for the disappearance of GPDH–
NADH is expanded. The apparent rate constant calculated from this latter curve
is 134 s^{-1}. The rate constants used for the simulation were 9×10^7 M^{-1} s^{-1}, 60 s^{-1},
1×10^8 M^{-1} s^{-1} and 200 s^{-1} for k_{12}, k_{21}, k_{34}, and k_{43}, respectively. The crucial
comparison is that of $k_{21} = 60$ s^{-1} with the apparent rate of reaching equilibrium
of 134 s^{-1}.

complex with E1. This is shown in figure 5.12 together with the data for the
concentration dependence of transfer in the other direction. Since the
affinity for NADH of lactate dehydrogenase is slightly greater than that of
glycerol-3-phosphate dehydrogenase, the slope of the plot is positive in that
case. Other evidence eliminating direct transfer of intermediates in the
glycolytic pathway is presented elsewhere in this chapter and by several
authors (Gutfreund & Chock, 1991; Pettersson, 1991; Wu *et al.*, 1991).
Gutfreund & Chock (1991) pointed out that the dependence of transfer rate
on receptor enzyme, E2, can have a positive or negative slope for transfer by
free diffusion, but for direct transfer (substrate channelling) via a complex
the slope must always be positive. Diagnostics taking into account the
direction of change with concentration parameters are much more reliable
for the exclusion of proposed models than arguments about small numerical differences in kinetic constants. The take home lesson is, as always:
don't confuse rates with rate constants.

While the precise numbers for the rate constants quoted above are very
sensitive to the conditions of the experiments and are therefore liable to

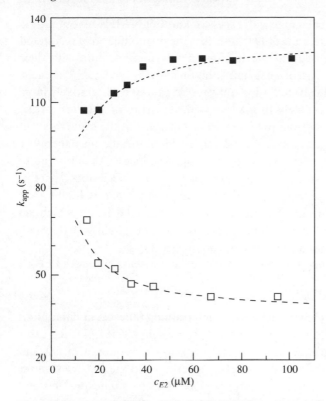

Figure 5.12   The rate of transfer of NADH from its complex with E1 to E2 when increasing concentrations of E2 are added to solutions of E1-NADH. In the upper curve E1 is lactate dehydrogenase and E2 is α-glycerol-phosphate dehydrogenase. In the lower curve E1 is α-glycerol-phosphate dehydrogenase and E2 is lactate dehydrogenase. (Data extracted from Wu *et al.*, 1991.)

revision, their concentration dependence provides an unambiguous inter-pretation of the mechanism of ligand exchange (see Wu *et al.*, 1991).

Since most of the NAD^+ linked dehydrogenases are oligomeric some interesting 'thought experiments' might be carried out with hybrids. A kinetic investigation of nucleotide transfer from one type of subunit to another could be illuminating.

It is important to point out that the values of the rate constant for the dissociation of a ligand from its specific binding site, obtained by the two above methods (displacement and transfer), are not necessarily the same. Wu *et al.* (1991) noted that the rate constants for NADH dissociation from its complexes with lactate dehydrogenase and glycerol-3-phosphate dehyd-rogenase are more than 50% higher when measured by ligand displacement than when acceptor exchange with excess of another NADH binding site is observed. Prinz & Striessnig (1993) describes the phenomenon of facilitated dissociation. They discussed this in some detail in connection with their studies of the interaction of a calcium antagonist for L-type Ca^{2+} channels. Their ideas can be illustrated using our present example. A relatively large ligand like NADH is likely to interact with an array of subsites on the enzyme. Not all the subsites are likely to be occupied at the same time. At high NAD^+ concentrations a second nucleotide molecule may therefore bind to a subsite not already occupied by the first ligand, thus leading to competition for individual subsites and accelerated dissociation. There is, of course, also always the possibility of a completely separate binding site being occupied at the high concentration of the displacing ligand, which can influence events at the specific binding site. All this goes to show that several methods should be used to investigate any process.

Calcium dissociation and exchange

The importance of calcium exchange in determining the rates and initiation of physiological processes has been referred to in several places in this volume. For instance, arguments persist about the sequence of calcium release/uptake and potential changes triggering and ending transmitter release in cells with excitable membranes (see Mulkey & Zucker, 1991 and references therein). Similarly, there is a long running dispute about the kinetic role of parvalbumin in the shuttle of calcium between the sarcoplasmic reticulum and troponin C (see for instance Permyakov, Ostrovsky & Kalinichenko, 1987). As another example of transfer kinetics we present some experiments on the exchange of calcium between two compounds in use in physiological investigations. The wide application of chromophoric and fluorescent compounds for the continuous monitoring of inter- and intra-cellular calcium concentrations made it necessary to obtain kinetic constants for their reactions. We shall take one of these systems as an example for the derivation of calcium dissociation rates.

Jackson *et al.* (1987) carried out stopped-flow experiments with solutions of the dye fura-2, called F, saturated with calcium. On mixing with a solution containing 100fold excess of EGTA or EDTA, called E, the rate of calcium dissociation from the dye can be monitored by observation of the decrease of the fluorescence signal. The resulting reaction can be described in a manner analogous to that given above for NADH transfer between enzymes

$$F + Ca^{2+} \underset{k_{21}}{\overset{k_{12}}{\rightleftharpoons}} F \cdot Ca^{2+}$$

$$E + Ca^{2+} \underset{k_{43}}{\overset{k_{34}}{\rightleftharpoons}} E \cdot Ca^{2+}$$

The equations describing the rates of change of the two complexes are [$c_E(0)$ is constant EGTA concentration]

$$\frac{dc_{ECa^{2+}}}{dt} = k_{34}c_E(0)c_{Ca^{2+}}(t) - k_{43}c_{ECa^{2+}}(t)$$

$$\frac{dc_{FCa^{2+}}}{dt} = k_{12}c_F(t)c_{Ca^{2+}}(t) - k_{21}c_{FCa^{2+}}(t)$$

The rate of change of free calcium is given by

$$\frac{dc_{Ca^{2+}}}{dt} = -k_{12}c_{Ca^{2+}}(t)c_F(t) + k_{21}c_{FCa^{2+}}(t) - k_{34}c_{Ca^{2+}}(t)c_E(0) + k_{43}c_{ECa^{2+}}(t)$$

If the chelator, E, is at high concentration and has a high affinity for calcium, the concentration of free calcium can be regarded as close to zero and at a steady state throughout the measurement (see figure 5.11 for NADH exchange). Clearly, there are transients at the beginning of the experiment, but they only last for a period which is insignificant compared to the total record. Following this argument we set $dc_{Ca^{2+}}/dt$ equal to zero and obtain an expression for calcium

$$c_{Ca^{2+}}(t) = \left[\frac{k_{21}c_{FCa^{2+}}(t) + k_{43}c_{ECa^{2+}}(t)}{k_{12}c_F(t) + k_{34}c_E(0)} \right]$$

This expression for $c_{Ca^{2+}}(t)$ and the approximation that the total calcium concentration $c_{Ca^{2+}}(0)$ is equal to the sum of the concentrations of the two calcium complexes can be used to derive an equation for the rate of transfer of calcium to EGTA

$$\frac{dc_{ECa^{2+}}}{dt} = \frac{k_{34}k_{21}c_E(0)c_{Ca^{2+}}(0)}{k_{12}c_F(t) + k_{34}c_E(0)} \left[\frac{k_{21}}{1 + [k_{12}c_F(t)/k_{34}c_E(0)]} \right]$$
$$- \frac{k_{43}}{1 + [k_{34}c_E(0)/k_{12}c_F(t)]}$$

This result shows that, under the prescribed conditions, the disappearance of the fura calcium complex or the formation of EGTA calcium complex is represented by a single exponential function with the observed rate constant

$$k_{obs} = \frac{k_{21}}{1 + [k_{12}c_F(t)/k_{34}c_E(0)]} + \frac{k_{43}}{1 + [k_{34}c_E(0)/k_{12}c_F(t)]}$$

A plot of k_{obs} against $c_E(0)$, obtained from the results of a set of experiments at different EGTA concentrations, will give a limiting value (at high EGTA concentrations) for k_{43} and an intercept at the ordinate, $c_E(\infty) \to 0$, approximating to k_{21}. Jackson *et al.* (1987) obtained the values $k_{21} = 84 \text{ s}^{-1}$ and $k_{43} = 0.3 \text{ s}^{-1}$. The association rate constants for calcium with fura-2 and EGTA respectively can be estimated from these experiments as $2.5 \times 10^{-8} \text{ M}^{-1}\text{s}^{-1}$ and $5 \times 10^7 \text{ M}^{-1}\text{s}^{-1}$ at pH 8. All these constants are very critically dependent on pH and ionic strength. Investigations on this system were also carried out by Kao & Tsien (1987) using the temperature jump method (see section 6.3).

6

Chemical relaxation phenomena

6.1 Applications of relaxation kinetics

Definition of relaxation

Apart from brief excursions into steady states all the kinetic phenomena discussed in this volume are, in principle, relaxations from one state to another. These relaxations can be between two equilibrium states or two steady states, or between an equilibrium and a steady state. They are the consequences of the perturbation (by a step or oscillations) of an intensive parameter (temperature, pressure or electric field), or a mechanical parameter, as for instance in the rapid length changes of muscle fibres (Huxley & Simmons, 1971). Concentration jumps can also be used to initiate relaxation processes. Other examples taken from common experience are the visco-elastic relaxation in response to a change in length or tension and the response of a capacitor to a change in the applied voltage (see section 2.1). There is no sharp distinction between the general use of the terms transients and relaxations. Relaxation analysis can be applied to the response to a perturbation of very complex systems, such as an engineering structure or a physiological system. In connection with the last example it is, of course, very interesting to study the range of response times and recoveries after sensory stimuli. However, this sort of problem has already been discussed in chapter 4 in terms of consecutive reactions and the enumeration of the number of steps of sequential processes and transients. Eigen (1968) used the term 'chemical relaxation spectrometry' for methods involving small perturbations (see below for definition of 'small') of chemical equilibria by rapid changes of an intensive parameter. Although this term could equally well be applied to many other relaxation phenomena, it will be used here for convenience in the sense just defined. It will be seen that the methods used to derive the kinetic equations for the response to small perturbations involve

197

the same algebra as the matrix methods described in section 4.2. The difference is that the linearity necessary for such solutions is achieved by buffered ligand concentration in the previous chapter and by negligible changes in reactant concentrations in the derivations presented in the next section.

Range of applications

For two reasons the subject of chemical relaxation, although of great importance in biology for the study of proteins and nucleic acids and their interactions, does not justify the same detailed treatment given to transient kinetics in the previous chapter. First, it is easy to present the general principles of relaxation theory from the lessons learned in chapters 2 and 4 and to derive from them the equations which describe the behaviour of any kinetic model. It may not be possible to obtain experimental data to make full use of the resulting set of equations, but that is another matter with which we shall concern ourselves later. In transient kinetics one tends to encounter a collection of special cases. This is at least partly due to the fact that a different assumption has to be made for each model to obtain analytical solutions for the equations. As will be seen in the next section on the restricted use of the term chemical relaxations, defined above, only one general assumption, that of small perturbations, has to be made to obtain a set of linear equations, which can be solved by standard algebraic methods. The second reason for a limited treatment of chemical relaxation methods is the availability of a number of excellent specialized texts which can be consulted for more detail of theory and examples (Eigen & de Maeyer, 1963; Eigen & de Maeyer,1973; Bernasconi, 1976). The development of experimental techniques for temperature jump measurements is discussed by Hammes (1974) and for pressure jump techniques by Knoche (1974) and Davis & Gutfreund (1976). It is of interest to note that Cattell (1934) had already suggested the use of pressure perturbation for the study of complex biological processes. He drew special attention to the effects of pressure on muscle contraction; an observation which was used very effectively 50 years later in the elucidation of some of the molecular events in the contractile process (see section 6.2).

The work of Eigen (1954) brought the potentialities of chemical relaxation studies to the wider notice of physical biochemists. As described by Eigen & de Maeyer (1973) in their historical introduction, the development of relaxation techniques was associated with the study of the rates of ionic reactions. These processes were previously considered too fast for resolution by existing techniques. The application of relaxation techniques to the study of ionic equilibria was originally stimulated by the discovery that the absorption of certain frequencies of sound by sea water, was due to association–dissociation reactions of electrolytes (Eigen, Kurtze & Tamm,1953). Sound waves result in periodic pressure perturbations and measurements of energy absorption and sound velocity dispersion by chemical reactions are still important for the study of very fast processes. However, the wavelength of sound in water is 3 cm at 50 kHz. Since one dimension of the sample has to be large compared to the wavelength, the useful frequency range for kinetic experiments is above 1 MHz. Acoustic and electric field perturbation are the methods of choice when reactions with time constants of less than microseconds are under investigation. Another method specifically applicable to periodic pressure perturbations is discussed in the next section. Both the experimental and theoretical aspects (spectral analysis) of periodic perturbation techniques segregate them into the category of a rather specialized occupation. In the present volume we tend to concentrate on kinetic methods which can be used as one of several to be applied to a particular problem. This is one of the reasons why, for instance, nuclear and electron magnetic resonance techniques, which can also be classed as relaxation methods, are not included in the present discussions.

The restricted treatment of relaxation in this chapter concentrates on the analysis of responses to very small perturbations of equilibria, quasi first order or linear conditions. However, it will be demonstrated that sequences of real first order reactions can be treated by relaxation equations regardless of the size of the perturbation. From a number of examples presented in the following sections it will be seen that relaxation techniques can be applied to study reactions in systems at various levels of organization. It is clearly of great value that, for instance, steps in reactions studied in solutions of isolated muscle proteins can also be identified in contracting muscle fibres. This has been achieved by the use of the pressure relaxation technique for the investigation of the interaction of myosin with actin, which can be correlated with pressure effects on the tension of muscle fibres (section 6.3). Results of temperature and pressure jump experiments on muscle fibres (Geeves & Ranatunga, 1987; Davis & Harrington, 1993) complement the mechanical relaxation studies of Huxley & Simmons (1971). Similarly, the interconversion of intermediates of the bleaching of rhodopsin can be investigated by temperature or pressure relaxation of the purified pigment and in subcellular preparations. However, the theory, practice and interpretation of data are best explained using systems at the

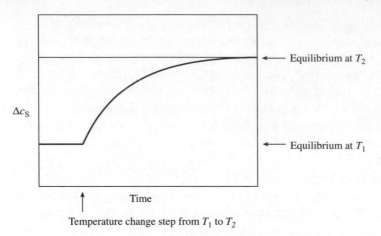

Figure 6.1   Diagram representing the relaxation between two equilibrium concentrations after a step function temperature perturbation.

molecular level as examples. The basic principles used to derive relaxation equations are described in the next section. In sections 6.3 and 6.4 examples are described of investigations on ligand binding, protein–protein interaction and ionic processes. These topics represent the main contributions of relaxation experiments to biological processes at the molecular level. Some other important investigations, such as base pairing in nucleic acids and intramolecular processes of protein and nucleic acid conformations, can be interpreted using analogous approaches.

Principle of the method

Methods involving single step perturbations, such as temperature jump by Joule or laser heating and pressure jump by rapid release of oil pressure, can be more readily used in non-specialist laboratories, than those involving periodic perturbations and spectral analysis. They have a time resolution down to about 1 μs and 50 μs respectively. Figure 6.1 shows schematically the single exponential response to a step change in temperature of the equilibrium concentration of a first order transition. In more complex reactions the response to a step perturbation can be the sum of a number of exponentials. In principle the number of exponentials (relaxation times) will be equal to the number of independent rate equations describing the reaction. In linear systems with n states there will be $n-1$ relaxations corresponding to the number of steps in the reaction. In parallel or cyclic reactions there are fewer relaxation times than steps, because the equations for different steps are not all independent. In practice, the number of relaxation times which can be resolved depends also on their degeneracy (similar values), on the position of the perturbed step in relation to the intermediate monitored during the relaxation, the position of the slowest step in the sequence and last, but not least, on the statistical resolution at the pertaining signal to noise ratio. The number of relaxation times observed will increase as the time and signal resolution improves. This is probably true of all kinetic experiments, more steps can always be elucidated as one or other aspect of the technique is improved. More detail about a kinetic mechanism will also be revealed when studies are extended to a wide range of conditions and concentrations. The use of a variety of perturbations and different monitors which record the concentrations of different intermediates, will also provide additional information.

6.2 Basic equations for the analysis of chemical relaxations

Relaxation amplitudes

The forcing functions used to initiate chemical relaxations are temperature, pressure and electric field. Equilibrium perturbations can be achieved by the application of a step change or, in the case of the last two parameters, of a periodic change. Stopped-flow techniques (see section 5.1) and the photochemical release of caged compounds (see section 8.4) can also be used to introduce small concentration jumps, which can be interpreted with the linear equations discussed in this chapter. The amplitudes of perturbations and, consequently of the observed relaxations, are determined by thermodynamic relations. The following three equations define the dependence of equilibrium constants on temperature, pressure and electric field respectively, in terms of partial differential equations and the difference equations, which are convenient approximations for small perturbations:

Responses of equilibrium constants to forcing parameters

Approximations for finite small changes

$$\left(\frac{\delta \ln K}{\delta T}\right)_P = \frac{\Delta H^0}{RT^2} \qquad\qquad \frac{\Delta K}{K} = \frac{\Delta H}{RT^2}\Delta T \qquad (6.2.1)$$

$$\left(\frac{\delta \ln K}{\delta P}\right)_T = -\frac{\Delta V^0}{RT} \qquad\qquad \frac{\Delta K}{K} = -\frac{\Delta V}{RT}\Delta P \qquad (6.2.2)$$

$$\left(\frac{\delta \ln K}{\delta E}\right)_{T,P} = \frac{\Delta M}{R} \qquad\qquad \frac{\Delta K}{K} = \frac{\Delta M}{R}\Delta E \qquad (6.2.3)$$

A second term, which might be included in equation (6.2.2) to describe the temperature change during pressure perturbation, can be neglected for aqueous systems and the relatively small pressure changes involved. Adiabatic heating in aqueous solutions results in a temperature change of $0.0015 \text{ K atm}^{-1}$.

Perusal of tables 6.1 (a) and (b) of the temperature and pressure dependence of equilibrium constants of a number of reactions of importance in biological processes illustrates one aspect of the relative merits and complementarity of the two parameters for the initiation of relaxations. Edsall & Gutfreund (1983) discuss the effects of temperature and pressure on equilibria in relation to potential relaxation amplitudes. Many temperature and pressure induced perturbations of equilibria of reactions involving proteins, nucleic acids and their ligands, are coupled to changes of one or more of the pK values of the many ionizing groups of such reactants. In turn, rapid pH perturbations can be achieved with pressure jumps on solutions containing, for instance, phosphate as a buffer, or with temperature jump using, for instance, Tris (2-amino-2-hydroxymethyl propane-1,3-diol) buffer. The use of more than one forcing function to initiate reactions on a system greatly extends the information which can be obtained from kinetic investigations. Different steps in a reaction will be perturbed by different forcing functions. The dependence of the individual equilibria on temperature, pressure, etc., provides information about the nature of the reaction steps. Interactions involving changes in the number of charged species are always very pressure dependent, while hydrophobic interactions between and within protein surfaces tend to be more temperature than pressure dependent. This is illustrated in section 6.3.

The number of intermediates which can be detected in a system will depend not only on the relation between the relaxation times, but also on the sensitivity of steps to different forcing functions. This can be illustrated with a model of a two step ligand binding reaction (three states of the receptor R):

$$\underset{}{\overset{1}{A}} + \underset{\text{SLOW}}{\overset{k_{12}}{\underset{}{R \rightleftharpoons}}} \underset{}{\overset{2}{AR}} \underset{\text{FAST}}{\overset{k_{23}}{\rightleftharpoons}} \overset{3}{[AR]}$$

Let us assume that $(k_{12}+k_{21}) \ll (k_{23}+k_{32})$ and that the first step is pressure independent, but temperature dependent, while the second step is pressure dependent but relatively temperature independent. If the observed concentration variable is [AR], pressure perturbation will result in two observed relaxations, while temperature perturbation will allow only the slow relaxation to be observed. There are variations on this theme when different concentration variables are observed. A monitor for the change in concentration of AR will also display two relaxations after pressure perturbation, but observation of R will only show the slow one. This should serve as a reminder that the use of different monitors, like that of different perturbation methods, will result in increased information from kinetic experiments. In the case of the above example relaxation times can also be separated through changes in concentration variables A and R; this is analysed below. By definition, systems that can be usefully studied with the temperature jump technique contain at least one step with significant temperature dependence. Experiments at different initial temperatures often result in the separation of relaxations which might be degenerate under some conditions and thus not observed (see section 7.2).

Numerical examples of relaxation amplitudes

Some numerical illustrations of the magnitudes of changes in equilibrium constants in response to step changes in temperature or pressure will lead us to the evaluation of relaxation amplitudes, that is in the total changes in reactant concentrations due to a perturbation. Whether a particular amplitude is observable depends, of course, on the availability of monitors and their sensitivity. Fluorescence measurements often permit a total change in concentration of less than $0.1 \mu M$ to be observed. Let us assume a temperature jump from 298 to 303 K on a reaction with $\Delta H^0 = 45$ kJ mole^{-1}. Substituting numbers into equation (6.2.1) we obtain a change in equilibrium constant of 30%:

$$\Delta K / K \approx (45\,000 \times 5)/(8.134 \times 90\,000) \approx 0.3$$

If we use a pH jump, as an example for the consequence of a temperature perturbation of an equilibrium, one can easily derive from the above that, in a solution in Tris buffer ($pK = 8$ and $\Delta H = 45.6$ kJ mole^{-1}), an increase of 5 degrees will reduce the pK by 0.12. The corresponding shift of a solution at pH 8 would be to pH 7.88. The effect of a temperature jump, or of any other perturbation of the equilibrium constant, is maximal when the ratio of the reactants is unity. In a simple first order isomerization one has little control over this, but most of the reactions under discussion are, at least distantly, linked to a pseudo first order process with equilibrium positions dependent on reactant concentrations. Unfortunately the choice of experimental conditions (concentrations) is often limited by solubilities and the optical properties used for monitoring the reactants, a problem also encountered in the displacement reactions discussed in section 5.3. During the discussion of relaxations of different model processes in the following sections, relaxation amplitudes will be used to distinguish between possible alternate mechanisms. Detailed analyses of expected amplitudes are presented by Bernasconi (1976). Relaxation experiments can also be used to determine the equilibrium constants and enthalpies of individual steps of sequential reactions (Bernasconi, 1976, p. 85; Winkler-Oswatitch & Eigen, 1979).

Pressure jump and periodic perturbations

Some additional comments must be made specifically about pressure perturbation. As can be seen in tables 6.1 and 6.2 volume changes for charge–charge interactions in an aqueous medium are approximately 25 $cm^3 mole^{-1}$. A pressure perturbation of 300 atm will result in a 32% change in equilibrium constant (see equation (6.2.2)). The technique normally used for step changes in pressure (see Davis & Gutfreund, 1976) involves the rapid release from an elevated level to the ambient atmosphere, resulting in kinetic data appertaining to the latter conditions. After a temperature jump relaxation times are observed at a higher temperature and re-equilibration to the chosen base temperature requires a few minutes. Also different methods have to be used to maintain a constant temperature at the elevated level when the time range of temperature jump experiments is extended to relaxation times longer than a few seconds. Pressure jump experiments can be carried out over time scales from 50 μs to the limit of stability of the system with the same equipment.

The easy rapid repetition of pressure perturbations makes it possible to average many records of small amplitudes to improve the signal to noise ratio. Recently (Pryse *et al.*, 1992) obtained some interesting results with a repetitive pressure jump technique, developed some time ago (Clegg & Maxfield, 1976). This involves the periodic perturbation with relatively small pressure pulses (about 10 atm) with a piezoelectric stack driven by a signal generator. The method can be used in two modes. If the system is subjected to square wave perturbations with a frequency several orders of magnitude less than the reciprocal relaxation time, the chemical response to both the 'on' and 'off' of the pulse can be observed. Records can be averaged over a large number (several thousand) perturbations, until satisfactory signal to noise ratios are obtained. In this case the analysis of the data is equivalent to single step perturbations, but both signals, those from rising and from falling pressure, are available.

The alternative approach is to 'tune in on' the frequency of the

Table 6.1 *Enthalpy changes for some ionizations*

Substance	pK	$\Delta G°/$ kJ mol^{-1}	$\Delta H°/$ kJ mol^{-1}
Acetic acid	4.756	27.14	−0.385
Lactic acid	3.860	22.04	−0.414
Succinic acid	4.207	24.02	3.188
	5.636	32.19	−0.452
Phosphoric acid	2.148	12.26	−7.648
	7.198	41.10	4.109
Ammonium ion	9.245	52.78	52.216
Tris(hydroxymethyl)aminomethane	8.075	46.10	45.606
Glycine	2.350	13.41	4.837
	9.780	55.81	55.815
Carbonic acid	6.352	36.26	9.372
	10.329	58.96	15.075
Amino acid and derivatives			
Lysine pK_3 (ε-ammonium)	10.79		
Lysyllysine (ε-ammonium)	10.05		
	11.01		
Arginine pK_3 (guanidinium)	12.5		
Tyrosine pK_3 (hydroxyl)	10.0		
Glycylglycine	3.148	17.96	3.62
	8.252	47.09	44.3
Ala-Ala-Ala-Ala	3.42		
	7.94		
Carbobenzoxy-Pro-His-Gly(imidazole)	6.42	36.64	33
N-acetyl-L-isoasparagine	4.08		
N-acetyl-L-isoglutamine	4.50		
Methylthioglycolate	7.8†		
	9.5†		
Cysteine	8.3†		
	10.8†		

Note: † These are hybrid constants involving the $-NH_3^+$ and the $-SH$ groups.

perturbation to the reciprocal relaxation time, as shown in figure 6.2. If the frequency of the perturbation is much greater than the reciprocal of the relaxation time of the process under investigation then the system cannot readjust itself and no concentration changes will be observed. As the frequency is reduced to the range $\omega\tau \approx 1$ the concentration change lags behind and conclusions about the relaxation time can be drawn from the phase shift (with respect to the perturbation) and amplitude dependence of the chemical response to changing perturbation frequency. A full analysis

Table 6.2. *Volume changes for some ionizations and selected protein structure changes*

Reaction	$\Delta V/\text{cm}^3\,\text{mol}^{-1}$
Ionization of buffers	
Phosphate (pK 2.1)	-15.7
Phosphate (pK 7.0)	-24.0
Acetate	-10.7
Bicarbonate	-25.4
Carbonate	-25.6
Borate	-32.1
Tris	$+1.0$
$\text{MgATP}^{2-} \rightleftharpoons \text{Mg}^{2+} + \text{ATP}^{4-}$	-23
Hydrophobic effects	
CH_4 in C_4H_{14}: CH_4 in H_2O	-22.7
C_2H_4 in C_4H_{14}: C_2H_4 in H_2O	-18.2
Hydrogen bonding	
α-Helical poly-L-glutamate: random coil	$+1.0$
Poly(A + U) nucleotide helix: random coil	-1.0
Protein self-assembly per monomer	
Myosin filament propagation	-280
Tubulin propagation at 35 °C	-26
Tubulin propagation at 15 °C	-50
Protein denaturation	
Chymotrypsin	-43
Myoglobin	-100

of this is presented by Clegg & Maxfield (1976). As the frequency is reduced to $\omega\tau \ll 1$ the full amplitude of the chemical relaxation will be observed, but no conclusion can be drawn about τ; the system will just follow the pressure change. In the related method of sound absorption the energy absorption per period of oscillation is measured by varying the frequency. The maximum occurs when $\omega = \tau^{-1}$. In dispersion methods the velocity of sound is measured as a function of frequency. These rather specialized methods are dealt with elsewhere in considerable detail (Stuehr, 1974).

Relaxation equation for a single reversible step

A reversible structural isomerization between two forms of a molecule

$$R \rightleftharpoons [R]$$

serves as a simple example for some of the terms used in the derivation of equations describing chemical relaxations. In this case, the rate equations,

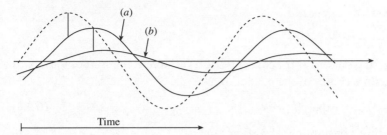

Figure 6.2   Oscillations of concentrations (solid lines) in response to a periodic perturbation of pressure with frequency ω are represented by broken lines. The periodic response of the reactant concentration is a function of the relaxation time τ. If the frequency of the perturbation is high ($\omega \gg 1/\tau$) the reaction cannot follow the pressure change and will not be observed. The traces (a) and (b) correspond to reactions with $1/\tau = \omega$ and $1/\tau = 4\omega$ respectively.

first order in both directions, are linear without the introduction of any other conditions (such as buffering or small perturbation). The linearization assumptions are introduced in the subsequent example.

As illustrated in figure 6.1 the concentrations of the reactants at time t are:

$$c_R(t) = c_R(\infty) + \Delta c_R(t) \qquad c_{[R]}(t) = c_{[R]}(\infty) + \Delta c_{[R]}(t) \tag{6.2.4}$$

where $c(\infty)$ indicates concentrations at final equilibrium and $\Delta c(t)$ the time dependent concentration variables during the approach to equilibrium, the latter, which can be positive or negative, is the distance from equilibrium. The rate equation for $c_R(t)$ is

$$\frac{dc_R(t)}{dt} = \frac{dc_R(\infty)}{dt} + \frac{d\Delta c_R(t)}{dt} = -k_{12}[c_R(\infty) + \Delta c_R(t)] + k_{21}[c_{[R]}(\infty) + \Delta c_{[R]}(t)]$$

$$\tag{6.2.5}$$

Equilibrium conditions and conservation of mass, respectively, specify that:

$$k_{12}c_R(\infty) = k_{21}c_{[R]}(\infty) \quad \text{and} \quad dc_R(\infty)/dt = 0 \tag{6.2.6}$$

and

$$-\Delta c_R(t) = \Delta c_{[R]}(t)$$

Hence

$$d\Delta c_R/dt = -\Delta c_R(t)[k_{12} + k_{21}] \tag{6.2.7}$$

which gives on integration

$$\ln\Delta c_R(t) = -[k_{12}+k_{21}]t + \text{constant}$$

If $\Delta c_R(0)$ is defined as the concentration variable at the start of the record then the time course of the relaxation is

$$\Delta c_R(t) = \Delta c_R(0)\exp[-\{k_{12}+k_{21}\}t] \tag{6.2.8}$$

and the time constant or relaxation time is

$$\tau = [k_{12}+k_{21}]^{-1} \tag{6.2.9}$$

Equations (6.2.5) and (6.2.9) should be compared with those derived for the rate of approach to equilibrium in section 3.2, where the evaluation of single time constants are discussed. Although the relaxation time can be obtained from a signal proportional to reactant concentration, for the evaluation of individual rate constants the equilibrium constant, k_{12}/k_{21}, has to be determined from the absolute concentrations.

Problems involved in ascertaining whether the record represents a single time constant and in the resolution of several relaxations will be discussed during the treatment of more complex reaction pathways. If the above simple reversible reaction is an adequate description of the results, then experiments carried out at different concentrations of the reactants at equilibrium must result in identical relaxation times. They will have different relaxation amplitudes, but the fractional change will remain the same.

In the case of a pseudo first order reaction like

$$A + R \rightleftharpoons AR \quad [\text{with } c_R(0) \ll c_A(0)] \tag{6.2.10}$$

the relaxation time will be

$$\tau = [k_{12}c_A(0)+k_{21}]^{-1}$$

and the relaxation time and amplitude will depend on $c_A(0)$. A plot of τ^{-1} against $c_A(0)$ will have the slope k_{12} and the ordinate intercept k_{21}. At a given concentration the relaxation time will still be independent of the perturbation amplitude.

Linearization of rate equations: small steps

When the pseudo first order condition of equation (6.2.10) does not hold, another limiting assumption has to be introduced to obtain linear equations for the derivation of the relaxation time. If the perturbation amplitude is small such that

$$c_A(0) \gg \Delta c_A(0) \quad \text{and} \quad c_R(0) \gg \Delta c_R(0) \qquad (6.2.11)$$

Then the terms in the rate equations which are the products of two small quantities, for instance $\Delta c_R(t)\Delta c_A(t)$, can be neglected. We shall return below to the question of '*how small is small?*'.

The rate equation analogous to (6.2.5) modified for the relaxation of a second order association reaction is

$$\frac{dc_R}{dt} = \frac{dc_R(\infty)}{dt} + \frac{d\Delta c_R(t)}{dt}$$
$$= -k_{12}[\{c_R(\infty) + \Delta c_R(t)\}\{c_A(\infty) + \Delta c_A(t)\}] + k_{21}[c_{AR}(\infty) + \Delta c_{AR}(t)] \qquad (6.2.12)$$

If the right hand side of (6.2.12) is rearranged and we use expression (6.2.11) and, as before, the conservation of mass

$$-\Delta c_{AR}(t) = \Delta c_R(t) = \Delta c_A(t)$$

and the equilibrium condition

$$-k_{12}c_R(\infty)c_A(\infty) + k_{21}c_{AR}(\infty) = 0$$

we obtain

$$\frac{d\Delta c_R(t)}{dt} = -[k_{21} + k_{12}\{c_R(\infty) + c_A(\infty)\}] \, \Delta c_R(t) \qquad (6.2.13)$$

Proceeding as for equations (6.2.7) and (6.2.8) we obtain

$$\Delta c_R(t) = \Delta c_R(0) \exp\left[-[k_{21} + k_{12}\{c_R(\infty) + c_A(\infty)\}]t\right] \qquad (6.2.14)$$

The single relaxation time for this reaction scheme is

$$\tau = [k_{21} + k_{12}\{c_R(\infty) + c_A(\infty)\}]^{-1} \qquad (6.4.15)$$

Analogous to the case of the pseudo first order relaxation, if experiments are carried out at different reactant concentrations, the two rate constants can be obtained from a plot of the reciprocal relaxation time against the sum of the equilibrium concentrations of R and A. In this case, if the equilibrium constant is not known accurately, so that $c_R(\infty)$ and $c_A(\infty)$ can only be estimated, the following iterative procedure can be used. A plot of the reciprocal relaxation times against estimated equilibrium concentrations is used to obtain a value for the equilibrium constant, $K = k_{12}/k_{21}$. The value obtained for K is then used to calculate the next estimate of equilibrium concentrations for another plot to obtain a new value for K. This can be continued until all values remain constant. A computer program can easily be devised to achieve this by iteration without the necessity of producing a sequence of plot. This also serves as a diagnostic for the simple one step mechanism. If that assumption is incorrect, the iteration will not converge.

Data compatible with single step mechanisms discussed in this section have to be treated with caution. Insufficient time resolution and degeneracy of time constants often hide additional parameters. It is also important to test proposed mechanisms by carrying out measurements over the widest possible range of concentrations. As will be seen in subsequent examples, more complex mechanisms will give data corresponding to a single step at low concentrations and will diverge as the range is increased.

A few comments are in order here on the assumption of linearity when small perturbations are applied to non-linear systems. The pragmatic approach is to vary the step size and ascertain the range over which essentially constant relaxation times are obtained. If increases and decreases in the amplitude of perturbation do not affect the relaxation times, then one can say that the perturbation is small enough. Numerical modelling of perturbations can also be informative. An example of this approach has been presented by Wu *et al.* (1992) for the relaxation of ligand exchange between two receptors (see also section 5.3). Clearly, if the changes in the concentrations of all the reactants during the relaxation are less than 5% of their individual totals, the effect of the linearity assumption on the results will, in most cases, be below the overall limits of error. The problem is somewhat similar to the steady state assumption used in enzyme kinetics (see section 3.3) and to deviations from pseudo first order conditions as illustrated in section 3.2. Another criterion for linearity is that it holds for the condition $\Delta c_A \Delta c_R \ll (\Delta c_A + \Delta c_R)$. Several other definitions are given in a chapter devoted to the problem by Bernasconi (1976). The assumption of small changes is identical to that required for the equations of non-equilibrium thermodynamics.

Multiple relaxation times

In the following paragraphs the treatment of a system of relaxations will be found to be closely related to the matrix method (described in section 4.2) for obtaining exponents and amplitudes of consecutive reactions in terms of eigenvalues and eigenvectors. While there is considerable duplication, it is convenient to present the derivations of relaxation times in the present section in the form generally found in the literature on this subject. To clarify the comparison of the two treatments the reader must be reminded that reciprocal relaxation times correspond to the eigenvalues or exponents ($\tau^{-1} = -\lambda$; see section 4.2).

We have shown above that a single isolated transition has one relaxation time, which is a function of two rate constants and, under certain conditions, also of the reactant concentrations. A reaction with two steps (three states) has two relaxation times. As explained in section 4.1, a system with n states will have $n-1$ relaxation times. However, each of the two relaxation times in a three state system, are, in principle, functions of all four rate constants. There are two possible extremes, which affect the interpretation of these relaxations. The first is that, within the concentration range allowed by solubility and size of signal, they are within a factor of two or three of each other and cannot be resolved in practice. The second limiting condition is that they are sufficiently far apart (more than an order of magnitude) and can be regarded as uncoupled. This phenomenon has already been discussed in connection with consecutive reactions (section 4.2). The time constants of steps of reactions which are significantly slower than anything that happened before and are faster than the following step, can be interpreted in terms of a single pair of rate constants (forward and reverse). Similarly, the first step of a reaction can be treated in isolation if it is significantly faster than the following one. As the above comments indicate, additional complexities arise from multiple relaxation times when one wishes to resolve the rate constants of coupled reaction steps; let us say under conditions when relaxation times are separated by a factor of 3 (just resolvable) and 30 (uncoupled and therefore separable; see section 4.1).

Some comments are necessary about the meaning of multiple relaxation times in terms of data analysis. The situation is analogous to that described in section 4.2 where consecutive reactions were described in terms of sums of exponentials. In the present context the approach to equilibrium of a reactant can be described by the time course of the total signal amplitude $A_0(t)$:

$$A_0(t) = A_1 \exp(-t/\tau_1) + A_2 \exp(-t/\tau_2) + \ldots + A_n \exp(-t/\tau_n)$$

where A_i is the amplitude of the signal for an exponential with time constant τ_i. The technique for resolving the record of a relaxation experiment into individual exponentials is common to most of the topics discussed in this volume and was given detailed attention in section 1.5. This is one of the two principal algebraic problems of kinetic investigations. The other is the prediction of the record of the relaxation spectrum by solving the kinetic equations for a proposed model by one of the methods discussed for different examples.

Before analysing a model requiring linearization of the equations for a two step reaction we shall derive the relaxation equation for the condition that there is a pseudo first order initial binding step. Using the scheme:

$$A + R \rightleftharpoons AR \rightleftharpoons [AR] \qquad (6.2.16)$$

pseudo first order conditions for the first step assume that $c_A(0) \gg c_R(0)$ and we define $k_{obs} = k_{12}c_A(0) + k_{21}$, so that the reaction can be described by the following equations:

$$\frac{dc_R}{dt} = \frac{dc_R(\infty)}{dt} + \frac{d\Delta c_R(t)}{dt} = -k_{obs}[c_R(\infty) + \Delta c_R(t)] + k_{21}[c_{AR}(\infty) + \Delta c_{AR}(t)]$$

$$\frac{d\Delta c_R}{dt} = -k_{obs}\Delta c_R(t) + k_{21}\Delta c_{AR}(t) \qquad (6.2.17)$$

$$\frac{d\Delta c_{[AR]}}{dt} = k_{23}\Delta c_{AR}(t) - k_{32}\Delta c_{[AR]}$$

Any two rate equations can be used to describe a three state system together with the conservation of mass to eliminate, for example, $\Delta c_{AR}(t)$:

$$\Delta c_{AR}(t) = -[\Delta c_R(t) + \Delta c_{[AR]}(t)]$$

and introduce the reciprocal time constants to write

$$\frac{d\Delta c_R}{dt} = \tau^{-1}\Delta c_R(t) = -\Delta c_R(t)[k_{obs} + k_{21}] - \Delta c_{[AR]}(t)k_{21}$$

$$\frac{dc_{[AR]}}{dt} = \tau^{-1}\Delta c_{[AR]}(t) = -\Delta c_R(t)k_{23} - \Delta c_{[AR]}(t)[k_{23} + k_{32}] \qquad (6.2.18)$$

The above two equations can be written in the matrix form:

$$\begin{bmatrix} -[k_{obs} + k_{21}] - \tau^{-1} & -k_{21} \\ -k_{23} & -[k_{23} + k_{32}] - \tau^{-1} \end{bmatrix} \begin{bmatrix} \Delta c_R(t) \\ \Delta c_{[AR]}(t) \end{bmatrix} = 0 \qquad (6.2.19)$$

Since $\Delta c_R(t)$ and $\Delta c_{[AR]}(t)$ are not zero the determinant of the matrix of coefficients must be equal to zero and we obtain the quadratic equation:

$$\tau^{-2} + \tau^{-1}[k_{obs} + k_{21} + k_{23} + k_{32}] + k_{obs}k_{23} + k_{obs}k_{32} + k_{21}k_{32} = 0$$

Clearly there are two solutions for τ in terms of rate constants. From the theory of quadratic equations we obtain:

$$\left.\begin{aligned} \tau_1^{-1} + \tau_2^{-1} &= k_{obs} + k_{21} + k_{23} + k_{32} \\ \tau_1^{-1}\tau_2^{-1} &= k_{obs}[k_{23} + k_{32}] + k_{21}k_{32} \end{aligned}\right\}$$

(6.2.20)

For a sequence of first order reactions the relaxation times are clearly independent of reactant concentrations and the equations apply equally to the interpretation of large transients. The effects of changing the concentration, for instance of the ligand in the pseudo first order system, will be discussed later. Without such additional diagnostics, which are available in the case of concentration dependent systems, the four rate constants can only be estimated by numerical fitting procedures. If signals in terms of absolute concentrations for A, R and [AR] are available, the equilibrium constants can be evaluated and serve as a useful restriction for the numerical solutions. If the two relaxations are uncoupled, $\tau_1 \ll \tau_2$, then we can simplify from equations (6.2.20):

$$\left.\begin{aligned} \tau_1^{-1} &= k_{obs} + k_{21} \\ &\text{and} \\ \tau_2^{-1} &= [k_{obs}k_{23} + k_{obs}k_{32} + k_{21}k_{32}]/[k_{obs} + k_{21}] \end{aligned}\right\}$$

(6.2.21)

The slow relaxation, τ_2, is dependent on the rapid pre-equilibration due to τ_1 (see also section 3.2).

Temperature and pressure jump techniques find their widest application in the study of the sequence of events following ligand binding and protein–protein interaction, both molecular recognition processes. In many cases one wishes to use a wide range of concentrations spanning from pseudo first order to second order conditions for the first step. We therefore derive the equations describing scheme (6.2.16) replacing k_{obs} with the general term $k_{12}c_A$ and include the linearizing conditions introduced in (6.2.11), which result in neglecting the product term $\Delta c_A \Delta c_R$. As before only equations for the relaxation of two out of the three states of R are required:

$$\frac{d\Delta c_R(t)}{dt} = -k_{12}[c_R(\infty) + c_A(\infty)]\Delta c_R(t) + k_{21}\Delta c_{AR}(t)$$

$$\frac{d\Delta c_{AR}(t)}{dt} = -[k_{12}\{c_R(\infty) + c_A(\infty)\} - k_{32}]\Delta c_R(t) + [k_{21} + k_{23} + k_{32}]\Delta c_{AR}(t)$$

These two equations can be written in the form:

$$\tau^{-1}\Delta c_R(t) = a_{11}\Delta c_R(t) + a_{12}\Delta c_{AR}(t)$$
$$\tau^{-1}\Delta c_{AR}(t) = a_{21}\Delta c_R(t) + a_{22}\Delta c_{AR}(t)$$

Using the same procedure as on p. 213 we obtain

$$\tau^{-2} - [a_{11} + a_{22}]\tau^{-1} + [a_{11}a_{22} - a_{12}a_{21}] = 0$$

and hence

$$\tau_1^{-1} + \tau_2^{-1} = a_{11} + a_{22} = k_{12}[c_R(t) + c_A(\infty)] + k_{21} + k_{23} + k_{32}$$
$$\tau_1^{-1}\tau_2^{-1} = a_{11}a_{22} - a_{12}a_{21} = k_{12}[k_{23} + k_{32}][c_R(t) + c_A(\infty)] + k_{21}k_{32}$$

If the two relaxation times are determined over a range of reactant concentrations, the following linear plots should be obtained for the above three state mechanism. The sum of the two reciprocal relaxation times plotted against $[c_R(t) + c_A(\infty)]$ will have slope k_{12} and ordinate intercept $[k_{21} + k_{23} + k_{32}]$, while the product of the two reciprocal relaxation times plotted against the sum of the equilibrium concentrations will have the slope $k_{12}[k_{23} + k_{32}]$ and the ordinate intercept $k_{21}k_{32}$. These plots provide all the information needed to evaluate the four rate constants from the two relaxations. The same criterion as stated above applies to the linearity of the plots. If they are not linear, either the equilibrium constants used for the calculations of $c_R(\infty)$ and $c_A(\infty)$ need revision by iteration or the assumed mechanism is incorrect.

The argument about simplified interpretation again applies to the present mechanism. If

$$k_{12}[c_R(\infty) + c_A(\infty)] + k_{21} \gg [k_{23} + k_{32}]$$

then k_{12} and k_{21} can be evaluated directly from a plot of τ_1^{-1} against the sum of the equilibrium concentrations, giving slope k_{12} and intercept k_{21}:

$$\left.\begin{array}{l} \tau_1^{-1} = k_{12}[c_R(t) + c_A(\infty)] + k_{21} \\ \text{and} \\ \tau_2^{-1} = k_{32} + k_{23}/\{1 + K^{-1}[c_R(t) + c_A(\infty)]\} \end{array}\right\} \qquad (6.2.22)$$

where $K = k_{12}/k_{21}$, the equilibrium constant for the first step. It follows that τ_2^{-1} approaches $[k_{23} + k_{32}]$ at saturating concentrations of reactants, $[c_R(\infty) + c_A(\infty)] \gg K^{-1}$, while as the equilibrium concentrations approach zero τ_2^{-1} tends to k_{32}. Figure 6.3 shows such a derivative plot from experiments on the pressure relaxation of the association of NAD$^+$ and liver alcohol dehydrogenase. A conformation change, after binding of the

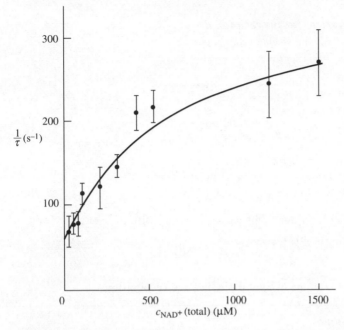

Figure 6.3 The reciprocal relaxation time for protein fluorescence of a solution (pH 7.7) containing 11 μM horse liver alcohol dehydrogenase (E) as a function of NAD^+ (total) concentration. The smooth curve is a fit to a hyperbola. The upper limit as $c_{NAD^+} \to \infty$ is $k_{23} + k_{32}$ and the ordinate intercept is k_{32}. The data are interpreted in terms of the mechanism:

$$E^H + NAD^+ \underset{k_{21}}{\overset{k_{12}}{\rightleftarrows}} E^H_{NAD^+} \underset{k_{32}}{\overset{k_{23}}{\rightleftarrows}} \overset{*}{E}^H_{NAD^+} \underset{k_{43}}{\overset{k_{34}}{\rightleftarrows}} \overset{*}{E}^-_{NAD^+} + H^+$$

$\overset{*}{E}$ indicates an enzyme species with quenched fluorescence.

nucleotide to the enzyme, results in the quenching of fluorescence, which is due to a change in the environment of one tryptophan residue.

As emphasized in the introduction to relaxation kinetics, the methods described in this section can, in principle, be extended to derive equations for mechanisms with any number of relaxation times. Clearly these become progressively more complex as the number of roots increases. Assumptions have to be made in terms of limiting conditions, to extract useful information from them. The practical difficulties of resolving multiple exponentials from noisy experimental records have been alluded to before and helpful hints on this topic are presented in section 2.3. The discussion of examples of investigations by temperature and pressure jump techniques in

Table 6.3. *Expressions for the reciprocal relaxation times of some simple reactions*

Reaction	τ^{-1}
$A \rightleftharpoons B$	$k_{12} + k_{21}$
$A + B \rightleftharpoons C$	$k_{12}(c_A + c_B) + k_{21}$
$A + B \rightleftharpoons C + D$	$k_{12}(c_A + c_B) + k_{21}(c_C + c_D)$
$A + B \rightleftharpoons C$ (B buffered)	$k_{12}^* c_A + k_{21}$ (where $k^* = k_{12} c_B$)
$2A \rightleftharpoons A_2$	$4k_{12} c_A + k_{21}$

Note: All concentrations at $t = \infty$.

the following two chapters will provide extensions of methods for the interpretation of relaxation.

6.3 Chemical relaxations of ligand binding and recognition

Range of applications

The description of a range of investigations with temperature and pressure relaxation techniques, which is presented in this section, is intended to illustrate what can be achieved with these methods. The major contributions of equilibrium perturbation techniques to our knowledge of enzyme mechanisms come from studies of partial reactions such as substrate, product and effector binding with the associated protein conformation changes. Initially there was considerable optimism about the potentialities of the temperature jump technique; it was thought that all other kinetic techniques would be replaced by chemical relaxation methods (Eigen, 1968). As pointed out in section 6.1, in theory, the spectrum of relaxation constants of a system contains all the kinetic information one needs. In practice because of degeneracy and lack of resolution, especially if the system cannot be poised at an equilibrium with all components at significant concentrations, relatively little information has been obtained from application to overall enzyme reactions. Information about the steps involving chemical transformation is largely dependent on observations of transients by flow techniques, as illustrated in section 5.1. Table 6.3 summarizes the relaxation times for some single step mechanisms.

Extending the range of examples of expressions for relaxation times for alternative and more complex pathways is easily carried out by the algebraic methods presented above. The difficulties arise in the experimental resolution of multiple relaxations and their use for the evaluation of individual rate constants. While expressions for relaxation amplitudes are of great value in identifying complex mechanisms (see Coates, Criddle & Geeves, 1985) they are more difficult to derive than relaxation times. Bernasconi (1976) devotes two chapters to the problem of amplitudes. Only a small selection of mechanisms is discussed below.

Ligand induced conformation changes versus pre-equilibration

A problem of considerable interest in ligand binding is the possibility to distinguish between the classical induced fit mechanism and the pre-equilibration between open and closed binding sites (see also section 3.3), that is clockwise or anti-clockwise progress in the following scheme.

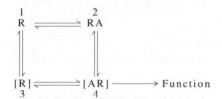

Here 'Function' can stand for enzyme catalysis, signal transmission, etc. Halford (1972) used temperature and concentration jump techniques to distinguish between the two possible reaction paths for the interaction between the active site of alkaline phosphatase (*E. coli*) and a substrate analogue. The catalytically active complex is formed via a collision complex and a protein conformation change. The question was, which comes first? Is the enzyme, in the absence of ligand, largely in a closed state in equilibrium with an open state, which has to form prior to binding, or does binding in the collision complex induce the conformation change to the active state? While the particular problem of alkaline phosphatase turned out to be much more complex, due to very tight binding of one of the products, Halford's analysis of the ligand concentration dependence of relaxation times predicted for the two mechanisms is of general importance. The kinetic mechanism of this enzyme was subsequently well documented from ^{31}P nuclear magnetic resonance studies (see Hull *et al.*, 1976).

The clockwise pathway and its concentration dependence is described at the end of the last section. The anti-clockwise pathway, which involves the pre-equilibrium of a protein conformation change was treated by Halford (1972) in terms of the simplified scheme:

$$R \underset{k_{31}}{\overset{k_{13}}{\rightleftharpoons}} [R] \underset{k_{43} \;\; \pm A}{\overset{k_{34}}{\rightleftharpoons}} [AR]$$
                                                                (6.3.1)

Using the procedure of section 6.2 to derive the relaxation equations, the following expressions for the sum and the product of the reciprocal relaxation times are obtained.

$$\tau_1^{-1} + \tau_2^{-1} = k_{13} + k_{31} + k_{43} + k_{34}[c_{[R]}(\infty) + c_A(\infty)]$$
                                                                (6.3.2)

$$\tau_1^{-1}\tau_2^{-1} = k_{34}k_{13}[c_{[R]}(\infty) + c_A(\infty)] + k_{32}[k_{13} + k_{31}]$$
                                                                (6.3.3)

If the relaxation times are sufficiently well separated (uncoupled) we can again derive simplified expressions:

If $\tau_2^{-1} \gg \tau_1^{-1}$ (ligand binding is fast compared to isomerization) then

$$\tau_2^{-1} = k_{43} + k_{34}[c_{R^*}(\infty) + c_A(\infty)]$$
                                                                (6.3.4)

and

$$\frac{\tau_1^{-1}\tau_2^{-1}}{\tau_2^{-1}} = k_{13} + \frac{k_{31}}{1 + K[c_{[R]}(\infty) + c_A(\infty)]} \quad \text{(where } K = k_{34}/k_{43}) \quad (6.3.5)$$

Inspection of equations (6.3.4) and (6.3.5) shows that changes in the equilibrium concentrations of protein and ligand serve as a diagnostic for this mechanism. The following limiting conditions can be used:

$$\tau_1^{-1} \longrightarrow k_{13} + k_{31} \quad \text{when } [c_R(\infty) + c_A(\infty)] \ll K^{-1}$$

and

$$\tau_1^{-1} \longrightarrow k_{31} \quad \text{when } [c_R(\infty) + c_A(\infty)] \gg K^{-1}$$

These relationships show that as the ligand concentration, $c_A(\infty)$, is increased τ_1^{-1} decreases. This is in direct contrast to the response of the system in which isomerization occurs after ligand binding. In the latter case rapid pre-equilibration of the collision complex responds to increases in ligand concentration so as to increase the rate of the subsequent protein conformation change.

The most common occurrence in ligand binding processes is the clockwise mechanism with an initial binding step (collision complex formation) which often does not provide a convenient signal. In section 7.4 we discuss the rates of these diffusion controlled initial steps. The second order rate constants ($k \cong 10^8 \, M^{-1} s^{-1}$) would result in relaxation times of the order of 100 μs at reactant concentration of 0.1 mM (clearly the value for k_{21} has to be added since $\tau^{-1} = k_{12} + k_{21}$). This can be easily resolved if it results in a direct optical response of sufficient magnitude. Of course in the planning of the concentration range for an equilibrium perturbation experiment one has to take into account not only the signal and time resolution, but also the equilibrium constant. As stated before, to obtain a significant perturbation the ratio of the perturbed concentration variables has to be near unity.

In many cases all the kinetic information about ligand binding has to be obtained from τ_2 and its concentration dependence, which represents the relaxation of the response of the receptor subsequent to initial ligand contact. This was illustrated, with an example, in the previous section. Among enzymologists this response is commonly referred to as a substrate induced conformation change. This phenomenon has been illustrated in section 5.1 as part of the sequence of a number of enzyme reactions. Similar phenomena are encountered as a consequence of agonist or antagonist binding to channel proteins and other receptors. There are, so far, no good examples in the literature of the obviously potentially fruitful application of relaxation methods to the exploration of such conformation changes using local labels introduced into enzymes with site directed mutagenesis, although one would predict that such studies will not be long delayed.

Lysozyme substrate binding

The first elucidation of a three-dimensional structure for an enzyme was that for lysozyme, which was soon followed by that of its complex with polymers of N-acetylglucosamine (see Blake *et al.*, 1967). In subsequent years this enzyme–substrate interaction was the subject of kinetic investigations in a number of laboratories and the results present several interesting general features. Fluorescence changes of two tryptophan residues, as well as pK changes of two carboxylic acid groups at the active site, provided a monitor for a variety of local events at the active site.

Relaxation experiments on the interaction of the saccharide with lysozyme were carried out by Chipman & Schimmel (1968). Their results, at that early time in the history of the application of the temperature jump method, should serve as a warning. The fact that reciprocal relaxation times increase linearly with ligand concentration should not be taken as evidence that one is observing a simple second order process, unless the rate constant derived from the slope corresponds to diffusion control ($k = \geq 10^8\,\mathrm{M}^{-1}\mathrm{s}^{-1}$). Chipman & Schimmel deduced a rate constant two orders of magnitude below that expected for a diffusion controlled reaction (see section 7.4). The complexity of the system was subsequently exposed by Holler, Rupley & Hess (1969) and Halford (1975) by carrying out experiments at higher

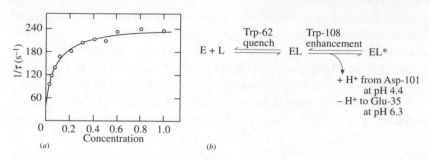

Figure 6.4   Data from temperature jump relaxation experiments (Halford, 1975) on the interaction of the β-1-4-linked trimer of N-acetyl-D-glucosamine with lysozyme at pH 4.4 monitoring H^+ release. The concentration scale corresponds to the sum of the equilibrium concentrations of enzyme and ligand. (*a*) shows that the concentration dependence of the reciprocal relaxation time reaches a limit of 240 s^{-1}, which is similar to the fluorescence data of Holler *et al.*, 1969. (*b*) illustrates a mechanism proposed by Halford (1975) from his relaxation and stopped-flow investigation of fluorescence changes and proton release over a range of pH.

ligand concentration, when a limiting relaxation time is observed (as in figure 6.4), and by utilizing the different signals obtained from the two tryptophans and from changes in protonation of glutamic acid residues situated in the binding cleft of the protein.

Ligand exchange reactions: calcium exchange

The importance of the rates of reactions involving the regulation of physiological processes by calcium binding and exchange has already been emphasized in section 5.3. Kao & Tsien (1987) used the temperature jump method to investigate whether the reactions of Ca^{2+} with the dye fura-2 are fast enough for its use as an indicator of the rates of change of concentration of the ion. This paper presents a good example of a clean second order binding process and demonstrates the evaluation of k_{12} from the slope and k_{21} from the intercept of a plot of τ against the sum of the equilibrium concentrations (see equation (4.4.15)). This complements the stopped-flow investigation of Jackson *et al.* (1987), who measured the rate of dissociation by the ligand exchange method described in section 5.3. The value for a dissociation rate constant obtained by this latter method is likely to be more accurate than that obtained from an intercept, provided it does not exceed a value of ≈ 500 s^{-1}. Association rates of 10^8 to 10^9 M^{-1} s^{-1} were, until recently, only accessible with relaxation techniques. The improved use of fluorescent ligands now makes it possible to resolve such rapid reactions at low concentrations (considerably below the micromolar range). Reasonable agreement of the value for the dissociation constant obtained from equilibrium titration with that obtained from the ratio of the rate constants obtained from kinetic experiments is of great value in the diagnosis of a correctly postulated mechanism.

Eigen & de Maeyer (1963; p. 903) have derived general relaxation equations for the exchange of ligands between two sites via two possible pathways:

$$RX + Y \underset{k_2}{\overset{k_1}{\rightleftharpoons}} RY + X$$

$$k_{13} \quad k_{31} \quad k_{32} \quad k_{23}$$

$$R + X + Y$$

Under steady state conditions ($dc_R/dt = 0$) they obtain an expression for the reciprocal relaxation time:

$$\tau^{-1} = \left[\frac{k_{12} + k_{13}k_{32}}{(k_{31}c_X + k_{21}c_Y)} \right] [c_{RX} + c_Y] + \left[\frac{k_{21} + k_{23}k_{31}}{(k_{31}c_X + k_{21}c_Y)} \right] [c_{RY} + c_X] \qquad (6.3.6)$$

(all concentrations are steady state concentrations).

Chock & Gutfreund (1988) used the steady state assumption for the interpretation of stopped flow experiments of ligand exchange and Wu *et al.* (1992) demonstrated how the ligand concentration dependence of the relaxation time can be used as a diagnostic to decide between ligand exchange via free diffusion or via ternary complex formation. This is discussed in some detail in section 5.3 and it has wider implication for the distinction between dissociative and associative mechanisms in chemical reactions.

Some other applications of the temperature jump method to the kinetics of the association of calcium to control sites are discussed by Tsuruta & Sano (1990), who use fluorescent dyes as well as the tyrosine fluorescence of calmodulin as an indicator of conformation changes resulting from calcium binding.

A classic temperature jump investigation from Eigen's laboratory resulted in suggestions that the specificity of carriers for membrane translocation of certain ions rested with the rate of dissociation of the complex formed. Several antibiotics act as specific carriers of alkali metal ions across membranes. The cyclic polyether nonactin, for instance, which can form complexes with either Na^+ or K^+, will selectively transport only the latter. Diebler *et al.* (1969) found that the association rate constants for both of the two ions were nearly diffusion controlled (3×10^8 M^{-1} s^{-1}) although desolvation will have had some effect on the rates. However, the two ions are distinguished by the rate constants for dissociation from their complexes with nonactin, which are $10^3 s^{-1}$ for K^+ and $3 \times 10^5 s^{-1}$ for Na^+. This means that the average lifetime for the potassium complex is 700 μs and 2 μs for the sodium complex. A value of 100 μs is a reasonable estimate for the time taken for the complex to diffuse through a lipid membrane of 10 nm thickness. Consequently the K^+-nonactin complex has a reasonable chance to survive a passage, while the Na^+-nonactin complex will break down before it can be transported. This is not intended to be a proposal for a definitive mechanism, but a potential example for the importance of lifetimes of complexes.

Polymerization equilibria

Pressure jump experiments have been particularly informative in the exploration of protein–protein interaction and its regulation by ligands. Both the assembly of myosin molecules into the thick filament and the interaction of myosin with actin are very pressure sensitive and serve as good examples for discussion of detailed investigations by appropriate relaxation techniques. The first of these problems will only be dealt with in a qualitative manner to describe the potential of the method for the study of polymerization reactions. The pressure dependence of the aggregation of myosin molecules was first discovered during attempts to determine the molecular weight of this protein using the ultracentrifuge. The variation of pressure throughout the centrifuge cell and at different speeds of rotation, results in a distribution of sedimentation velocities (Josephs & Harrington, 1968). In vivo myosin molecules are assembled into filaments with a precise length of 1.57 μm. On extraction the filament length becomes markedly dependent on ionic strength and pH, as well as on pressure and temperature. A detailed review of filament structure by Davis (1988) also contains a summary of the kinetic studies designed to elucidate the assembly mechanism.

The pressure perturbation technique can be applied to this problem in a variety of ways by following the mass of filaments by light scattering: (1) small perturbations of fully assembled filament solutions; (2) release of pressure from completely dissociated myosin solutions (initial rate measurements); and (3) essentially unidirectional dissociation rates obtained by rapid increases in pressure. This shows the value of the pressure perturbation technique, which, unlike rapid temperature changes, can operate in both directions. The conclusions from these experiments were as follows. Assembly depends on a second order reaction, dimerization of myosin prior to addition to filaments, and is also dependent on filament concentration. Thus it is really third order and it is independent of filament length. In contrast the dissociation rate of myosin from the filaments is length dependent, the rate decreases of the order of 1000fold as the thick filament grows (in both directions) from the central zone to its full length. Consequently there is kinetic control of length at the point of equilibrium between rate of addition and removal. The conditions used for the kinetic studies in vitro do not result in the formation of filaments of exactly the size found in vivo and it remains to be seen what other factors contribute to the precise limitation. The presence of other proteins as well as the composition of cytoplasm must influence the process.

Pressure perturbation of actin–myosin interaction in solution and fibres

The molecular basis for the production of mechanical force during muscle contraction should, in principle, be clarified by an understanding of the interaction between the myosin molecules of the thick filament and the actin molecules of the thin filament and the consequent structure changes. Reactions of regulatory proteins are also of importance (McKillop & Geeves, 1993) but these need not concern us in the present discussion. The investigations by pressure perturbation of solutions of F-actin and the myosin subfragment S1 (see ATPase mechanism of S1, section 5.1) serve as a good example of several aspects of the potentialities of relaxation techniques (Coates *et al.*, 1985; Geeves, 1991). The three steps of the interaction are

$$A + S \underset{\text{fast}}{\overset{K_0}{\rightleftharpoons}} A...S \underset{\text{low}}{\overset{K_1}{\rightleftharpoons}} A—S \underset{\text{fast}}{\overset{K_2}{\rightleftharpoons}} AS$$

collision   weak   light
complex   binding   binding

The relaxations were observed by changes in light scattering and by the quenching of fluorescence of a pyrene label covalently attached to cys-374 of actin. Light scattering monitored the combined changes in the concentrations of all three complexes and resulted in a single relaxation determined by the equilibration of K_1 and the rapid pre-equilibrium (K_0). The first relaxation time is thus defined by:

$$\tau_1^{-1} = k_{12} \frac{1}{1 + 1/(K_0[c_A(\infty) + c_S(\infty)])} + \frac{k_{21}}{1 + K_2} \qquad K_0 = \frac{k_{12}}{k_{21}} \qquad K_2 = \frac{k_{23}}{k_{32}} \qquad (6.3.7)$$

Fluorescence changes showed two relaxations, a very fast one

$$\tau_1^{-1} = k_{23} + k_{32} \tag{6.3.8}$$

and a slow one in which the equilibration of K_1 was affected both by pre- and post-equilibrium. Under all experimental conditions $K_0^1 \gg c_A$ or c_S and the collision complex is, therefore, always a sparsely populated state.

From the arguments developed above one can conclude that the transition from weak to strong binding is the pressure dependent one. On correlating these results with those obtained from studies of the pressure dependence of tension in active muscle fibres, one can reach the conclusion that the fast, pressure dependent relaxation observed in solutions of muscle proteins, is connected with force development. The large volume change ($100 \text{ cm}^3 \text{ mole}^{-1}$) is indicative of large changes in charge during the transition of weak to strong complex, the putative force development. The equilibrium of this last transition is controlled by the stage in ATP hydrolysis, that is by the particular myosin bound components of that reaction. Geeves (1991) should be consulted for a review of the application of different kinetic methods to these problems. The exciting results obtained from X-ray and electron microscope studies of the structure of actin, myosin and their complexes will be usefully complemented by the information about the changes during interactions, which are deduced from the kinetic experiments.

For more detail of temperature and pressure relaxation experiments on muscle fibres the papers of Davis & Harrington (1993) and Geeves (1991) respectively should be consulted.

6.4 Relaxations of proton equilibria

The role of protons in structure and function

We have already emphasized in chapter 1 that the study of electrolytes has been a bridge between physical chemists and physiologists formed by common interests. Many of the latter researchers have made considerable contributions to the subject and some of the former owe their introduction to and interest in biological systems to the important role of ionic equilibria in physiological phenomena. In the historical introduction (section 6.1) to relaxation kinetics it was pointed out that this field originated in studies of ionic equilibria and in the elucidation of their chemical mechanisms. Electric field and ultrasound investigations on solutions of electrolytes established that ionic association reactions involve two step mechanisms: encounter and union of their respective solvation shells. Rates of interaction depend critically on the properties of the hydration shells of the ions. The hydrogen ion (proton), through its effects on the reactivities of amino acid side chains, plays a key role in the reactions of proteins and thus in the regulation of most physiological functions. As pointed out in section 3.4, protons are substrates, products and modifiers (inhibitors and activators) of enzyme reactions. Furthermore, there are reciprocal relations between protein conformation changes and pK changes, resulting in proton uptake and release during most functions.

For the above reasons the present section is devoted to a discussion of the rates of relaxation of ionic equilibria, and protonic equilibria in particular, as well as of conformational changes linked to them. The rates of many functions are controlled by the uptake and release of protons by functional groups of proteins. Some attention needs to be given, in connection with this topic, to rates of response of indicator dyes used to monitor the changes in concentrations of protons; the discussion below of the relation between pK values and proton transfer rates applies to this problem. The interaction of calcium ions with indicators has already been referred to in the previous section. Reference to discussions of the chemical mechanisms of enzyme action (see sections 2.6 and 3.3 as well as Gutfreund, 1972; Fersht, 1985) will make clear the importance of the state of ionization of particular amino acid residues during specific elementary steps of binding and catalysis.

Protons as substrates and products

Alcohol dehydrogenase serves as an example of an enzyme reaction which involves proton uptake and release during a substrate induced conformation change, and also as part of the stoichiometry of the reaction catalysed. The present discussion is only intended to deal with certain elementary steps without going into the detail of the whole reaction mechanism (see section 5.1). The reaction catalysed by this enzyme

$$\text{ethanol} + \text{NAD}^+ \rightleftharpoons \text{aldehyde} + \text{NADH} + \text{H}^+$$

can be subdivided into the partial reaction

$$\text{EH}^+ + \text{NAD}^+ \rightleftharpoons \overset{*}{\text{E}}\text{NAD}^+ + \text{H}^+$$

Binding NAD^+ to the enzyme, E, involves a conformation change which can be monitored by concomitant fluorescence quenching:

$$EH^+ + NAD^+ \underset{k_{21}}{\overset{k_{12}}{\rightleftharpoons}} EH^+ \cdot NAD^+ \underset{k_{32}}{\overset{k_{23}}{\rightleftharpoons}} \overset{*}{E}H^+ \cdot NAD^+ \underset{k_{43}}{\overset{k_{34}}{\rightleftharpoons}} \overset{*}{E} \cdot NAD^+ \\ + H^+$$

The pH dependence of the binding equilibrium shows that the pK of a group adjacent to the binding site shifts from 9.8 to 7.6 as the complex between enzyme and NAD^+ is formed. Consequently the pH independent equation for the relaxation time (equation (6.3.7)) for the conformation change should be expanded to

$$\tau^{-1} = \frac{k_{23}}{1 + k_{21}/\{k_{12}[c_{EH^+} + c_{NAD^+}]\}} + \frac{k_{32}}{1 + [K/c_{H^+}]} \tag{6.4.1}$$

where $K = k_{34}/k_{43}$ (a dissociation constant).

Clearly, equation (6.4.1) has to be further expanded at pH significantly above neutrality when the dissociation of EH^+ to $E + H^+$ has to be taken into account. This is a typical example of a relaxation linked to a protonic equilibrium. Such systems can be studied with pH jumps coupled to pressure or temperature dependent buffers (see pp. 205–6). In the particular case of the reaction of alcohol dehydrogenase the cycle must be completed during the catalytic reaction when a proton is taken up by the enzyme during product dissociation.

A more complex and very interesting situation arises in the dark reactions of rhodopsin. More detail of the whole cascade of reactions involving rhodopsin is presented in section 4.2. There are at least two, but probably three, proton linked steps in the transition from metarhodopsin-I (MR-I) to metarhodopsin-II (MR-II) and to the state in which metarhodopsin interacts with the G-protein, transducin, for its activation. The chromophore of rhodopsin, 11-*cis*-retinal, is bound to the protein via a Schiff's base linked to the ε-amino group of a lysine residue. After the photochemical isomerization of 11-*cis* to all *trans*-retinal, deprotonation of the Schiff's base (which may be an intramolecular proton transfer and not result in a free H^+ ion) results in the spectral change (approximately 100 nm red shift of absorption peak) associated with the transition MR-I to MR-II. The very large temperature dependence of the equilibrium constant for this transition, discussed in section 7.2 permits its adjustment to near unity at 5 °C and pH 7. This makes it possible to study the effects of pH and thus of proton uptake, on the relaxation time of the transition. The equilibrium constant is also very pressure dependent and pressure relaxation experiments on disk and rod outer segments from bovine retinas established that the transition monitored at 380 or 480 nm is truly reversible (Attwood & Gutfreund, 1980). The pH dependence of relaxations can be studied both by pressure perturbation and with small light flashes, which are essentially concentration jumps of MR-I. Determination of the equilibrium constant (K) and of the reciprocal relaxation time, which gives the sum of the forward (k_{12}) and reverse (k_{21}) rate constants, allows the evaluation of the individual rate constants as follows:

$$K = k_{12}/k_{21}; \quad \tau^{-1} = k_{12} + k_{21}; \quad k_{12} = K\tau^{-1}/(1+K); \quad k_{21} = \tau^{-1} - k_{12}$$

Actually the assignment of the larger and smaller numerical values to the rate constants k_{12} and k_{21} respectively has to be decided from the equilibrium constant. Clearly if $K > 1$ then $k_{12} > k_{21}$ and vice versa. The relaxation times observed over the range pH 5 to 8.5 show an increase in k_{12}, with $pK_1 \cong 5$ and a decrease in k_{21} with $pK_2 \cong 8$ (King & Gutfreund, 1984; Parkes & Liebman, 1984). The uptake of protons, before and after the MR-I → MR-II transition, can be described by the following alternative mechanisms:

$$(1) \quad \text{MR-I} \underset{K_1}{\overset{\pm H^+}{\rightleftharpoons}} \text{H}^+\text{MR-I} \underset{k_{21}}{\overset{k_{12}}{\rightleftharpoons}} \text{H}^+\text{MR-II} \underset{K_2}{\overset{\pm H^+}{\rightleftharpoons}} \text{H}^+\text{MR-II}\cdot\text{H}^+$$

$$(2) \quad \begin{array}{ccc} \text{MR-I} & \overset{K}{\rightleftharpoons} & \text{MR-II} \longrightarrow \\ K_1 \Updownarrow & & \Updownarrow K_2 \\ \text{H}^+\text{MR-I} & \underset{}{\overset{K_{H^+}}{\rightleftharpoons}} & \text{H}^+\text{MR-II} \longrightarrow \end{array} \Big\} \begin{array}{l} \text{cascade} \\ \text{(see figure 4.3)} \end{array}$$

The pH dependence of the rate constants calculated from the relaxation times for mechanism (1) can be expressed in terms of the two dissociation constants K_1 and K_2 corresponding to the two pKs quoted above

$$k_{12}(\text{observed}) = k_{12}/[1 + K_1/c_{H^+}]$$
$$k_{21}(\text{observed}) = k_{21}/[1 + c_{H^+}/K_2]$$

As c_{H^+} increases, the observed forward and reverse rate constants increase and decrease respectively. Their sum, that is the reciprocal relaxation time, is at a minimum near neutrality. For mechanism (2), when two pathways are allowed for the interconversion of MR-I to MR-II, the pH dependence of the individual rate constants is more complex since they will each be the sum of two rate constants. An essential difference between the two mechanisms is that (1) involves the uptake of two protons, while (2) involves the uptake of only one proton; the internal deprotonation of the Schiff's base is not taken into account here. These problems are still under active investigation and more accurate kinetic data should be able to distinguish between the two mechanisms. In mechanism (1) the two observed rate constants must tend to zero at low and high pH respectively, while in mechanism (2) this is not required. The studies of Arnis & Hofmann (1993), who investigated proton uptake directly with indicators, provides more direct evidence for an additional protonated MR-II state.

Rates of proton transfer in buffered systems

As has been emphasized before, the linkage of conformation changes and perturbations of pK values of ionizing groups on, or in, the vicinity of binding sites is a frequent occurrence. In connection with this phenomenon and with the uptake or release of protons as substrates or products, questions are often asked about the speed of proton equilibration in aqueous solutions at neutral pH and whether, for enzymes with turnover numbers larger than 1000 s^{-1}, the rate of change of protonic equilibria is fast enough. Proton dissociation from relevant side chains could, in a pure aqueous environment, be as slow as 10^3 to 10^4 s^{-1}. In a classic paper Eigen (1963) discussed elementary processes of proton transfer and their role in acid–base as well as enzyme catalysis. Eigen *et al.* (1964) presented a survey of rate constants of protolytic reactions. Most of these processes are very fast (see table 6.3) and were measured by electric field relaxation and sound absorption methods (see section 6.1). The mechanism proposed for proton exchange between a donor A and an acceptor B is illustrated by the following scheme (for generality charges are omitted):

$$AH + B \underset{}{\overset{k}{\rightleftharpoons}} (AH \ldots B \underset{}{\overset{K}{\rightleftharpoons}} A \ldots HB) \underset{}{\overset{k}{\rightleftharpoons}} A + HB$$

The proton exchange occurs between the two encounter complexes shown within the brackets. The interconversion between these encounter complexes is a very rapid redistribution, apparently via quantum mechanical tunnelling.

The association rates are diffusion controlled with $k > 10^{10} \text{ M}^{-1} \text{s}^{-1}$. The equilibrium constant, K, for $(A \ldots HB)/(AH \ldots B)$ is defined by

$$\log K = pK_B - pK_A = \Delta pK$$

and the fraction of $A \ldots BH$ is

$$K/(K+1) = 10^{\Delta pK}/(10^{\Delta pK}+1) \tag{6.4.2}$$

This allows one to write for the second order rate constant, k_{tr}, for transfer

$$k_{tr} = k(10^{\Delta pK}/(10^{\Delta pK}+1))$$

and for the pseudo first order rate constant, k_{ex}, for proton exchange between A and B

$$k_{ex} = \sum k_{tr} c_{cat} \tag{6.4.3}$$

where c_{cat} are the concentrations of the proton donors/acceptors. The summation is for the general case of several buffer components in the reaction mixture. The above derivation includes the same argument as that developed in section 3.2 for the role of rapid pre-equilibration in pseudo first order reactions.

We can now consider proton exchange rates during the steady state of enzyme catalysis or other reactions of proteins. These considerations also apply to studies of hydrogen ion exchange used to monitor the mobility of different parts of protein molecules (see section 7.3). In all these cases the question is often asked, how fast can proton transfer occur between a group on a protein and the aqueous medium at a specified pH? For example, the imidazole residue of histidine fulfils the function of a proton acceptor or donor during the catalytic cycle of many hydrolytic and other enzymes (for a review of this topic see Fersht, 1985). Transient kinetic studies of several enzymes have explored mechanisms by following the release and uptake of protons with indicators (Gutfreund, 1955; Whitaker *et al.*, 1974). Is the potential rate of proton transfer fast enough to be involved in the mechanism of enzymes with really fast turnover? For instance the k_{cat} for carbonic anhydrase is $> 10^6 \, s^{-1}$ and the reaction depends on proton release for dehydration of carbonic acid and proton uptake for the hydration of CO_2 by a group with $pK = 7$. If the rate of proton transfer from that group depended on exchange with hydroxyl groups ($pK = 15.7$) in the solvent, the following calculation would apply: $c_{OH} = 10^{-7}$ M, $pK = 8.7$, hence from equation (6.4.3) the rate constant for the exchange $k_{ex} = 10^3 \, s^{-1}$.

This discrepancy between enzyme turnover and proton exchange presented a puzzle. Enzymes with turnover rates larger than $10^4 \, s^{-1}$ like carbonic anhydrase and acetylcholinesterase, were dubbed 'impossible enzymes'! This is still sometimes reiterated with respect to these and other systems. However most biochemical reactions are studied in the presence of buffers. If the reaction of carbonic anhydrase is studied in the presence of 10 mM buffer with $pK = 8$, then the rate constant for proton exchange would approach $k_{ex} \geq 10^{10}$ s^{-1}. In fact the experiments of Silvermann & Tu (1975) show that the reaction of carbonic anhydrase is critically dependent on buffer concentrations. A similar explanation could account for the apparent exceptionally fast proton transfer rates to a channel protein deduced by Prod'hom, Pietrobon & Hess (1987) who assumed the hydrogen ion concentration to be defined by pH. The authors neglected the concentration of potential proton donors available from the buffers in the solution. It should also be pointed out that charge fields on the surface of proteins can give exceptionally high rates of interactions with ions.

Other examples of the dependence of proton exchange rates on the pKs and on the sum of the concentrations of bases present in the reaction mixture are described by Englander & Kallenbach (1983). These authors apply the above rationale to the interpretation of exchange rates of solvent components with different bases in protein and nucleic acid molecules.

7

Factors affecting rates of reactions

7.1 Concepts and their history

Introduction

In homogeneous systems the major factors which control the rates of reactions are the concentrations, the steric requirements and the energy barriers. The first of these is described by the law of mass action (section 3.1) and the other contributions by the rate constants. In organized systems spatial distributions of the components add to the complexities. In this chapter we are mainly concerned with the chemical dynamics of single reaction steps which determine the individual rate constants, as distinct from the phenomenological description of the time course of a sequence of events; the latter being the main concern of most of the rest of this volume. The frontiers between these two subjects are diffuse and the limits of their semantic distinction are raised in several sections. Another fundamental aspect of rate processes, related to their chemical dynamics, is the thermodynamic description of the extent to which a reaction will proceed (see for instance Edsall & Gutfreund, 1983, for biochemical reactions). As will be seen below, thermodynamic considerations are involved in the description of energy barriers as well as in the description of the final outcome of reactions. Diffusion is an important part of molecular dynamics and different problems arise at different levels of molecular complexity and cellular organization. Spatial and steric problems, including conformational mobility, are of special importance in biological processes which involve the reactions of large and complex molecules. The additional problems, arising as we consider more organized systems, are due to the fact that at the cellular level the production or release of a substance may be at some distance from the location of its subsequent utilization, either in a chemical reaction or for interaction at a receptor site. At the molecular level the considerable complexities of protein or nucleic acid molecules and their assemblies, for instance in muscle fibres, introduce factors requiring considerations additional to those needed in the interpretation of reactions of small molecules in a homogeneous system.

It must be emphasized that the theories originally developed for the interpretation of rate processes in simple gaseous reactions are not realistic when applied to reactions of large molecules in aqueous solutions. Furthermore the interaction of water with all components of biological systems plays a much bigger role in their reactivity than the solvents used for most reactions featuring as examples in textbooks on chemical kinetics in solutions.

Biotechnology in the form of reactors with enzymes bound to a matrix requires considerations similar to reactions localized on membranes of cells or intracellular organelles. When one reaction partner is localized the other reactants also develop concentration gradients through the depletion near the reaction site, even if they are initially homogeneously distributed. Some techniques for kinetic studies can themselves introduce heterogeneity into the volume under observation. Flash photolysis for the initiation of reactions through, for instance, photodissociation of ligands from haem proteins or the liberation of reactants from caged compounds (see section 8.4) will usually only activate the reaction in a segment of the solution. Diffusion into and out of the field of observation has to be taken into account. However, this very phenomenon is utilized in the study of diffusion in cells and tissues by a variety of optical methods mentioned in chapter 8. In such techniques local bleaching of natural or artificially introduced photolabile compounds is followed by diffusion of unbleached material into the depleted volume segment. Optical observation of this process provides data for the calculation of intracellular and local rates of diffusion.

The diffusion of products away from localized reaction centres has received wide attention from investigators interested in pattern formation and in developmental biology. This is of particular interest when oscillations occur in initially homogeneous systems through the coupling of diffusion, itself a rate process, to a chemical reaction. This summary should give some indication of the importance of the study of diffusion phenomena both as a basis of some of the experimental techniques and of the behaviour of biological systems. It will be seen in section 7.4 that in homogeneous reactions and in spatially organized reactions different aspects of the principles of the laws of diffusion have to be taken into account. The differential equations describing diffusion phenomena in organized systems or for coupled reaction–diffusion processes require specialized mathematical techniques, which are not covered in this volume, but references are given in section 7.4.

One of the major problems encountered in writing a book on any particular physical approach to the study of biological processes is that every one of them is dependent on several others, yet not all of them can be treated in the same detail. Quite apart from any purely mathematical simplifications of the treatment given here, it is clearly not possible to provide enough detail of the molecular statistics of diffusion or of statistical thermodynamics, both topics seminal to rate processes, to satisfy a reader who wishes to find a thorough explanation of all concepts used. Hopefully, the references given to sources for treatments of basic principles will be found to be adequate.

Arrhenius energy

Most elementary treatments of factors controlling rates of chemical reactions are principally concerned with the energy barrier and an often ill defined pre-exponential factor. In the literature on biochemical and physiological problems the energy barrier is variously termed 'activation energy' or 'Arrhenius energy' and is sometimes analysed in terms of the thermodynamic parameters Gibbs energy, enthalpy and entropy (i.e. enthalpy of activation, etc.). In chapter 1 it was pointed out that rates of reactions in solutions were investigated systematically in the late nineteenth century by several authors who had considerable interest in the power of such physicochemical studies for the elucidation of the mechanisms of biological functions. Arrhenius was quoted as a prime example and it is of interest to reflect that in most presentations in the biochemical literature of results of kinetic measurements, the effects of temperature on reaction rates are still interpreted in the manner suggested by Arrhenius (1889). The only difference is that nowadays a logarithmic plot of the velocity, v, against reciprocal temperature is used (see, however, the historical comments below) to obtain the slope E/R, where E is the energy and R is the gas constant, and the intercept $\ln A$ of the Arrhenius equation, given here in its modern form:

$$v = A\exp(-E/RT)$$
$$\ln v = \ln A - E/RT$$

$$(7.1.1)$$

Alas authors, including the present one, often only consult reviews of reviews of classic papers. It is, however, nice to see sometimes how the

Table 7.1.

Q_{10}	E_a kcal^{-1} mole^{-1}	E_a kJ^{-1} mole^{-1}
1.2	3.1	13
1.5	6.9	29
2	11.8	49
3	18.8	79
4	23.7	99
5	27.5	115
6	30.6	128

originator of a concept expressed his ideas. The actual equation which Arrhenius (1889) used to fit the data from the temperature dependence of rates of a number of reactions was:

$$\Omega_{T_1} = \Omega_{T_0} B \exp[A(T_1 - T_0)/T_0 T_1] \tag{7.1.2}$$

Ω_T are the rates at specified temperatures, T_0 and T_1, while A and B correspond to the activation energy and pre-exponential factor respectively. During the subsequent hundred years only the nomenclature has changed in the basic formulation of the temperature dependence of reaction rates.

While we are on this historical note (see also section 1.2) it is of interest to point out that nowhere in his long paper, in which he presents a large amount of data on the dependence of the rates of different reactions on temperature and solvent composition, did Arrhenius show a graphical display of rates against temperature or log(rate) against $1/T$. The reliance on numerical tabulation is a reflection on the space available in journals, however the present-day practice of displaying data graphically has other advantages apart from economy in space. We shall, none the less, refer to the plot of log(rate) against $1/T$ as the Arrhenius plot. In the biological literature one often finds the temperature dependence of a process expressed in terms of a value for Q_{10}, that is the ratio of the rate constants measured at temperatures 10 degrees apart. Within the temperature range $20 \pm 10\,°C$ the approximate values for Q_{10} and the corresponding Arrhenius energies, expressed in modern nomenclature E_a, are shown in table 7.1.

As a rule of thumb one can say that, in biological systems, the majority of reactions has a Q_{10} near 2. Those with a lower Q_{10} are likely to be diffusion controlled and those with a larger Q_{10} involve entropy effects due to structural rearrangements. Many aspects of the temperature dependence of reactions and of the shapes of Arrhenius plots, are discussed in detail in section 7.2.

Theory of absolute reaction rates

The pre-exponential factor given by the ordinate intercept contains the obscurities of the fruitful collisions with energies larger than the barrier. Before we turn, in the next section, to examples of Arrhenius plots of biochemical reactions which exhibit interesting features, comments are required on somewhat more elaborate theoretical analyses of activation energies. Most of the developments in the theories of reaction rates which are found in standard texts today (for instance Atkins, 1982) have their foundation in the work of Wigner, Eyring, Hinshelwood and their collaborators in the 1930s. A Discussion Meeting on 'Reaction Kinetics' held by the Faraday Society in 1937 (see *Transactions of the Faraday Society*, volume 33) contains contributions from these and other founders of kinetic theory. Unlike the proposals of Arrhenius, which were based on the phenomenological interpretation of studies on reactions in solutions, the later work was more ambitious and intended to interpret and predict reactions in the gaseous phase by the application of the, at that time new, theories of quantum mechanics and statistical mechanics. This resulted in the name 'theory of absolute reaction rates'. This approach had little useful impact on studies of complex biological molecules in aqueous solutions. The recent interest in the molecular dynamics of proteins and nucleic acids has opened new theoretical approaches to the study of the mechanisms involved in their reactions. The availability of computers with ever increasing storage capacity and speeds has made it possible to simulate the molecular dynamics of the mobility of proteins and of the approach of substrate to their binding sites (McCammon & Harvey, 1987). However, at this time Newtonian rather than quantum mechanics are still used for most of such molecular dynamics calculations.

At the same time as the interest in molecular dynamics came a wave of enthusiasm for the use of the theory of Kramer (1940) for the interpretation of reaction rates. This approach, which is stimulated by the attention now given to the internal mobility of macromolecules, is discussed in section 7.3. Some more background to absolute reaction rates has to be given since, rightly or wrongly, it is often encountered in the biological literature. Hirschfelder (1983) presented an interesting and lively personal account of the development of the ideas on absolute reaction rates in the 1930s. As intimated above, the principal motivation was the application of quantum mechanics to the calculation of the potential energy diagrams during the approach of two reaction partners. These calculations of gas phase reactions were intended to describe such processes as the interconversion of *para-* to *ortho*-hydrogen and the interaction between hydrogen and fluoride. While the mathematical and numerical techniques developed during the last 50 years made it possible to predict rates for more complex reactions, only the empirical formalism, and not the quantum mechanical basis, of the theory of absolute reaction rates is found to be relevant when the complexities of biological systems are being considered. There are, of course, exceptions in the field of photobiology and in the application of photochemical methods, when more fundamental calculations on reaction profiles have been fruitful. In section 7.3 some reference will be made to so-called trajectory calculations for the approach of reactants, such as a substrate into an active site of an enzyme. Electrostatic and Newtonian theory is used for such predictions of reaction rates, usually comparative ones. Eyring and his colleagues (Johnson, Eyring & Stover,1974) attempted to popularize among biologists the idea that statistical mechanics and thermodynamics could be used to describe the formation of activated complexes as transition states in equilibrium with the reactants. This leads to the drawing of energy profiles with a peak at the activation energy. Thermodynamic equations for the transition state pictured in a well at the top of the barrier (see figure 7.1) are discussed in section 7.2. The alternative interpretation of the pathway across the energy barrier, mentioned above (Kramer, 1940), depends on a diffusion model. It is more appropriate to discuss this in the section on the effects of viscosity on molecular dynamics (see section 7.3), when the differences between the two theories will become more apparent.

7.2 Temperature dependence of reactions and activation parameters

Uses and abuses of Arrhenius plots

In the last section it was stated that most biologists wisely restrict themselves to the pragmatic approach of just finding the temperature dependence of a process as 'useful information'. However, it is necessary for us to give some attention to the interpretation of Arrhenius plots and its pitfalls. Besides the purely descriptive 'how much faster is the reaction when the temperature is raised by ten degrees' (see the previous section), there are questions which can be answered relating to reaction mechanisms, to the

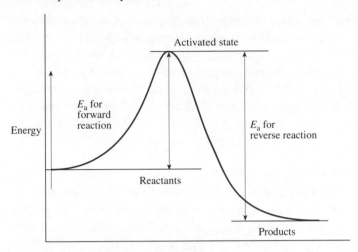

Figure 7.1 Diagrammatic presentation of the potential energy barrier for forward and reverse reactions.

identification of particular molecular events within a complex physiological process (see for instance Anson, 1992) and to the adaptation to the environment. Recently interest in the temperature stability of proteins, which is measured by the effect of temperature on the rate of denaturation, has become of great interest in connection with attempts to 'engineer' enzymes of superior stability for industrial processes. Relatively small energy changes in the stability of a structure introduced by site directed mutagenesis, cause considerable changes in rates of denaturation (see for instance Matouschek *et al.*, 1990). From the relation between the Gibbs energy of activation and the reaction velocity (see equation (7.2.3)) it can be calculated that, at 25 °C the latter changes by a factor of 10 for every 5.7 kJ mole^{-1} change in the energy barrier. This corresponds to the energy of the formation of a rather weak hydrogen bond and does show, incidentally, how difficult it is to predict changes in rates from information about structure. Conversely, it shows how sensitive kinetic parameters are to energy changes.

In subsequent sections we shall find that the heat of activation, derived from Arrhenius plots, gives pointers to the character of the reaction. For instance, reactions which are controlled by the rate of diffusion have very low energy barriers (30 to 40 kJ mole^{-1}), as have the elastic processes in muscle fibres (see Davis & Harrington, 1993), while reactions involving protein conformation changes characteristically have values three to four times as large. The major difficulties in the interpretation of many reports on the temperature dependence of biological processes are that, even the rates of a single enzyme reaction or of the binding of a ligand to a receptor, are rarely determined by a single step of the complex sequence of events involved. We shall now concern ourselves with the consequences of the complexities which result in 'apparent' heats of activation. These complexities demonstrate themselves in multiple linear sections or in non-linear Arrhenius plots.

Arrhenius energies of coupled reactions

If a reaction depends on the state of ionization of a reactant then the temperature dependence of the process will have, as one of its components, the effect of temperature on the ionization equilibrium. Analogous to the rapid pre-equilibrium mechanism of ligand binding (see section 3.2) in the reaction

$$AH^+ \underset{}{\overset{K}{\rightleftharpoons}} A + H^+; \quad A \xrightarrow{k} B$$

(where K is the dissociation constant and k the true rate constant for the interconversion of A to B) the observed rate constant for the formation of B is given by

$$k_{obs} = \frac{k}{[c_{H^+}/K] + 1} \tag{7.2.1}$$

The opposite situation, involving proton association prior to the reaction, can also occur and contribute to composite Arrhenius plots. Examples are found in the differences between the apparent activation energies for an enzyme reaction measured at pH values near or well above the pK of a group on the active site, which has to be in the basic form for catalysis. Experiments carried out on the trypsin catalysed hydrolysis of amino acid esters illustrate this point (see figure 7.2). A histidine in the basic form is an essential part of the catalytic site of this enzyme (Gutfreund, 1955). The activation energy at pH 6.0 (near the pK of the histidine) is 76.15 kJ mole^{-1} and at pH 7.5 (when the histidine is predominantly in the basic form) it is 46.44 kJ mole^{-1}. This is only one of many ways in which the coupling of two reactions can give misleading information deduced from Arrhenius plots. In the particular example described above the effect of pH on the apparent activation energy does throw some light on the mechanism of the coupled reaction.

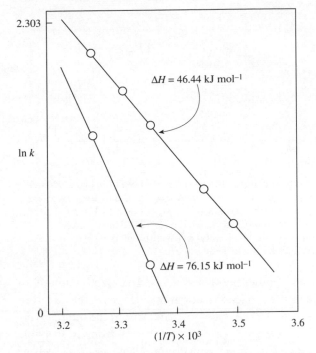

Figure 7.2   Arrhenius plots for the trypsin catalysed hydrolysis of benzoyl
L-arginine ethylester at pH 6.0 and 7.5. The difference in the activation energies
of the two reactions is 30 kJ mole^{-1}. This corresponds to the extra energy
required to dissociate a proton from an imidazole group (pK = 6.25) which has to
be deprotonated in the active form of the enzyme.

Non-linearity of Arrhenius plots (sequential reactions)

Another feature of the analysis of the temperature dependence of reaction
velocities, which has received much attention and frequently dubious
interpretation, is the non-linearity of the Arrhenius plot. Moore & Pearson
(1981) use the term anti-Arrhenius plots for some of their graphical
representations of rates which first increase and then decrease as the
temperature is raised. They state, correctly, that such graphs usually result
from multi-step mechanisms or a change in mechanism as the temperature
increases.

One has to distinguish between results indicating a decrease in rate as
temperature is increased and the conclusion that the rate would increase as
the temperature is decreased. Tests of reversibility of the effect of any
change in condition on the rates of reactions are of particular importance in

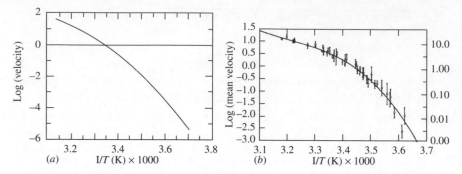

Figure 7.3   Non-linear Arrhenius plots due to sequential reactions with change in rate limiting step as temperature is varied. (*a*) A simulated plot for a two step essentially irreversible reaction with activation energies of 50 and 150 kJ mole^{-1}, respectively. (*b*) Arrhenius plot of data from Anson (1992) of the average velocities of actin filaments moving on skeletal myosin (in vitro assays). A cubic curve is fitted through the data by linear least square.

the study of enzymes and of other protein systems. The inactivation of an enzyme at higher temperature can lead to apparent changes in the direction of an Arrhenius plot. These effects can be reversible due to the interconversion of two forms of the enzyme, or they can result from irreversible denaturation. In the latter case rates will be time dependent as well as temperature dependent. The effects of changes in the specific heat of reactants on the temperature dependence of Arrhenius constants will be discussed below.

Changes in mechanism, shifts of rate limiting steps or even in pathway, are a frequent cause of changes in slope of the Arrhenius plot. In the following minimal enzyme mechanism

$$E + S \underset{k_{21}}{\overset{k_{12}}{\rightleftharpoons}} ES \underset{k_{32}}{\overset{k_{23}}{\rightleftharpoons}} EP \underset{k_{43}}{\overset{k_{34}}{\rightleftharpoons}} E + P$$

at substrate saturation one of two constants (k_{23} or k_{34}) or a function of both, can be rate limiting for the overall reaction at a specified temperature. If the energy of activation of step 2 is much higher than that of step 3, then the former can change from being slower to being faster than the latter as the temperature is increased. This is illustrated in figure 7.3 which shows an Arrhenius plot for a steady state process with successive steps with activation energies of 50 and 150 kJ mole^{-1}. It can be seen that a distinct change in slope occurs in this simulated plot. Investigations of the temperature dependence of a process can be used as a diagnostic for the involvement of additional steps. Two recent examples are studies of reactions involved in muscle contraction and phototransduction in the retina. Levy, Sharon & Koshland (1959) compared the temperature dependence of ATP hydrolysis by myosin with that of the 'walking rates of ants'. While the available data were not detailed enough for distinctions between possible mechanisms, their paper was one of the first in which reactions at the molecular and the physiological level were compared. Since that time many other studies of the temperature dependence of reactions of myosin, actomyosin and of muscle fibre preparations have been carried out. These are reviewed in a paper by Anson (1992) who studied the Arrhenius activation energy of the velocity of actin filaments moving on a myosin surface. Curved Arrhenius plots were obtained corresponding to large activation energies (about 150 kJ mole^{-1}) at the lower end of the temperature scale and small activation energies (< 50 kJ mole^{-1}) at higher temperatures. The process studied by Anson is, of course, a very complex one. However, even for apparently simpler reactions, curved plots were obtained by Kahlert & Hofmann (1991) from studies of the temperature dependence of the activation of G-protein by interaction with rhodopsin R* (see section 4.2). These results can be interpreted by a change in rate limit from a conformation change on interaction (at low temperature) to diffusion control (at higher temperature). Both the Arrhenius plots of Anson (1992) and of Kahlert & Hofmann (1991) are very much like the theoretical one of figure 7.3 corresponding to two activation energies. However, such changes in slope of Arrhenius plots have sometimes given rise to an erroneous interpretation in terms of a temperature dependent transformation (structure change or phase change) of enzyme or substrate. An interesting discussion of this point is given in connection with Brahm's (1977) studies of anion transport through human red blood cell membranes. The temperature dependence of chloride transport gives an activation energy of 125 kJ mole^{-1} below 15 °C and of 96 kJ mole^{-1} above that temperature. The observed break point for the Arrhenius plot for bromide transport is at 25 °C. The activation energies for the two ions are the same below and above their respective break points and furthermore the rate of transport for chloride at 15 °C is the same as that for bromide at 25 °C. The conclusion is that there is a change of rate limit to a step which is common for the two ions.

Changes in specific heat and structure of reactants

Changes in the heat of activation with temperature can, at times, be attributed to changes in the specific heat of a reactant and this argument leads to the proposals of structure changes or phase transitions. The relation of the specific heat of a reaction to its enthalpy and entropy change is given by

$$C_P = \left(\frac{\delta H}{\delta T}\right)_P = T\left(\frac{\delta S}{\delta T}\right)$$

This indicates that changes in specific heat with temperature will change the slope of the Arrhenius plot. The relation between enthalpy of activation, $H^{\downarrow}$ and Arrhenius energy, E_a, is discussed below.

The distinction that changes in mechanism, such as those displayed in figure 7.3, will give a curved plot while structure transitions will give a sharp kink, is not very reliable. Basically, curved Arrhenius plots are just an indication that there is an interesting phenomenon to be investigated by other methods, such as the transient kinetic analysis of individual steps and differential scanning calorimetry of reactants to detect phase transitions. The latter gives information about changes in specific heat (see Edsall & Gutfreund, 1983, chapter 6). There are some examples of breaks in the Arrhenius plot which do indicate temperature dependent transitions in proteins. In theory, if there is something like a phase transition in the protein, there should not only be a change in slope but also a step displacing the two parts of the plot. This type of behaviour is seen with reactions of myosin (see figure 7.4) studied by Biosca, Travers & Barman (1983), who give references for other examples such as lipoproteins (see also Biosca *et al.*, 1984).

Some comments on the theory of absolute reaction rates

Before we turn to the connection between the temperature dependence of reaction rates, structure changes and the temperature of the environment, some comments are required on the relation of the Arrhenius parameters (slope and intercept of the plot) to those of the theory of transition states (or activated states). There is assumed to be a quasi equilibrium between such a state and the ground state of a reactant. The relative concentration of the transition state to that of the ground state is characterized by an equilibrium constant $K^{\downarrow}$. Eyring's theory led him to postulate the following relation between the rate constant of a reaction, k, and $K^{\downarrow}$:

$$k = K^{\downarrow}(k_B T/h) \tag{7.2.2}$$

where k_B is the Boltzmann constant and h is the Planck constant.

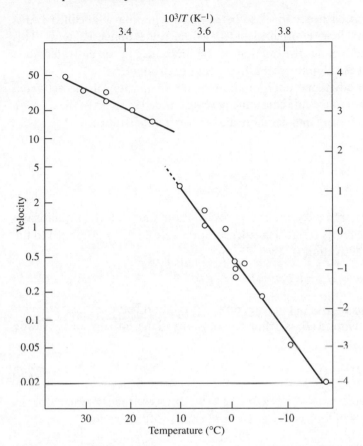

Figure 7.4 Arrhenius plot of the data from Biosca *et al.* (1983). ATP hydrolysis by myosin S1 at pH 8 in 5 mM KCl and 40% ethylene glycol. A distinct step function rather than a smooth curve indicates a type of phase transition involving a change in specific heat.

In terms of the Gibbs energy of activation, $G^{\downarrow}$, we can write

$$k = (k_B T/h) \exp(-G_0^{\downarrow}/RT) \qquad (7.2.3)$$

On substituting numerical values for the Planck and Boltzmann constants ($h = 6.626 \times 10^{-27}$ ergs and $k_B = 1.381 \times 10^{-16}$ erg K^{-1}) one obtains for the pre-exponential factor at 25 °C $(k_B T/h) = 6.24 \times 10^{12}$ s^{-1}. This is the frequency of vibration and thus has the units of a first order rate constant. If $G^{\downarrow} = 0$ this frequency determines the maximum rate with which a reaction can occur. In the case of a second, or higher, order reaction a factor to account for the collision frequency, has to be added to the basic equation.

This is discussed in section 7.3. In Eyring's equation a so-called transmission coefficient, is often included in the pre-exponential term. This (fudge factor) defines the fraction of the molecules in the activated state which is successful in their transformation to product.

The thermodynamic interpretation of the formation of the activated state, with that of the rate constant, can be extended in terms of the division of the Gibbs energy into enthalpy and entropy of activation

$$G^{\downarrow} = -RT \ln K^{\downarrow}$$
$$G^{\downarrow} = H^{\downarrow} - TS^{\downarrow} \tag{7.2.4}$$
$$H^{\downarrow} = RT^2 (d \ln K^{\downarrow}/dT)$$

If one takes into account the concentration units of the equilibrium constant, the temperature dependence of the rate constant is related to the enthalpy of activation $H^{\downarrow}$ by

$$H^{\downarrow} = RT^2 (d \ln k/dT) - RT(n-1) \tag{7.2.5}$$

(n is the molecularity of the reaction). The relation between the Arrhenius energy of activation (E_a as defined on p. 234) and the enthalpy of activation is given by

$$H^{\downarrow} = E_a - RT(n-1) \tag{7.2.6}$$

If we use equations (7.2.3) and (7.2.4) we can write the expression for the rate constants in terms of enthalpy and entropy of activation

$$k = \frac{k_B T}{h} \exp(S^{\downarrow}/R) \exp(-H^{\downarrow}/RT) \tag{7.2.7}$$

On comparing this to the Arrhenius equation

$$k = A \exp(-E_a/RT) \tag{7.2.8}$$

we obtain for Arrhenius's pre-exponential and temperature independent factor

$$A = \frac{k_B T}{h} \exp(S^{\downarrow}) \quad \text{or} \quad \ln A = \ln(k_B T/h) + S^{\downarrow} \tag{7.2.9}$$

A plot of $\ln k$ against $1/T$ will have the slope $-H^{\downarrow}/R$ and the ordinate intercept extrapolated to $1/T = 0$ will be $S^{\downarrow}/R + \ln(k_B T/h)$.

In an analogous manner the volume change from ground state to transition state (volume of activation), $V^{\downarrow}$, can be derived from the pressure dependence of the rate constant. The relation between the rate constants at two pressures P_0 and P is

$$k_P/k_{P_0} = \exp(-\Delta PV^{\downarrow}/RT)$$
$$\ln(k_P/k_{P_0}) = -\Delta PV^{\downarrow}/RT$$

If the above interpretation is taken literally it means that the activation parameters represent values for the change in Gibbs energy, entropy, enthalpy and volume during the transition of the reactants from the ground state to the activated state. Organic chemists dealing with relatively simple reactions, compared to those involving proteins, postulate specific structures for transition state intermediates. This endeavour has spread into the field of enzyme catalysis and will be given a little attention in connection with isotope effects in section 7.5. It is not clear whether these transition states are comparable to the ones suggested by Eyring, who was really concerned with reactions in the gas phase. However, he and his colleagues (Johnson *et al.*, 1974) have applied the theory to steps of quite complex biological functions. This highlights the problem that what is to a bio-chemist a single step in a reaction is a sequence of events to a physical-organic chemist and in turn the latter's single step is a sequence of events to a physicist.

If one accepts that the transition state is in true equilibrium with the ground state one can draw the same conclusions from $G^{\downarrow}$, $S^{\downarrow}$, $H^{\downarrow}$ and $V^{\downarrow}$ about structure changes during its formation of as one can from thermody-namic parameters of overall reactions. It must be emphasized that biologi-cal reactions taking place in an aqueous environment are under the major influence of changes in solvation. Entropy changes are dominated by the binding and release of water molecules and of ionic components of the solution. These are of particular importance during assembly or folding of macromolecular structures and the interaction of ligands with binding sites. Interactions with solvent components can overshadow any entropy changes expected from the structure changes in the protein or nucleic acid molecules themselves. Activation parameters become more meaningful in comparative studies of reactions of a series of related compounds. In the case of protein folding, comparisons between molecules with amino acids selectively replaced by site directed mutagenesis have been interpreted in terms of thermodynamic activation parameters by Fersht and his collea-gues (Matouschek *et al.*, 1989). These studies are part of an increasing effort to study the effects of site directed mutations on protein stability and eventually to improve on nature. The use of enzymes for bio-engineering calls for stability under demanding conditions.

Some examples of the effects of temperature changes on rates

In many cases the assembly of oligomeric enzymes, viruses and ribosomes is favoured by an increase in temperature and can be shown to be accompanied by an increase in entropy (decrease in order). This is due to the liberation of water and ions bound to the areas of contact. Cold lability of structures is due to this phenomenon since reduction in temperature can, under some circumstances, favour dissociation and unfolding. If a negative entropy change, due to immobilization of solvent, is the dominant factor in the Gibbs energy change ($\Delta G = \Delta H - T\Delta S$) during a reaction involving the dissociation of an oligomer into its component units, then a decrease in temperature will stabilize the dissociated state by a reduction of ΔG. It must be emphasized that this is not due to an increase in the rate of dissociation with decreasing temperature, but is the result of a decrease in the rate of association. An example of this is presented in the study of the rate of attachment of actin to fragments of myosin by Geeves & Gutfreund (1982). The conclusion from their experiments was that the observed rate described a reaction step subsequent to the formation of a diffusion controlled collision complex. Apparent negative heats of activation, that is positive slopes in the Arrhenius plot, must always be due to complexities and cannot be characteristic of a single rate constant. Other anomalies in the behaviour of Arrhenius plots are due to factors connected with alternative interpretations of reaction rate theory, which involve diffusion across the energy barrier and include a viscosity term in the rate equation (see section 7.3).

The many functional proteins which are embedded in a lipid matrix in membranes present an additional problem in connection with the temperature dependence of their action. Membrane lipids usually have solid to liquid phase transitions at temperatures well below that of the normal environment of the particular organisms. However substitution of lipids with different chain length or degree of saturation, and consequently melting points, is often used to investigate their effects on the action of intrinsic membrane proteins. The consequences of phase changes of the bulk lipid on the function of, for instance, the calcium pump has been controversial. In the present context some experiments on the transition of metarhodopsin-I (MR-I) to metarhodopsin-II (MR-II) are of interest. In the case of the pigment from bovine retinal rods this reaction has a very large (150 kJ mole^{-1}) energy of activation. The natural lipids of mammalian retinas have a melting point at about $-10\,°C$. If the lipids in rhodopsin preparations are replaced by DMPC (dimyristoyl phosphatidylcholine), which has a melting point in the region of 25 to 28 °C, the Arrhenius plot shows the following features of interest. While the rate of the reaction is reduced by a factor of about 10, the energy of activation is essentially unchanged by the altered lipid environment. The most interesting point is the linearity of the plot over the temperature range of the phase transition of DMPC. The bulk property of the lipid in terms of the solid or liquid state appears to have no effect on the mechanism of the reaction, although the actual structure of the lipid does.

Returning to the question of what is of interest to the physiologist who wants information about the effect of temperature on an organism or a specialized cell, we come to, what L. J. Henderson (1958) called 'The Fitness of the Environment'. Let us again consider the reaction of rhodopsin MR-I to MR-II, which has a very large temperature dependence of that transition in bovine rhodopsin. This transition has a time constant of about 600 μs at 37 °C and 1 s at 0 °C. The question arises as to how cold blooded animals can react to visual responses; could a frog react at low temperature (near 0 °C) to escape from a predator or to catch a fly? The answer is given by the results obtained by Liebman *et al.* (1974) from their studies of frog retinas. In this system the time constants of the transition only increase by a factor of approximately 6 over the same temperature range.

A similar phenomenon is observed in the adaptation of myosin in fishes living under widely different conditions of temperature. Johnston & Goldspink (1975) reported that the energy of activation of the myofibrillar ATPase of fishes living in hot springs is as high as 137 kJ mole^{-1}, while the same reaction in material obtained from Antarctic species has an energy of activation of 30 kJ mole^{-1}, which is exceptionally low for an enzyme reaction. Johnson & Goldspink's (1975) data on the temperature dependence of the ATPase of muscle from species living in different temperature environments also provide a good example for the phenomenon of enthalpy/entropy compensation (see figure 7.5). The Gibbs energy of activation is maintained constant during adaptation of the function to different temperatures by compensating changes in the activation parameters $H^{\downarrow}$ and $TS^{\downarrow}$.

The use of kinetic analysis for the study of the chemical detail of evolutionary changes in proteins during adaptation of organisms to different temperature environments provides the most sensitive measure of the effects of amino acid replacements. As pointed out in various contexts, this is not only of academic interest. Site directed mutagenesis is used to study adaptation of protein structures for use in enzyme reactors, where

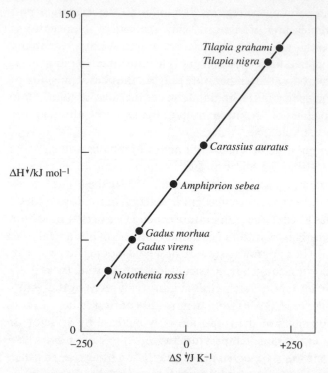

Figure 7.5   Compensation of enthalpy and entropy of activation to maintain a constant Gibbs energy of activation during adaptation of myosin ATPase to different temperature environments. The study was carried out on species of fish living at temperatures ranging from hot springs to arctic conditions. The plot is recalculated from the data of Johnson & Goldspink (1975).

stability and optimum activity under a wide range of conditions is of paramount importance.

7.3 Viscosity and molecular dynamics

Introduction

In recent years several developments of ideas about reaction mechanisms have encouraged a considerable increase in interest in diffusion control of substrate binding and in the internal mobility of protein molecules. The well-established relation between viscosity and bimolecular collision processes has been the subject of many investigations and will be discussed in the next section. Here we shall concern ourselves with a reassessment of the application of the theory of absolute reaction rates and alternative proposals for the interpretation of molecular dynamics in solutions, especially of proteins. Interest in this subject has been stimulated by the observed effects of solvent viscosity on unimolecular reaction rates and has led to the appraisal of this variable as a means of studying the consequences of ligand binding and protein dynamics. In the subsequent discussion the terms 'theory of absolute reaction rates' and 'transition state theory' will be used interchangeably. A distinction is sometimes made between the terms 'activated state' and 'transition state' when used in different contexts. The former tends to be used in discussions of activation energies by physical chemists, while the latter is used whenever the detailed chemical mechanism of a reaction is discussed by organic chemists. The latter often includes speculations about the structure of the transition state intermediate. These, not very precise, distinctions will be clarified later in this section and in section 7.5.

Kramer's theory: an alternative to Eyring

Frauenfelder and his colleagues (see summaries in Frauenfelder & Wolynes, 1985; Frauenfelder, Parak & Young, 1988) have stimulated the use of an approach to the interpretation of reaction rates of proteins which has its origin in the theoretical treatments of Kramer (1940) and Chandrasekhar (1943). As pointed out by Hynes (1985), the transition state theory gave a good approximation to the behaviour of reactions in the gas phase, but solvent dynamics had no part in it. Gavish (1986) expressed surprise that, since the presentation of the transition state theory by Glasstone, Laidler & Eyring (1941) (see section 7.1) this remained the only one mentioned in textbooks. Hynes (1985) also comments on the neglect of Kramer's theory, but quotes some examples of its application in the physicochemical literature. Gavish surmises that this neglect may be due to the mathematical complexity of Kramer's theory. The apparent relative simplicity of the Eyring theory is probably due to an illusion created by the usual diagrammatic presentation (see figure 7.1). Kramer (1940) does also write in terms of a transition state in the summary of his paper: 'A particle which is caught in a potential hole and which, through the shuttling action of Brownian motion, can escape over a potential barrier, yields a suitable model for elucidating the applicability of the transition state method for calculating the rates of chemical reactions.'

As stated before, the present text is not concerned with the theoretical treatment of chemical dynamics but only with kinetic profiles and with the effects of conditions on them. This leads to the principal aim, namely to distinguish and characterize the steps of a reaction sequence. Basically this means that we are stopping short of the detail which requires quantum mechanical treatment and the consideration of molecular orbitals. The examples used to illustrate the topics presented in this and other sections should make it clear to the reader what specific questions we are attempting to answer.

Some re-iteration and expansion of the difference between the approaches of Eyring and of Kramer to the theory of reaction rates should help to explain why the latter has stimulated new experiments. Whatever the merits of Kramer's theory, it has certainly helped to promote interest in the contribution of intra- as well as inter-molecular mobility of reactants, and consequently of solvent viscosity, to reaction parameters. As we have seen in section 5.2, Eyring's theory is based on the hypothetical thermodynamic properties of an activated state in a potential well above the ground states of the reactants. Kramer's theory is based on a process of diffusion over a potential barrier. The resulting equation (7.3.1), just like Eyring's (7.2.2), has the same form as the Arrhenius equation (7.1.1) and one might wonder what all the argument is about. The article by Gavish (1986) is recommended for a good insight into the theory. From an experimental point of view the introduction of viscosity as a component of the pre-exponential factor in the equation for the rate constants of both bimolecular and unimolecular reactions, is of great interest. Gavish (1986) writes for Kramer's equation for the reaction

$$S \longrightarrow P$$
$$k = \tau^{-1} \exp(-U/RT) \tag{7.3.1}$$

where τ, the time constant of structural fluctuations, is proportional to viscosity of the medium, and U is defined as the height of the potential energy barrier. It is important to distinguish the viscosity term in the above equation for a unimolecular event, from the effect of viscosity on the diffusion of the substrate to the reaction site. We have to consider separately the effect of the viscosity of the medium on the frequency of diffusion controlled collisions and of its influence on the structure fluctuations of the complex during its path over the energy barrier U. We shall emphasize this distinction again when we interpret the role of diffusion in the second order reaction of ligand binding and in the reaction of the complex, for instance of enzymes with substrates (see p. 268).

Microscopic and macroscopic viscosity

Before any further discussion of the effects of viscosity on any reaction in solution, whether collision frequencies or structure fluctuations, we have to consider the distinction between macroscopic and microscopic viscosity. A practical definition of viscosity is

{energy × time (required to maintain flow)/volume of liquid}

The data for viscosities of solutions are usually given in the literature in terms of poise or centipoise (cp). For practical reasons cgs units are generally used and poise has the units dyne s cm^{-2} (the dimensions of viscosity are $ML^{-1}t^{-1}$). Comprehensive tables of viscosities are found in the *Handbook of Physics and Chemistry*. In terms of cp the viscosity of water is 0.890, 1.002 and 1.787 at 25, 20 and 0 °C respectively. An example of the large temperature dependence of the viscosity of water is presented by the decrease in collisional fluorescence quenching on cooling of a solution (see section 8.2).

The viscosity of a solution measured in a capillary flow viscometer is the macroscopic viscosity. However the viscosity encountered by a diffusing solute, that is the viscosity term in the equation for the diffusion constant (see p. 267) or Kramer's equation (7.3.1), is the microscopic viscosity. It is the microscopic viscosity which therefore concerns us, both in diffusion control of collision frequency and in the pre-exponential factor of equation (7.3.1) for unimolecular reaction rates. The difference between these two types of viscosities is only of significance when the macroscopic viscosity of the solution is due to components of molecular weights considerably higher than that of the diffusing reactant or mobile part of the reaction site. Graham (1862) showed that the presence of gelatine in an aqueous medium had little influence on the rate of diffusion of ions. Since that time a number of investigations have shown that the solvent medium can no longer be considered a continuum for a diffusing solute which has considerably smaller dimensions than components of the solvent (for key references see Biacheria & Kegeles, 1957; Phillips *et al.*, 1977). Phillips *et al.*, for instance, have shown that the diffusion constant for Na$^+$ ions was only decreased by 30% for a 20fold increase in viscosity due to the addition of polyethylene glycol ($M_r = 20\,000$) to the aqueous medium. In contrast to these findings the addition of ethylene glycol or sucrose results in a change in the diffusion constant which is approximately inversely proportional to the increase in viscosity. We shall return later in this section to further discussions of viscosity effects on intramolecular transitions; the effect on collision frequency is discussed in section 7.4.

Conformational mobility of protein molecules

The dynamics of protein molecules in solution and in crystals has become of ever greater interest in recent years (see Daggett & Levitt, 1993). Two types of conformational mobility make important contributions to the kinetic behaviour of protein molecules. The first involves distinct structure changes on the direct pathway of the overall reaction and the second results in a distribution of many substates of structures in rapid equilibrium with each other; these substates having different reactivities. The observation of an apparent distinct change could be due to the freezing (perhaps through ligand binding) of a selected conformation from the fluctuating distribution. The success of X-ray crystallography in producing atomic models of protein molecules and even of enzyme–substrate complexes (for instance that of lysozyme; Blake *et al.*, 1967) has made an enormous contribution to our understanding of enzyme catalysis and other biological reactions. Structure changes on substrate binding have long been inferred from kinetic observations (see section 5.1). However, the difference between the structures of free enzymes and their complexes with substrates or other ligands, deduced from X-ray analysis, made it more obviously convincing that specific conformation changes occurred during such interactions. Similarly the tertiary and quaternary structure changes, which occur during the cooperative transitions of oligomeric proteins responsible for the regulation of metabolism and transport, are indicative of the flexibility of protein molecules in solution. For a detailed and authoritative review of the crystallographic evidence for distinct structure changes see Perutz (1989). It was also noted in the early days of the elucidation of protein structures that some parts of the molecule, as for instance a few residues at the $-NH_2$ terminal of myoglobin, were ill defined. This is even more striking when one inspects the results of the recent X-ray structure determination of the major fragment of myosin, which contains both the actin binding and ATPase sites (Rayment *et al.*, 1993).

About the time when the first three-dimensional structures were published, studies of deuterium exchange into groups in the interior of protein molecules indicated flexibility (for review see Englander & Kallenbach, 1983). However, in the 1960s X-ray crystallography was the dominant technique for the study of proteins. Although Perutz & Matthews (1966) realized that the haemoglobin molecule must be flexible for oxygen to reach its bonding position at the haem iron, the conclusions of generally rigid structures were extrapolated from crystals to solutions. The picture of unique structures with the potential to undergo some specific transitions were generally accepted at that time. To quote Karplus & McCammon (1986) 'a curious case in which a method that yielded dramatic advances also led to a misconception. . . . The intrinsic beauty and remarkable detail of structures obtained from X-ray crystallography resulted in the view that proteins are rigid.' The attitude among crystallographers changed dramatically when the recent development of Laue techniques made it possible to follow structure changes during reactions of enzymes with substrates (Moffat, 1989; Hajdu & Johnson, 1990). The application of this technique has, in the case of enzymes with slow turnover, such as the hydrolysis of GTP by Ha-Ras p21, displayed the reaction mechanism by the changing Laue pattern (Schlichting *et al.*, 1990). As NMR techniques developed to the stage of being applicable to the determination of structures of small protein molecules in solution, the views on rigidity versus flexibility became more balanced. NMR provides structure information as well as a dynamic picture and should really be part of a discussion on kinetics. However, a meaningful presentation of this method would require half the space of this volume.

It must be pointed out that, even before time resolved crystallographic methods appeared on the horizon, structural mobility was beginning to be accepted. The experiments of Lakowicz & Weber (1973a,b) on the diffusion of oxygen into protein molecules, which were interpreted in terms of 'breathing' of the whole structure, had considerable influence on opinion towards a belief in the internal mobility of protein molecules. The experiments showed that oxygen quenches the fluorescence of tryptophan residues internal to protein molecules. The very low activation energy of this process, 12 to 17 kJ mole^{-1}, is indicative of a diffusion controlled process. Lakowicz & Weber also estimated (with some considerable assumptions) that the diffusion constant for oxygen inside protein molecules was only two- to fivefold reduced over that in water.

To a physical chemist an idealized picture of rigid protein molecules always did appear improbable. Secondary and tertiary protein structure is maintained by a system of hydrogen bonds, complementary charge pairs and non-polar interactions. The equilibria between intramolecular and intermolecular (with water) pairing of the first and second of these are not far from unity and are, therefore, readily perturbed. Two types of flexibility with consequences for the reaction profile of protein molecules were to be expected, and indeed have been found to occur. These have already been alluded to above. The first is induced by the changes in the local environment when a substrate binds. This results in conformation changes which are detected by the kinetic techniques described in section 5.1 and confirmed by structure analysis. For example well-defined changes have been observed when glucose binds to hexokinase (Shoham & Steitz, 1982) and the time course of the movement of a polypeptide chain embracing the substrates of lactate dehydrogenase and in triosephosphate isomerase has been recorded (Clarke *et al.*, 1985; Pompliano, Peyman & Knowles, 1990). These conformation changes have been accommodated in the complete elucidation of the respective reaction mechanisms of, for instance, lactate dehydrogenase (see section 5.1).

A second type of flexibility of protein molecules has been gradually documented in more and more detail as new experimental techniques were devoted for its study, that is, the existence of proteins in an ensemble of conformational substates. This depends on the degeneracy of the ground states of protein structures. At room temperature these substates are in rapid equilibrium with each other and the observed kinetic parameters are a weighted average for the population. The observed rate constant, k, for the formation of all n substates of B from all substates of A (in rapid equilibrium with each other) is given by

$$k = \sum_{i=1}^{n} k_i p_i$$

where k_i is the rate constant for the transition from substate A_i to B and p_i the probability of A to be in substate A_i. As the temperature is lowered kinetic and spectroscopic techniques that can sample the characteristics of the system faster than the transitions occur can detect the heterogeneity of protein structures. There are clearly a number of factors connecting ligand induced conformation changes between two well-defined states and a wide distribution of substates. As suggested by Karplus & McCammon (1983), a conformation change will occur via a pathway of fast searches along a sequence of energy troughs. If the binding of a substantial ligand freezes a selected local conformation of previously mobile segments, or if the ligand will only bind to such a selected conformation, this would be kinetically observed as a pre-equilibrium of the kind described in section 3.4. For instance there is evidence that avidin, transferrins (Anderson *et al.*, 1990) and other binding proteins have an open and very flexible structure in the absence of their specific ligand and close up like a clam when the ligand has entered the site. Extensive studies on some periplasmic chemotactic binding proteins in Quiocho's laboratory (see for instance Sack, Saper & Quiocho,

1989) led the authors to use the term 'flytrap model'. In addition to structural evidence, kinetic studies on the biotin–avidin system showed that the enthalpy of activation for entry is very small, while it is very large for exit. For entry through a flickering gate the rate constant will depend on diffusion reduced by the equilibrium constant of the positions of the gate (see p. 69 for rapid pre-equilibria in ligand binding). Neither the equilibrium constant nor diffusion will have a large temperature dependence. Fixed gate opening for exit will have the large enthalpy of activation of a conformation change.

In addition to the early suggestions for protein flexibility mentioned above, the evidence for conformational substates of proteins comes principally from three areas of experimental investigations. The results of kinetic studies by Frauenfelder and his colleagues suggested to them that steps of reactions of proteins are controlled by a distribution of energy barriers resulting from an ensemble of reactant conformations. From the same laboratory came reports of the asymmetry in the spectral changes during the time course of reactions. This was interpreted as the difference in reaction rates between different substates. Inhomogeneous broadening of visual pigment spectra has also been explained in terms of conformational substates (Cooper, 1983). However, care has to be taken if protein samples are not obtained from a single animal. Polymorphism could account for the differences in spectra from different individuals (Mollon, Bowmaker & Jacobs, 1984). Collaboration of kineticists with X-ray crystallographers encouraged the extended investigations of the so-called temperature factor of crystal structures to the interpretation of structural heterogeneity. Frauenfelder (1989) describes how the Debye–Waller effect, which at one time was thought to make structure determination difficult, changed from a villain to a hero. Crystal structure determinations at low temperatures made it possible to distinguish between thermal vibrations and protein flexibility (Frauenfelder, Petsko & Tsernoglou, 1979; Ringe & Petsko, 1984). In addition to these experimental contributions, theoretical investigations by Karplus and his colleagues (see Karplus & McCammon, 1983) have been used to predict structural mobility and its involvement in reactions of proteins. An interesting historical account, albeit a very personal one, of the development of these ideas is given by Frauenfelder (1989). Debrunner & Frauenfelder (1982) also refer to the application of Mossbauer spectroscopy to the study of the structural environment of the haem iron in connection with the Debye–Waller effect. Mossbauer measurements are sensitive to the mean square displacement of the iron atom in myoglobin. The present discussion of these problems has been strongly influenced by the extensive publications from Frauenfelder's laboratory (see reviews by Debrunner & Frauenfelder, 1982; Frauenfelder, Parak & Young, 1988). Investigations by Petrich *et al.* (1994) led these authors to complementary conclusions about non-exponential geminate recombination of ligands with haem proteins. They suggest that at low temperatures this could be due to distributions of states, as suggested by Frauenfelder's group, while at high temperatures structural relaxations on the time scale of the recombination could be the cause.

Just as for the development of many other techniques and ideas in protein chemistry the haem proteins, haemoglobin and myoglobin, have been used as a paradigm for the exploration of molecular dynamics. In particular myoglobin, the first protein to have its three-dimensional structure elucidated, has the advantages of small size (17 000 daltons) and a single haem group which provides an excellent spectral monitor. During the classic kinetic studies, using flow techniques to investigate the reaction of haem proteins with oxygen and carbon monoxide, the overall bimolecular process was observed on the time scale of milliseconds. Gibson (1956) initiated the use of flash photolysis to resolve individual steps of this process on a microsecond time scale. Austin *et al.* (1975) presented the first of many detailed accounts of the experiments on flash photolysis in Frauenfelder's laboratory. Their measurements were carried out over a wide range in time (100 ns to ks) and temperature (10 to 320 K). In their experiments they were able to resolve at least three phases in the recombination of CO with myoglobin after photodissociation. One bimolecular process of ligand entering the binding site from the solvent is followed by events within the haem pocket leading to binding to the ferrous iron. The steps within the haem pocket were studied in terms of geminate recombination, the rebinding of photodissociated ligand which had not left the pocket into the solvent.

At temperatures above about 200 K geminate recombination of CO follows an exponential time course. The kinetic evidence for conformational substates comes from investigations at lower temperatures. Experiments in glycerol–water mixtures at 200 K and below indicate that each protein molecule is frozen into its conformation; the substates are no longer in rapid equilibrium. Pressure perturbation experiments indicate relaxation times of 10^5 seconds. Under these conditions the recombination of ligand is no longer exponential but follows a power law (see Steinbach, 1991). This could be due either to heterogeneity in protein structure and hence in reactivity (a distribution of activation enthalpies) or it could be caused by multiple pathways within each protein molecule of a homogeneous population. These two cases can be distinguished by multiple flash experiments with the frequency of the flashes in the middle of the range of time constants of the final CO–Fe docking step. The response to a single flash is indistinguishable in the two cases. However, repetitive flashing with a suitable frequency results in a constant amplitude for each flash in the heterogeneous case, while in the homogeneous case each flash will commit more and more ligand molecules to the slower pathways and the observed recombination amplitude will decrease. Frauenfelder *et al.* (1988) discuss the hierarchical distribution of substates in at least three tiers. The problem of a detailed kinetic analysis of a sequence of distinct reaction steps becomes enormous, even for a 'simple' process like ligand binding, with a distribution of reactivities among the reactants. Much more will have to be done with the armoury of new kinetic techniques to determine which features of protein flexibility control and discriminate in biological processes. Although most of the detailed photochemical studies on ligand binding were done with myoglobin, similar evidence on conformational substates was obtained with a number of other haem proteins.

An increasing number of observations point to an important transition in the state of proteins as the temperature is lowered from 220 to 200 K. Results from neutron scattering, reviewed by Smith (1991), provide information about the mean square displacement of hydrogen atoms in proteins. It is found that proteins undergo a transition to a glasslike structure at 200 K and only harmonic motion remains below that temperature. This is of interest in connection with the change in kinetic behaviour from that of an apparent (rapidly averaged) single state to a frozen distribution of states, at temperatures below 200 K. Furthermore this phenomenon has been used by Petsko and his colleagues to investigate the involvement of flexibility in catalysis. They studied binding of substrate to pancreatic ribonuclease crystals, monitoring X-ray reflections sensitive to productive interaction at the active site. Using a flow cell/cryostat system, they found no evidence of binding at 212 K while on raising the temperature to 228 K binding occurred (see Rasmussen *et al.*, 1992). This, combined with observations from the same laboratory (Tilton, Dewan & Petsko, 1992) on the more rapid increase above 220 K in the Debye–Waller factor, indicates that, for substrate binding, more complex motions are required than the simple harmonic ones allowed below 220 K.

Viscosity and the transition over the energy barrier

If energy barriers are no longer considered static, as in Eyring's theory of absolute reaction rates, but are treated in terms of dynamic, fluctuating amplitudes, then one can account for the effects of solvent viscosities. This parameter is included in Kramer's (1940) treatment of reaction rates but not in Eyring's. In turn, it enables one to use viscosity as a variable for the exploration of some aspects of reaction mechanisms. Only a very small number of first order reactions has been studied with respect to their viscosity dependence. For second order reactions this parameter has been used extensively as a diagnostic for diffusion control (see section 7.4). However, it will be emphasized there, that the viscosity dependence of a reaction does not necessarily identify it as a diffusion controlled collision process. Some examples can be given where viscosity dependent reactions of proteins have been ascribed to conformation changes.

We again refer to work from Frauenfelder's laboratory (Steinbach, 1991) where it was shown that during recombination of CO with myoglobin, a step at the entrance to the haem cleft was viscosity dependent while the final step close to the haem iron was not accessible to solvent. Movements of parts of a protein molecule are just as susceptible to changes in solvent viscosity as diffusion of the whole molecule, provided the moving parts are accessible to solvent. The restriction on the size of the reagent added to increase the viscosity is more stringent than that discussed above (p. 251). Of the reagents frequently used to control viscosity, glycerol for instance might have access to a moving hinge while sucrose does not. This point was investigated in connection with the study of the movement of the polypeptide chain which embraces the substrates of lactate dehydrogenase after their initial contact. The viscosity dependence of the reactions of this enzyme had previously been studied by steady state kinetics (Demchenko *et al.*, 1993). However, transient kinetic studies could provide clearer evidence for the viscosity dependence of a single step, which could be identified with the movement of a segment with a tryptophan residue. It could also be identified with the rate limiting step of the enzyme reaction (Clarke *et al.*, 1985; and section 5.1). Gavish (1986) reported steady state studies of viscosity effects on the reaction of carboxypeptidase. The data from this paper were also analysed in a good theoretical treatment of viscosity and Kramer's theory (Gavish, 1986).

Studies of the effects of viscosity on hydrogen exchange into proteins could be very informative, but few reports are available. Somogyi *et al.* (1988) studied the exchange of deuterium for a hydrogen in the indole ring of tryptophan-63 of lysozyme. This exchange reaction, which can be followed by a change in fluorescence, is diffusion controlled with free indole but not so for that group within the protein. The effects of increasing viscosity were studied by the addition of glycerol or ethylene glycol to the solvent. The reaction mechanism was identified in terms of a transient hydration of the buried residue with a high activation barrier. The viscosity gradient was 0.6.

The kinetics of quaternary structure changes of proteins undergoing cooperative transition has, so far, only been studied with two systems. For haemoglobin the rate constants are in the 100 μs range (Eaton *et al.*, 1991; Jones *et al.*, 1992) and for aspartyl transcarbamylase they are about 50 ms (Kihara *et al.*, 1987). In both cases these rates will be critically dependent on the pre-equilibria with different ligands and their concentrations. In a very interesting application of time resolved X-ray scattering Tsuruta *et al.* (1990) followed changes in the quaternary structure of aspartyl transcarbamylase and in an attempt to slow down the reaction established its viscosity dependence.

Rates of protein conformation changes

To conclude this section on controlling factors in first order transitions of protein molecules we must discuss a question frequently posed: how fast are protein conformation changes? One must remember that one person's rapid conformation change is another's slow motion picture of a multitude of events. In connection with the last statement one is concerned with major changes observed as distinct conformations on the reaction pathway. It is, however, analogous to asking 'how long is a piece of string?'. Table 7.2 gives some observed rates of such gross conformation changes. In reality the rate constants quoted should be called the frequency of transitions from one conformation to another. This is analogous to the terms 'open times' and 'closed times', which are used in connection with the direct observations of states of ion channels. In that connection it is emphasized that the lifetimes of states do not represent the rate constants for opening the gate in the membrane. As pointed out above, only observations on a single molecule, or channel, can give information about rates of movement in terms of distance/time. The time resolution of patch clamp measurements on single channels results in step functions between different states. Lifetimes can give minimum rates. Recently exchanges of ideas and terminology have occurred between electrophysiologists and protein chemists. Diffusion over energy barriers and power law kinetics have been invoked for ion channel gating (Millhauser *et al.*, 1988). On the other hand, in connection with the approach of ligands to their docking sites, the terms 'gates' and 'channels' are being applied outside the field of excitable membranes. A theoretical treatment of reaction rates controlled by gating is presented by Northrup & McCammon (1984). The most concrete picture for this is again drawn for

Table 7.2. *Rates of some protein-conformation changes*

Protein	Transition	Rate(s)	Reference
Lysozyme	Tryptophan fluorescence and proton release	120	Halford (1975)
Lactate dehydrogenase	Tryptophan fluorescence	150	Clarke et al. (1987)
Alkaline phosphatase	Substrate chromophore	170	Halford et al. (1969)
Myosin	ATP analogues fluorescence	340 (20 °C)	Millar & Greeves (1988)
		20 (1 °C)	
Aspartyl trancarbamylase	T to R transition	50 (4 °C)	Kihara et al. (1987)
Haemoglobin unliganded	T to R transition	4×10^4	
Haemoglobin triliganded	T to R transition	1×10^3	Eaton et al. (1991)
Acetylcholine receptor	Channel opening (τ^{-1} closed state)	9.4×10^3	Matsubara et al. (1992)
Troponin C, Ca^{2+} binding	Tryptophan fluorescence	700	Rosenfeld & Taylor (1985)

myoglobin where a motion of the E helix is required to open a channel for the approach of ligand to the haem iron from the distal side (see for instance studies from Olson's and Gibson's laboratories; Gibson, 1990). Several of the other examples mentioned above, might involve a gate closing for ligand binding and opening for ligand release as the major energy barrier. Special cases involve the protection of a reactive intermediate. The protective function of enzymes was discussed by Gutfreund & Knowles, 1967. A more recent example comes from Knowles's work on triosephosphate isomerase. He has shown that the flexible loop of this enzyme is required to prevent the loss of a reaction intermediate, enediol phosphate, prior to reprotonation to product (Pompliano *et al.*, 1990). Whether gating mechanisms are viscosity controlled, even if the moving hinge is readily accessible to solvent, depends on the mechanism. The analysis of transients (section 5.1) and of relaxations (section 6.3) of sequential processes shows that for some mechanisms only the equilibrium constant would affect the observed rate. The position of a gating process in the sequence of the reaction will determine whether the probability of its 'open' state or its rate of opening is of importance for the overall rate. The equilibrium may not be affected by viscosity even if the rate of movement is.

The rate constants of conformation changes listed in table 7.2 are all 'observed' values under special conditions and should perhaps not be honoured with the term constant. The values are likely to be variables depending on pre- and post-equilibria. These in turn will depend on many conditions. Furthermore, the molecular dynamics simulations reviewed by Karplus & McCammon (1983) showed that conformation changes on the millisecond time scale are likely to be a sequence of much faster steps.

7.4 Diffusion control of reactions

Brownian motion: random walk

As pointed out in section 7.1 diffusion influences the velocities of biological processes in a number of ways in addition to its effect on homogeneous reactions studied in a chemical laboratory. Chemical engineers have considerable interest in transport processes associated with production in large reactors (Danckwerts, 1969). These processes are related to problems encountered in attempts to understand the functioning of organized cellular systems when reactants have to come together from different locations. The two problems, diffusion control in homogeneous reactions and diffusion in organized systems, require some separate discussion. Many

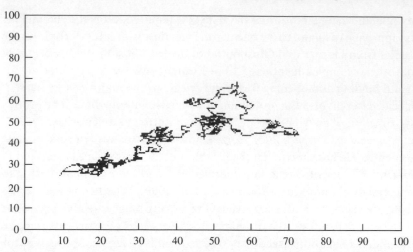

Figure 7.6   Simulation of Random Walk showing tendency of local search of space, with occasional excursions into a new environment.

biochemical reactions are first order transformations but they are always part of a sequence initiated by a bimolecular (second order) encounter, which depends on two reactants diffusing to within their reaction distance. Exceptions to this are the processes initiated by light or electrical impulses. In the rest of this section we are principally concerned with the relation between rates of diffusion and 'target finding' of a ligand onto the binding site of a large molecule, with both of them in free solution.

Diffusion is a consequence of the random Brownian motion of molecules. The mean free path of a solute molecule in a liquid is very small, compared to events in a gas, since its momentum is lost rapidly to solvent molecules. If one were to observe a single solute molecule over an extended time one would find that it moves at random within a small volume element not much larger than its own. Occasionally it will shoot off some distance and then again explore a small volume (see figure 7.6). The random motion of solute molecules can be observed by light scattering (Schurr, 1976). Indeed this is how Brownian motion was first detected for large particles. Robert Brown (1773–1858), a botanist, reported seeing the random motion of various kinds of particles suspended in water. He was observing, under a microscope, the light scattered by pollen and dust particles. The physical explanation of the random movement is found in the fluctuations of solvent collisions. Pais (1982) presents a good historical account of investigations into the physical basis of Brownian motion in a chapter entitled 'The reality of molecules'. Einstein (1905) showed that the kinetic energy per particle with velocity v is, in each direction (for instance x)

$$k_B T/2 = m v_x^2 /2$$

or, since the velocity fluctuates

$$k_B T/2 = \langle m v_x^2 \rangle /2$$

(throughout this volume $\langle\ \rangle$ is used to indicate averaging, k_B is the Boltzmann constant, m is the mass of the particle, v_x is the velocity in one direction). The record of the scattering of a laser beam from a small volume element of a solution allows the calculation of the rate of diffusion. This involves the principle of the correlation of density fluctuations of the system over a range of successive time intervals. Other methods for measuring the rates of diffusion (diffusion constants) involve the observation of the rate of flux of the solute across an artificial boundary between solution and solvent. These two methods each have their counterpart in systems of interest to us. Random motion in a homogeneous system is of importance in the control of collision limited reactions, which are the subject of this section, while diffusion across boundaries influences reactions in organized systems. The basic theory of diffusion applies to both these problems and a definition of the diffusion constant D is required for an understanding of both of them. The volume by H. C. Berg (1983) *Random Walks in Biology* and the review by O. Berg & von Hippel (1985) will be found a stimulating addition to much of this section. The latter review either covers or gives references to most of the field of diffusion controlled reactions. The present discussion is certainly influenced by and owes much to these authors.

The random movement of a molecule in solution results in the spherical extension, with time, of its probability distribution in space from its original position. The distribution is an expanding spherical Gaussian with a half width of approximately $(4Dt)$ at time t for a molecule with diffusion constant D.

Derivation of diffusion equations

Berg (1983) shows very clearly how the equations for the macroscopic diffusion of an ensemble of molecules can be derived starting with the random motion of a single particle in one dimension. Just as in other statistical problems when the mean is zero, the important parameter for the distribution is the mean square displacement of the particle, $\langle x^2 \rangle$, which is related to the diffusion constant by $\langle x^2 \rangle / 2t = D$. The factor $1/2$ comes from the definition of the diffusion constant as the flux across a boundary in one direction.

Berg gives the following interesting example for the movement of a protein molecule, lysozyme, with a mass of $14\,000/6 \times 10^{23} = 2.33 \times 10^{-20}$ g per molecule. Using the Einstein equation at 300 K ($k_B T = 4.4 \times 10^{-14}$ g cm^2 s^{-2}) one obtains the velocity in one dimension $\langle v^2 \rangle^{1/2} = 1.3 \times 10^3$ cm s^{-1}. The macroscopic diffusion constant of lysozyme is $D = 10^{-6}$ cm^2 s^{-1} (a summary discussion of the dimensions of the diffusion constant is given below). With a velocity, $v = \delta / u$, corresponding to a step length δ per unit of time u, and $D = \delta^2 / 2u = 10^{-6}$ cm^2 s^{-1} the step length is $2D/v \approx 10^{-9}$ cm and the step rate is $v/u = 10^{12}$ s^{-1}. The random walk argument proceeds from this. Still thinking in terms of one dimension, half the steps will be taken to the right and half to the left. Using the binomial distribution one can calculate the probability of k out of n steps occurring in the same direction. This reaches a Gaussian distribution in the limit of large n. Actually the probability of a particle having moved a distance to the right is half the probability density curve and movement to the left is given by the other half. For k steps to the right in n steps of length δ the distance moved $x = (2k - n)\delta$ in time nu.

The above is intended to give a physical picture of the diffusion process rather than a full mathematical derivation of the probability distribution. For a somewhat more complete algebraic treatment Berg (1983) and references therein should be consulted. There are ongoing discussions among specialists in statistical physics about the validity of the so-called ergodic hypothesis (see Margenau & Murphy, 1943; or most good texts on statistical mechanics) which is concerned with the equivalence of summing the behaviour of a single particle over time and the behaviour of an ensemble of particles at a point in time. In solution the statistical behaviour of a single particle is assured by collisions with the solvent, which gives a meaning to temperature. Here the description of the behaviour of a single particle is intended to give a picture of how it moves and explores one-, two- and three-dimensional space. This is of significance in the interpretation of the control of homogeneous reactions by diffusion.

Diffusion and flux

Before applying diffusion constants to the interpretation of collision frequencies and reactions let us consider the phenomenological interpretation of diffusion in terms of flux. This is also required if an attempt is

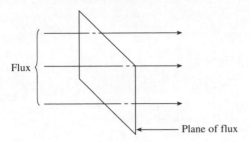

Figure 7.7

made to interpret reactions in organized systems (Maynard Smith, 1968). The flux ,J, is defined as the quantity (moles) which crosses a unit area (cm^2) in unit time (s^{-1}) and is, therefore, expressed in moles cm^2 s^{-1}. The factor which determines the relation between the flux (of diffusing solute with concentration c_S) across a boundary and the diffusion constant is the concentration gradient, dc_S/dx, (the driving force), which is expressed in moles cm^{-3} cm^{-1} at time t as Fick's first law which defines the flux δ

$$J = -D \left[\frac{\delta c_S}{\delta x} \right]_t$$

(see figure 7.7).

A check on units is always a useful assurance that they are consistent; the above argument assigns the correct units to the diffusion constant, cm^2 s^{-1}. It is significant that Fick (1829–1901), a physiologist, essentially founded and made the major contributions to the theory of diffusion. Fick's second law describes the rate of change

$$\left[\frac{\delta c_S}{\delta t} \right]_x^2 = D \left[\frac{\delta c_S}{\delta x^2} \right]_t \tag{7.4.1}$$

The simplest model for macroscopic diffusion from a source, and also the one widely used in the determination of diffusion constants, consists of a sharp boundary between solution and solvent in a tube of infinite length in both directions. It is assumed that at $t = 0$, dc_S/dx is a step function. As time progresses the boundary spreads and the concentration profile is represented by the Gaussian distribution function (see figure 7.8).

Fick's second equation and the diffusion constant

As indicated above, the diffusion constant is related to the variance (the mean square displacement) by $\langle x^2 \rangle = 2Dt$ and it can, therefore, be calcu-

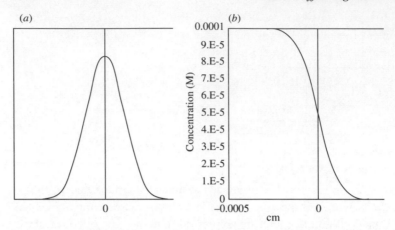

Figure 7.8   (*a*) Simulation of Gaussian distribution across a boundary as observed by refractive index changes. (*b*) Simulation of concentration across a diffusion boundary corresponding to a gradient 1 ms after initiation at 25 °C. Initial concentration at the left hand side of the perfect boundary 1 mM, $D = 4.59 \times 10^{-6}$ cm^2 s^{-1}.

lated from the time course of the change in solute distribution across a boundary. The optical signals used for the experimental determination of the diffusion constant result either in probability distribution curves, from light absorption, or in probability density curves from refractive index changes (see figure 7.8). The distribution curve can be derived in a number of different ways from basic diffusion models (see Berg, 1983). The solution of Fick's second equation (see equation (7.4.1)) is

$$c_S(t) = \frac{c_S(0)}{2} \left[1 - \frac{2}{\sqrt{\pi}} \int_0^{x/(4Dt)^{1/2}} \exp[-x^2/4Dt] \mathrm{d}x \right] \qquad (7.4.2)$$

The integral, which can be obtained from tables of error functions, has the value of 0 to 1/2 for $x = 0$ to ∞ and similarly for $x = 0$ to $-\infty$. This merely shows that whatever one has lost from one side of the boundary will have been gained on the other side. The plot of concentration gradient against x (obtained from refractive index changes) gives an integral probability density curve

$$\left[\frac{\mathrm{d}c_S}{\mathrm{d}x} \right]_t = \frac{c_S(0)}{\sqrt{[4\pi Dt)}} \exp[-(x^2/4Dt)] \qquad (7.4.3)$$

To obtain the diffusion constant one has to calculate the variance of the

Gaussian curve thus obtained. The converse is the use of the diffusion constant for the calculation of the change with time of the concentration/distance profile for a substance in an organized system.

It is of interest to tabulate the values for the diffusion constants of a number of solutes of different molecular weight:

Substance	Molecular weight	D (cm^{-2} s $\times 10^6$)
Glycine	72	9.34
Sucrose	342	4.59
Ribonuclease	13 683	1.07
Haemoglobin	67 000	0.60

The kinetic equation for the diffusion constant

$$D = k_B T/f \quad \text{(where } f = 6\pi r\eta, \text{ with } \eta \text{ viscosity)}$$

shows that it is approximately proportional to the cube root of the molecular weight. The term approximately is used because molecules are not necessarily spherical and the frictional constant, f, of an elongated particle will be a function of both axes (see Tanford, 1961). It is of interest to look at the change in concentration profile of glucose diffusing across a boundary. Figure 7.8(b) shows the status quo after 0.1 μs and 1 ms respectively. This has been simulated for the idealized system of a column of infinite length, which stipulates the constant concentration $c_S(0)$ at one end and zero at the other end.

Collision frequencies and productive collisions

Over the last 40 years the literature on diffusion as a controlling factor in rates of homogeneous reactions has been quite extensive. In the above references to H. C. Berg and O. Berg & von Hippel special attention is given to this problem in enzyme–substrate and other biological receptor–transmitter systems. Berg & von Hippel paid particular attention to protein–nucleic acid interaction. The problem of measuring binding rates and determining whether they are diffusion controlled will be separated from the theoretical one of calculating the rates of diffusion controlled reactions. The latter problem became acute when exceptionally rapid binding was observed for protein–DNA interaction by Riggs, Bourgeois & Cohn (1970). Let us first consider the experimental investigation of substrate binding in some selected enzyme systems.

Knowles and his colleagues (see for instance Knowles & Albery, 1977) were concerned with the idea that certain enzymes have evolved to perfect their catalytic mechanism (however, see also Pettersson, 1989). Clearly if the overall turnover of an enzyme reaction is rate limited, at the ambient substrate concentration, by the formation of the collision complex controlled by Brownian motion, then no further improvement in catalysis can speed up the pathway from substrate to product. Later in this section, experiments will be discussed which were carried out to test this hypothesis by investigation of the effect of viscosity on ligand binding and thus into its mechanism and maximum rates. Such speculations about diffusion limitation and evolutionary perfection of enzymes have been extended from enzymes to ion channels (Hille, 1992).

Arguments used to draw conclusions from the effect of viscosity on reaction rates, as to whether reactions are diffusion controlled, involve the distinction between encounter and collision. A reaction will only be affected by viscosity if the first collision in an encounter between reactants is productive, which means that it is diffusion controlled. If the reaction is not diffusion controlled, in terms of the above definition (i.e. not every collision is productive), an increase in viscosity will decrease the number of encounters, but it will also increase the number of collisions per encounter due to slow separation of the reactants. The longer lifetime of the encounter will balance the decrease in their frequency and the effects of viscosity are cancelled out. Among the earliest investigations into the effects of viscosity on ligand binding in biological systems were those of Ackerman and his colleagues (1963). They found that, when suitable controls are taken into account, the rates of binding oxygen or carbon monoxide to haemoglobin were not significantly affected by increases in solvent viscosity. Related results come from measurements of rates of quenching of porphyrin fluorescence which gave a second order rate constant for the collision of des-Fe myoglobin with oxygen $k = 4 \times 10^8 \, \mathrm{M}^{-1} \, \mathrm{s}^{-1}$. This is considerably faster ($40 \times$) than the rate of binding of oxygen to myoglobin or haemoglobin and it appears that the rate limit of this reaction is not the penetration into the haem pocket (see also section 7.3). Dunford & Hewson (1977) examined the dependence on viscosity of one of the classic reactions in transient kinetics of enzymes, the formation of compound I of horse-radish peroxidase. Using m-chlorobenzoic acid as the substrate they found that the rate constant of this reaction decreased from 8.2×10^7 to $3.1 \times 10^7 \, \mathrm{M}^{-1} \, \mathrm{s}^{-1}$ (a factor of 2.65) when the viscosity was increased by a factor of 3 by the addition of sucrose. With H_2O_2 as the substrate the rate constant for the formation of compound I was only decreased from 1.9 to $1.8 \times 10^7 \, \mathrm{M}^{-1} \, \mathrm{s}^{-1}$ for a 2.4-fold increase in viscosity.

Enzymologists often use the function k_{cat}/K_M as a measure of the catalytic efficiency of an enzyme. The justification is clearly the fact that k_{cat} is proportional to the maximum rate and K_M is a measure of the substrate concentration dependence of the rate. This function is however also used as a substitute for a measure of the rate of substrate binding. Reference to the steady state equations (p. 91) shows that k_{cat}/K_M is likely to be smaller than k_{12}, the rate constant for substrate binding, and contains other constants, which could be affected by viscosity (see below). However, useful indications about diffusion control have been obtained from this indirect estimation of substrate binding rates. Kirsch and his colleagues showed that when a range of substrates with different values for k_{cat} are investigated, the dependence of k_{cat}/K_M on viscosity decreased with the slow reacting compounds. For instance in the chymotrypsin catalysed hydrolysis of N-acyl-tryptophan nitrophenylester, with $k_{cat}/K_M = 3.7 \times 10^7 \, M^{-1} s^{-1}$ they found a significant viscosity dependence, while with substrates with smaller k_{cat}, the methylester and p-nitroanilide, they found no viscosity dependence (Brouwer & Kirsch, 1982). The conclusion from this series of experiments is that with slower substrates chemical catalysis is rate limiting, while with the faster ones the limit of diffusion control is almost reached.

Knowles and his colleagues (Blacklow *et al.*, 1988) extended their investigations on the evolutionary perfection of the enzyme triosephosphate isomerase by studying the effects of site directed mutagenesis on diffusion control. For the wild type enzyme, with $k_{cat}/K_M = 4 \times 10^8 \, M^{-1} s^{-1}$ at a viscosity of 0.01 poise, this second order constant is directly proportional to viscosity. With a mutant enzyme, which has k_{cat} decreased by a factor of 1000, no viscosity dependence is observed. They conclude that the viscosity effect is on diffusion controlled binding and not on catalysis.

In section 7.3 the effects of viscosity on reaction steps other than bimolecular (second order) interactions are discussed. The observed rate constant for ligand binding is often a function of the equilibrium constant for the formation of a collision complex and the rate constant for a ligand induced conformation change (see section 3.2). Either or both of these constants may be dependent on viscosity and their individual evaluation usually requires analysis by several techniques. Hinge bending to provide access for substrates to the active sites of lysozyme, hexokinase and other enzymes has been treated as a diffusion model in molecular dynamics calculations (McCammon & Harvey, 1987). These structure changes, which follow the transient diffusional encounter, can also be affected by changes in the viscosity of the medium. Unfortunately there is, at present, only limited information available on the effects of viscosity on the elementary steps of enzyme reactions (see section 7.3). Hopefully such information will become available on the kind of reaction sequences discussed in section 5.1. In summary, it must be emphasized that k_{cat}/K_M is a complex function of rate constants and corresponds only to a minimum value for k_{12}, the diffusion controlled rate constant; it can be affected by the viscosity dependence of subsequent steps in the enzyme–substrate reaction.

The difficulties due to the fact that changes in the medium will not only affect the viscosity can be overcome in a variety of ways. A good control is the use of different reagents to raise the viscosity, but with different effects on the dielectric constant. It also is informative if either poor substrates or damaged enzymes retain the viscosity dependence (see Blacklow *et al.*, 1988).

Models and theories for maximal bimolecular rates

Let us now consider the contribution of diffusion to the maximum rates of homogeneous reactions involving associations as a first step in the interactions of proteins with both small and large substrates. In the present context association to form a specific complex does not include any covalent bond formation. Three events are involved in the approach to equilibrium: collision, alignment to correct mutual orientation and for some processes separation after the event. These events depend on translational diffusion for approach and separation and on rotational diffusion for alignment. Earlier in this section the diffusion constant was defined using the principle of random Brownian motion and this will now be used in equations which have been derived to evaluate maximum collision rates.

The following equation is widely used to calculate the upper limit for the collision frequency in a homogeneous system of uncharged particles

$$k = 4\pi N(D_A + D_B)(r_A + r_B)/1000 \qquad (7.4.4)$$

The frequency, k, has the units of a second order rate constant, $M^{-1} s^{-1}$, N is Avogadro's number and D_A, D_B, r_A, r_B are the diffusion constants (in $cm^{-1} s^{-1}$) and the radii of reactions (in cm), respectively, of the two reactants A and B. In keeping with common practice we shall call this the Smoluchowski equation. It is difficult to ascertain the first use of the equation in this form. It has its origins in Smoluchowski's (1916) mathematical treatment of the kinetics of the coagulation of colloidal particles in solution. He derived the equation for the time course of the formation of aggregates of increasing size. His theoretical predictions were found to to fit the results obtained by colloid chemists at that time. Smoluchowski's

Table 7.3. *Second order rate constants for enzyme–substrate complex formation*[a]

Protein	Substrate	Method	Rate $k \times 10^{-8}$
Myoglobin	Oxygen	Direct (spectrum)	0.14
Haemoglobin (R state)	Oxygen	Direct (spectrum)	0.15
Des-Fe haemoglobin	Oxygen	Fluorescence quench	4.0
Free porphyrin	Oxygen	Fluorescence quench	10.0
Acetylcholine esterase	Acridinium	Direct (fluorescence)	up to 100^b
Acetylcholine esterase	Acetylcholine	k_{cat}/K_M	up to 40^b
Lactate dehydrogenase	NADH	Direct (fluorescence)	1.0
Glycerol-3-phosphate dehydrogenase	NADH	Direct (fluorescence)	0.9
Superoxide dismutase	Oxygen radical	k_{cat}/K_M	20.0
Triosephosphate isomerase	Glyceraldehyde 3-phosphate	k_{cat}/K_M	4.0
Horse-radish peroxidase	Chlorobenzoic acid	Direct (spectrum)	0.82

[a] For references see Gutfreund (1987).
[b] Ionic strength dependent.

derivation was based on the assumption that the rate of aggregation is inversely proportional to the diffusion constants and the radii of the particles. Earlier models for diffusion controlled processes were developed to explain the extent of collisional fluorescence quenching. This phenomenon will be discussed later in the present section and in section 8.2. Some data for rates of quenching are given in table 7.3 to illustrate maximum collision frequencies.

The above equation and other related classical treatments of the role of diffusion in solution kinetics (see references in Berg & von Hippel, 1985) have been used to calculate the upper limits of the rates of second order reactions. This limit is usually defined as the diffusion controlled collision rate. This equation is only strictly applicable to spherical molecules with uniform reactivity over the two surfaces at a centre to centre distance equal to $r_A + r_B$ and has been shown to have other theoretical limitations (Collins & Kimball, 1949).

A concept, discussed above, which appears in many models under different names is the distinction between collisions and encounters, or what H. C. Berg (1983) calls micro- and macro-collisions. This was originally introduced as particularly relevant to ionic reactions and applied to models when attractive forces and/or solvation shells affect the approach and separation of reactants. In such cases the tendency for reactants to stay close to each other during an encounter, which corresponds to local searches during random walks (see figure 7.6), will be prolonged by the solvent structure. Consequently the number of collisions per encounter will also be increased. This point has already been referred to during the discussion of the effects of viscosity on rates of reactions and its use in locating diffusion controlled steps in a sequence (see section 7.3).

The concept of encounters can be treated like the formation of a weak complex in rapid equilibrium with the free reactants. We shall return to this when complexes formed between large molecules are considered.

Before considering special mechanisms, it is instructive to make comparisons of values for collision frequencies calculated from the Smoluchowski equation with the fastest rates observed for ligand binding to proteins. If the diffusion constants (5.0×10^{-6} and 0.5×10^{-6} cm^2 s^{-1} for substrate and protein respectively) and the sum of the radii (1000×10^{-7} cm) are substituted into equation (7.4.4) we obtain the collision frequency $k = 2 \times 10^{10}$ M^{-1} s^{-1}. One can make the rough assumption that the target area on the surface of the protein corresponding to the size of the substrate is 1/100 of the total and this leads to the estimate that the maximum number of productive collisions is 2×10^8 M^{-1} s^{-1}. The fact that this crude derivation corresponds within a factor of two or three to many experimental determinations of diffusion controlled reactions (see table 7.2) indicates that a number of factors must cancel each other fortuitously. Such factors include the correct orientation of substrate to target on collision and the apparent increase of the target size due to the fact that the rotational relaxation time of the protein may be short (in the range of a few microseconds) compared to collision intervals.

Direct and indirect methods for the determination of rates of substrate binding are discussed extensively in chapters 5 and 6 and there are special methods for the study of transmitter mediated channel gating (Hille, 1992). As emphasized above, the indirect evaluation of substrate binding to enzymes, in single substrate reactions, relies on the determination of k_{cat}/K_M. As shown in section 5.1, this function, which has the units of a second order reaction, has a minimum value of k_{12}, the rate constant for the first step of substrate binding. Even so, as seen in table 7.2 it is often near 10^8 M^{-1} s^{-1}. Similar complications for the extraction of k_{12}, the rate constant for the first step, arise in the usual two step binding reactions of ligands.

The above calculations for maximum collision frequencies do not take into account the possible effects of attractive forces due to complementary charges or dipoles on ligand and binding site. Pertinent data are available from experiments on the rate of binding of acetylcholine and of a fluorescent analogue (*N*-methylacridinium) to the anionic site of acetylcholinesterase (Nolte, Rosenberry & Neumann, 1980). The rate constant for substrate binding was deduced from k_{cat}/K_M and the one for the analogue directly from fluorescence observation of temperature jump experiments. Measurements were carried out over a range of ionic strength (about 1 mM to 120 mM) and the derived rate constants extrapolated to zero ionic strength gave 10^{10} M^{-1} s^{-1} for analogue binding and 4×10^9 M^{-1} s^1 for k_{cat}/K_M for the reaction with acetylcholine. Even at physiological ionic strength the two constants are still as high at 7×10^8 M^{-1} s^{-1} and 1.5×10^8 M^{-1} s^{-1} respectively. Another interesting case is the enzyme superoxide dismutase, which holds a record with k_{cat}/K_M of 2×10^9 M^{-1}s^{-1}. The overall charge of both substrate and protein is negative. However, an analysis of the charge distribution over the protein molecule led Getzoff *et al.* (1983) to the conclusion that the substrate is guided to its docking site by a suitable electrostatic potential gradient (see also Sines, Allison & McCammon, 1990). Further investigations by Getzoff *et al.* (1992) included the alteration of the surface charge of the enzyme by substitution of amino acid residues, using site directed mutagenesis. Exchange of a lysine for a neutral residue resulted in increased reaction rate. At low ionic strength, rates around 10^{10} M^{-1} s^{-1} were observed.

The arguments developed above provide a pragmatic picture for the limits of diffusion controlled reactions, which is in reasonable agreement with data available for the binding of small substrates to proteins with molecular weights of the order of 10^5 daltons. Additional information can be obtained from the large literature of theoretical treatments that is reviewed by Berg & von Hippel (1985) prior to discussing their principal concern, which is the problem of the interaction of proteins with large substrates, namely nucleic acids. Interest in the mechanism of such binding processes arose from the original findings of Riggs *et al.* (1970) that the *E. coli* lac repressor protein locates the operator site on a long DNA molecule with rate constants approaching 10^{10} M^{-1} s^{-1}. Adam & Delbruck (1968) first made the suggestion that the phenomenon of 'reduction in dimensionality' can play a role both in such macromolecular interactions and in binding to membrane surfaces. In discussing this topic one is bound to lean heavily on the ideas and experiments reported by von Hippel *et al.* (1984) and the more theoretical article of Berg & von Hippel (1985).

Reflecting on the complex theoretical papers that have been published on the kinetics of binding to sites on macromolecules, one cannot help wondering whether any of these treatments would have predicted the remarkably efficient capturing rate. All the theories are very model dependent. The problem brings to mind Crick's (1988) distinction between models and demonstrations. He calls a demonstration a 'don't worry theory' which demonstrates that one can see at least one way in which nature might have solved the problem. A model is supposed to be the way one believes that something actually works and is contrasted to a demonstration, which does not pretend to approximate the right answer. This is in no way intended to denigrate the interesting contributions made by many of the authors referred to here, either directly or indirectly in connection with the present discussion.

Two related relatively simple concepts can play a role in accelerating diffusion controlled reactions over and above the Smoluchowski limit. The first of these is the tendency of a ligand attached anywhere on the surface of a large molecule to diffuse in one or two dimensions on the surface of that molecule until it finds the target site. Adam & Delbruck (1968) and Berg & Purcell (1977) in their penetrating approach to the problem of reduction in dimensionality, discuss the mean diffusion time in one, two and three dimensions. Reaching a small target with radius b in the middle of a large surface with radius L ($L \gg b$) can be expressed as:

$$\tau_3 = (L^2/3D_3)(L/b) \quad \text{in three dimensions}$$
$$\tau_2 = (L^2/2D_2)\ln(L/b) \quad \text{in two dimensions}$$
$$\tau_1 = (L^2/3D_1) \quad \text{in one dimension}$$

(Wittenberg & Wittenberg, 1990). D_1, D_2, D_3 are the respective one-, two-, three-dimensional diffusion constants. Inspection of these expressions shows that one- or two-dimensional search is more efficient than three-dimensional. Adam & Delbruck (1968) suggest that a messenger only has to find the whole structure, macromolecule or cell, by three-dimensional diffusion; then it can home in on the small target by two-dimensional (on a membrane surface) or one-dimensional (along a polymer) random walk. Detailed calculations on the one-dimensional sliding model, for the lac repressor protein binding to its specific site (the operon) on a long DNA strand, can be found in von Hippel *et al.* (1984). One would hope that, before long, experiments will be designed which help to distinguish the kinetics of initial attachment from the location of the target. This problem also arises in connection with another concept which can be used to speculate on accelerated target location.

The terms encounter and collision have already been referred to, as well as the equivalent terms macro- and microcollision. If a ligand, on its erratic random walk, comes near the surface of a macromolecule or membrane, it will stay close to it until the rare event occurs of a significant departure in one direction. After each first collision there will be many in rapid succession. However, due to rotational diffusion of the host molecule, each of these microcollisions can be on a different part of the surface. The rotational relaxation time is

$$\tau_0 = 4\pi r^3 \eta / 3k_B t$$

Let us consider a reaction in two steps, the first depending on the ligand coming within the orbit of the macromolecule and the second a mutual search for the docking area. Only the first step will be concentration dependent and can occur with a rate constant of 10^9 to 10^{10} M^{-1} s^{-1}. Whether a reaction will occur during this encounter will depend on whether the encounter is of sufficient duration for a successful search. One can set up a speculative scheme which is somewhat like a rapid pre-equilibrium prior to a reaction.

It might be possible to get some insight into the reality or otherwise of the above scenario if one studied the fluctuations of non-radiative energy transfer with a suitable pair of fluorescent markers on ligand and target molecules. This might make it possible to define the two positions: one in orbit, the other at the docking site. The fluorescence techniques involved in such experiments and in the measurement of rotational diffusion are discussed in section 8.2.

7.5 Some comments on the use of isotopes in kinetic investigations

Range of applications of isotope replacement

There are many uses for isotopes in the exploration of biological processes. Radioactive atoms, or deuterium for reactions involving hydrogen transfer, placed in specific positions of metabolites, provide information about stereochemical requirements of enzyme reactions. The classic work of Westheimer and Vennesland on the stereospecificity of hydrogen transfer by NAD^+ linked reactions was summarized by Gutfreund (1959a). Similarly the fate of groups during a sequence of reactions (for instance in the citric acid cycle) was determined. There is no doubt that the availability of carbon-14 was responsible for the rapid exploration of most metabolic pathways in the 1950s. Radioactive isotopic tracers are not only used for the qualitative exploration of pathways, but they are also widely applied to the study of rates of transfer of metabolites between their respective pools.

This topic, compartmental analysis, is not covered in this volume and the reader is referred to specialized literature (Zierler, 1981, and references therein). Stable isotopes of different weights can also be used for this purpose in conjunction with mass spectrometry. Deuterium in particular, because of its large ratio to the atomic weight of hydrogen, is a suitable tracer.

In some cases replacements of isotopes result in spectral and/or fluorescence changes. An interesting application of deuterium/hydrogen exchange is the study of the mobility of different domains of proteins (Englander & Kallenbach, 1983). This approach to the study of protein dynamics has already been referred to in section 5.3. The exchange of isotopes into specific positions (for instance deuterium and/or carbon-13) is also widely used to increase the potentialities of NMR investigations.

Last but not least is the application of isotopic substitutions for the determination of kinetic isotope effects, which is our main concern in this section. This technique depends on the fact that the activation energy of a bond breaking step is influenced by the mass of the atoms separated by this process. The interest in this phenomenon is due to its potentialities for the exploration of reaction mechanisms, the nature of transition state complexes and of the rate limiting step. We shall, however, restrict ourselves largely to the following aspect. In conjunction with the display of individual steps by transient kinetic analysis (see section 5.1), the presence or absence of kinetic isotope effects can enable one to make statements about a particular step of an enzyme reaction. For instance we shall discuss how a bond breaking step can be distinguished from a conformation change.

The isotope rate effect

The physicochemical basis, theoretical analysis and the results of some fundamental investigations on the kinetic isotope effect are found in texts on physical organic chemistry. Some of these have enzymological examples (see for instance Jencks, 1969; Bender, 1971). In investigations of reaction mechanisms the effect of substitution of an atom by a heavier isotope can produce a significant reduction in the rate of a reaction. The size of the effect depends on the mass ratio of the two isotopes. For deuterium replacement of hydrogen up to tenfold reduction in rate has been observed. For other isotopes used in such investigations the weight ratios are small and the effects on rates more difficult to determine. Expected ratios of rates for different replacements at bonds involved in rate limiting steps are given by:

$$\frac{k(^{12}C)}{k(^{13}C)}=1.25; \quad \frac{k(^{12}C)}{k(^{14}C)}=1.5; \quad \frac{k(^{14}N)}{k(^{15}N)}=1.14; \quad \frac{k(^{16}O)}{k(^{18}O)}=1.19$$

Secondary isotope effects can sometimes be observed when vibrations of the reactive bond are coupled to nearby bonding involving isotopically substituted atoms. They are always very small and require very precise kinetic analysis. Sometimes it is possible to carry out reactions with and without substitution, using a differential technique (for instance using the two cuvettes of a spectrophotometer). The following elementary description of the kinetic isotope effect will provide some insight in terms of the theory of absolute reaction rates. For more profound treatments and references to fundamentals the texts mentioned above must be consulted (see also Westheimer, 1961; Jencks, 1969).

As an example the discussion here is based on the discrimination between reactivity at a C—H and a C—D bond. There is no discrimination between the electronic, rotational or translational properties of H and D atoms. The vibrational frequencies observed by infrared spectroscopy are near the values of 2900 cm^{-1} for C—H and 2100 cm^{-1} for C—D. The zero point energy of vibration ε is given by

$$\varepsilon = -\tfrac{1}{2}hv$$

(h is the Planck constant and v the vibrational frequency).

From Hooke's law the vibrational frequency is

$$v=\tfrac{1}{2}\pi\sqrt{(c/m)} \tag{7.5.1}$$

(c is the force constant of the bond, m is the mass of the atom). It follows that

$$\varepsilon=h\tfrac{1}{4}\pi\sqrt{(c/m)} \tag{7.5.2}$$

The force constant of C—D and C—H bonds are essentially the same and the zero point energy will be inversely proportional to the square root of the mass. Accordingly the zero point energy of C—D should be $1/\sqrt{2}$ times that of C—H, which corresponds very closely to the relative energies calculated from the above infrared frequencies:

$$\varepsilon_{H-C}=17.4 \text{ kJ/mole} \quad \text{and} \quad \varepsilon=12.5 \text{ kJ/mole}$$

Figure 7.9 shows the relationship between zero point energies and energies of activation. In the transition state the vibration is transformed into translation and there is no difference in the transition state energy of C—H and C—D. From the difference in the energy of activation of the two

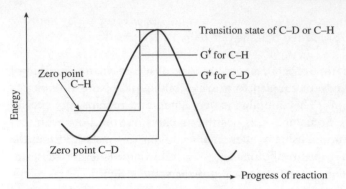

Figure 7.9   Transition state diagram illustrating the deuterium isotope effect: the lower zero point energy of the carbon–deuterium bond is responsible for a correspondingly larger activation energy and lower reaction rate for bond cleavage.

bonds (8 kJ) and equations (7.2.4) to (7.2.8) one can derive the ratio of the rate constants for reactions with their rate limiting step involving breaking the H or D substituted bonds: $k_H/k_D = 7$. Isotope effects approaching this value are often found in reactions involving a deuterium substituted bond. However, in many cases deuterium isotope effects are found to be considerably smaller than those calculated from the above simplified theory. Westheimer (1961) discusses the reasons for reduction in the primary isotope effect. When the ratio becomes smaller it becomes difficult to distinguish from a secondary effect due to substitutions at a step other than the bond breaking one (see Jencks, 1969).

Some examples of the application of kinetic isotope effects

Much of the interpretation of isotope effects falls within the realm of physical organic chemistry and some of the references quoted above should be consulted for such applications. Most of the biological processes discussed in this volume are dominated by non-covalent physical chemistry. Demarcations are notoriously vague, but the exploration of the detail of chemical mechanism is not part of our remit. We shall concern ourselves here principally with the application of this technique to the distinction between chemical events and conformation changes (non-bond-breaking). A number of transient kinetic studies have been carried out with dehydrogenases, which provide good examples for the type of evidence provided by isotopic substitutions in substrates. These systems have also been investigated by steady state isotope effects.

The transient kinetic studies of liver alcohol dehydrogenase by Shore & Gutfreund (1970) showed that the rate of hydride transfer from alcohol to NAD^+ appeared as an initial burst of enzyme bound NADH (see p. 155). This initial NADH production could be identified by a deuterium isotope effect of 5.2, when deuterated ethanol was used as a substrate. Using a series of alcohols as substrates we found that as k_{cat} decreased so did the isotope effect and the transient amplitude. In the reverse direction, when acetaldehyde together with NADH or NADD were used as substrates, no isotope effect was observed. The conclusion was that the rate limit is an isomerization of the ternary complex prior to hydride transfer. Similar experiments were performed by us (Whitaker et al., 1974) with heart lactate dehydrogenase. With this enzyme no deuterium isotope effect was found in either direction. This is to be expected from the experiments (see p. 181) which indicate that a protein conformation change in response to substrate binding is rate limiting both for the oxidation of lactate and the reduction of pyruvate. More detailed investigations with deuterium labelled substrates for this enzyme should make it possible to define conditions under which hydride transfer becomes rate limiting. Fierke, Johnson & Benkovic (1987) observed an isotope effect of 3 during the transient involving the oxidation of NADPH and NADPD by dihydrofolate reductase.

With flavoproteins there arises another problem in the interpretation of experiments involving deuterated substrates. The atoms on the flavin residue which shuttles between hydrogen donor and acceptor readily exchanges with protium or deuterium from solvent water. One can carry out experiments in D_2O and calculate the residence time of the shuttle hydrogens on the flavin from the relative transfer of protium or deuterium. Estimates of kinetic isotope effects under these conditions were obtained by Gutfreund & Sturtevant (1959) and Gutfreund (1959b). Kinetic isotope effects exhibited by another flavoprotein, glutathione reductase, were studied by steady state and transient kinetics (Vanoni et al., 1990). The complex chain of events of hydrogen transfer between glutathione and NADP, flavin and an $-S \cdot S-$ bond on the enzyme could be dissected by the latter technique. A full isotope effect was found in the steady state and single turnover experiments on deuterium transfer from NADPD to reduced glutathione. The comparison of reaction rates in H_2O and D_2O as solvents was used to study the mechanisms of a number of enzyme catalysed hydrolytic reactions. These mechanisms were among the first to be studied in detail by approaches developed in the field of organic reactions. Such investigations permit a distinction between general base catalysis and nucleophylic catalysis by, for instance, imidazole groups at the active site.

General base catalysis involves proton transfer from water to imidazole and exhibits deuterium isotope effects up to 3, while nucleophylic catalysis shows no such effect (Bender, 1971).

pH measurements in D_2O solutions

A practical point of importance involves the measurement of pD ($-\log c_{D^+}$). When a pH meter with a glass and calomel electrode is calibrated with standard solutions of known c_{H^+} made up in H_2O and it is then used for the measurement of pD in D_2O solution, the following relation holds

pD = pH meter reading + 0.4

Isotopes as tracers for rates of exchange

The preceding paragraphs provide a few isolated examples of additional information obtainable from kinetic studies when isotopic substitutions are used. The use of both stable and radioactive isotopes for tracing the fate of groups was mentioned at the beginning of this section, together with the comment that this also has kinetic applications. Similarly the exchange of oxygen-18 into phosphate groups during the enzymatic hydrolysis of phosphate esters has provided both stereochemical and kinetic information. The time profile of oxygen incorporation established the rate constants and equilibria of several steps of, for instance, the hydrolysis of ATP by myosin. This topic is discussed by Knowles (1980, 1989) and by Hibberd & Trentham (1986) and its interpretation in terms of transients is described in section 5.1. The use of kinetic techniques which can be applied to the study of systems at different levels of organization has been advocated throughout this volume. The measurement of rates of exchange of isotopes falls into this category. The rates of oxygen exchange during ATP hydrolysis by myosin can be followed in solution and in active muscle fibres. In this way steps of the reaction can be linked to mechanical events (Hibberd & Trentham, 1986).

The entry of many physical organic chemists into the field of enzyme mechanisms has resulted in extensive investigations of the detail of reaction mechanisms beyond the aims of a kineticist. The texts on enzyme mechanisms by Jencks (1969), Bender (1971), Walsh (1977) and Fersht (1985) should be consulted by readers with wider interests in that field.

8

The role of light in kinetic investigations

8.1 Applications of photochemistry

Introduction and early applications

The heading of this section will conjure up a range of different expectations in terms of subject matter. In fact the whole chapter is more method oriented than the rest of this volume. It is necessarily, because of the range of the topics, more in the nature of a review of available tools and does not treat any of them in detail. None the less, the emphasis is on the kinetic problems which can be solved by the use of light to initiate and/or monitor reactions. It will be seen that the methods themselves present some interesting kinetic problems. In this area, as elsewhere (see section 1.1), classic physical investigations were rapidly applied to biological problems. Early studies of the fundamental properties of light were associated with experiments on vision by Thomas Young and on spectral changes during reactions of haemoglobin by Sir George Stokes, who also contributed to the discovery of fluorescence (see below).

Biological responses to light

A meaningful discussion of the role of light for sending and receiving signals as well as for an energy source in biological systems would cover several volumes. The photochemistry of photosynthesis, of the activation of plant enzymes, the initial reactions of rhodopsins involved in vision and bacterial energy transduction and the emission of light by photoproteins (see section 4.2) all present challenging kinetic problems. The kinetic events consequent on the photochemical reaction of the visual pigment are used in section 4.2 to illustrate kinetic modelling of sequential reactions. As the time and signal resolution improved, with the development of laser and

281

detector technology, ever increasing detail was found in the spectra of the components of, for instance, the photosynthetic apparatus and rhodopsin. During the 20 years from 1952 to 1972 the time scale of photochemical experiments improved from millisecond to picosecond resolution. Witt (1991) described how these technical developments provided increasing detail of the reactions taking place in chloroplasts. Some results due to the further resolution into the femtosecond range are discussed below. In fact the exploration of the individual events during photosynthesis, from the absorption of light by antenna chlorophyll to the splitting of H_2O and the reduction of NAD^+, serves to illustrate the two principal aims of kinetic investigations: the analysis of the number of distinct components of a reaction pathway and their associated time resolved spectroscopic characterization. In the particular case of photosynthetic reaction centres, the elucidation of three-dimensional structures, combined with pico- and femtosecond time resolved spectroscopy, has given insight into some of the most exciting chemistry of recent years (see for instance Moser *et al.*, 1992). It is, of course, evident that in considering the role of light in kinetic investigations, one keeps switching between its function to initiate reactions and its role as a monitor for the observation of the time course of chemical change. For instance in the ingenious method of Hofrichter *et al.* (1983) one laser flash is used for the photodissociation of haemoglobin ligand complexes and a second flash for the excitation of the fluorescence of a dye. The fluorescence lifetime of the dye is sufficiently long, compared to the time constant of the reaction studied, so that the emitted light provides a stable source to monitor the early steps in ligand association. The combined functions of initiation and identification of intermediates will become evident during the discussion of various techniques.

An important characteristic of all the photochemical reactions is the *quantum yield*. In photochemistry it is defined as the ratio of the number of excitations to the number of molecules undergoing the reaction (e.g. photolysis). In fluorescence it is defined as the proportion of the excited molecules decaying by emission of radiation.

Light for the initiation of reactions

It was emphasized above that flash photolysis is a prime example which demonstrates that the number of intermediates resolved in a reaction is open ended; it is dependent on the time resolution of the technique. From the equation

$$k = (k_B T/h)\exp(-G^{\downarrow}/RT) \tag{8.1.1}$$

which is discussed in section 7.1, it is evident that the largest rate constant $k = 6.25 \times 10^{12}$ s^{-1} at 298 K, is found when the energy barrier to the transition state is zero. To get the time resolution to that level requires laser pulses in the femtosecond range. The state of the art is illustrated in a volume of *Annual Review of Physical Chemistry* by Khundar & Zewail (1990) for the study of transition states, by Birge (1990) for the resolution of intermediates in the photoreactions of rhodopsins and by Mukamel (1990) for femtosecond spectroscopy. This methodology is obviously of such a specialized and expensive nature that it is restricted to a few laboratories and detailed discussion is not justified here.

Fortunately most of what is of general interest for the study of biological processes can be discussed in terms of methods and reactions with lower limits in the micro- and millisecond ranges. Like many other areas of biological research, progress in flash photolysis benefited considerably from the developments of electronic techniques during the war of 1939–1945. This not only affected the flow and relaxation techniques described in chapters 5 and 6, but also the design of flash lamps and of recording equipment pioneered by Norrish & Porter (1954) for photochemical investigations. The latter methods helped Gibson (1956) to apply flash photolysis studies to reactions of haem proteins. The volumes describing the proceedings of two discussion meetings on rapid reactions in 1954 and 1967 provide surveys of such investigations on simple chemical reactions as well as on biological systems. The applications of photochemical techniques are illustrated side by side with rapid flow and relaxation methods. As reviewed above, photochemistry now works on such a different time scale that it will rarely be discussed in the same place as the other methods. As pointed out by Zewail (1990), resolution over milliseconds and microseconds detects reaction intermediates, while pico- and femtosecond techniques are required for the study of transition states.

Among the earliest attempts to use light to initiate reactions for kinetic investigations were those of Hartridge & Roughton (1923a,b). The complexes of haemoglobin with carbon monoxide and dioxygen can be dissociated by photolysis; the former more readily than the latter. Gibson (1990) gives a brief summary of the history of the use of flash photolysis for the study of haem proteins. The quantum yield for the dissociation of the ligand from CO–haemoglobin is approximately 20 times that for O_2–haemoglobin. It is, therefore, possible to change the equilibrium between different states of liganded haemoglobin and free ligand by exposing solutions to strong light. Hartridge & Roughton's experiments involved observations of the reaction after rapid shielding of the light source, which had been used to equilibrate the reaction mixture. Their attempts to get kinetic data with this method did not succeed with the instrumentation available at that time. However, they combined this procedure with the first rapid flow experiments. Solutions were equilibrated under strong illumination and made to flow rapidly (approximately 10 m s^{-1}) through a tube, which was kept in the dark, except for the low level of light required for the use of the Hartridge (1923) reversion spectroscope. The reaction was thus expanded on the time axis and observations of spectral changes were made at various distances from the point of exposure to light. This flow technique was the origin of the now extensively used methods involving the initiation of reactions by the injection of reaction partners into a flow tube through a mixing chamber. As electronic techniques developed, stationary observations on a millisecond time scale became possible (see section 5.1).

Properties of light sources

For the initiation of reactions a light source is required to produce the maximum number of photons during a time period which is short compared to the subsequent reaction under investigation. For observations of reactions by absorbence changes maximum stability of light intensity is the prime concern. Light intensity is rarely a problem for absorbence measurements and high intensity can result in photodegradation of reactants. When steady state fluorescence changes are used as a monitor for the progress of a reaction both intensity and stability are critical. However, the problem is that as the signal improves with increased intensity of the exciting light, radiation damage shortens the time over which reliable measurements can be carried out. For experiments involving lifetime records of fluorescence, phosphorescence and other excited states, the profile of the light flash is critical for the resolution of time constants of the excitation from those of the system under investigation. Some aspects of these specifications will become apparent when different types of experiments are discussed.

Some comments on light (from UV to IR) as a monitor

In biochemical kinetics by far the most important role for light within the wavelength range from ultraviolet (UV) to infrared (IR) is its use for the observation of the progress of reactions. This is coupled with the importance of changes in absorbence spectra in providing information about the nature of the changes occurring during the interconversion of states. The latter makes particular demands in connection with the analysis of rapid reactions. In the study, for instance, of enzyme reactions, the characterization of intermediates displayed along a time axis is often more informative than the rate constants for their interconversion. This applies not only to enzymes with chromophoric prosthetic groups, but also to systems where the absorbence or fluorescence spectrum of the substrate or the protein gives information about the nature of the reaction. Examples of such reactions are described in section 5.1.

The sequence of reactions of both the visual pigment and bacterial rhodopsin has been studied by a wide range of spectrophotometric techniques. The recent exciting developments in time resolved FTIR (Fourier transform infrared) spectroscopy are reviewed by Gerwert (1993). In the particular example of bacterial rhodopsin Hessling, Souvignier & Gerwert (1993) were able to characterize a series of changes in the protein (opsin) structure, which could be correlated with the reaction intermediates defined by absorbence changes in the visible spectrum of the chromophore (retinal). The spectra of some chromophores are very sensitive to structure changes, which affect their environment. Amino acid replacements in the protein opsin, in vivo or in vitro, are associated with changes in the spectral sensitivity of rhodopsin.

Among other widely used monitors of reactions are a variety of luminescence methods and other techniques involving excited states of chromophores. These are discussed in the next two sections. Light scattering also often provides information about changes occurring during the progress of a reaction. The importance of using a wide range of monitors for the study of any reaction is frequently demonstrated in the preceding chapters.

8.2 Luminescence and other excited states

Different forms of luminescence

Luminescence, which is the emission of light, can result from chemical reactions or excitation by light. The former can be artificial (for instance the reaction of luminol plus hydrogen peroxide) or the response of the stimulation of natural reactions of photoproteins (see the discussion of the kinetics of aequorin in section 4.2). The form of luminescence of most general interest for kinetic studies is that obtained in response to excitation by light (fluorescence and phosphorescence). All forms of luminescence are due to the emission of photons from electronically excited states. In the singlet excited state the electron in the higher orbital has opposite spin orientation from the electron in the lower orbital. In that case they emit a photon with spin $\hbar$ ($=h/2\pi$), so $\Delta s = 1$ (from $-\frac{1}{2}$ to $+\frac{1}{2}\hbar$ or vice versa) resulting in fluorescence. Such transitions are quantum mechanically allowed and are fast (about 0.1 to 10 ns). If the excited state of the molecule is a triplet state the electrons have the same spin, they are unpaired. In that case return to the ground state requires a change in spin orientation. This is not allowed and is, therefore, a rare event (see Atkins, 1984, p. 615). The lifetime of the excited triplet state is between milliseconds and seconds. The resulting light emission is called phosphorescence. The difference in the lifetimes of fluorescence and phosphorescence makes the two phenomena complementary for applications to the study of events which occur over a range of time scales. This will become apparent when the procedures for studying molecular motions are discussed below. Clearly, when events with relaxation times of micro- or milliseconds are investigated, fluorescence measurements can give little information, while phosphorescence and other methods using triplet states can be informative.

Both fluorescence and phosphorescence can be used in either the lifetime mode or the steady state mode. In the former the events after a flash of light are interpreted in terms of fluorescence decay (lifetime measurements) or molecular motion (rotational or translational). A phase modulation technique, briefly discussed in the next section, is now used by some specialists for fluorescence lifetime measurements in place of the single flash time resolved mode. In the rest of this section we are concerned with the steady state mode, when the time course of changes in light emission during a reaction are recorded with continuously applied illumination. These have applications to investigations of many systems discussed in previous chapters, when recording of fluorescence changes of various kinds was used to monitor the interconversion of intermediates. We are separating the steady state methods into those involving quenching of light emission and those involving effects on the ratio of polarized excitation and emission due to molecular motion.

Properties of some fluorophores

Fluorophores of interest in the study of the behaviour of biological molecules can be intrinsic or extrinsic. The principal intrinsic fluorophores are the aromatic amino acids tryptophan, tyrosine and phenylalanine (with decreasing quantum yields in that order) and the cofactors NADH and riboflavin. Extrinsic fluorophores are introduced as reporter groups either

Table 8.1. *A guide to fluorescence characteristics of some amino acids*

Amino acid	λ_{max}	$\varepsilon(M^{-1} s^{-1})$	λ_{emit}	ϕ
Phenylalanine	257.44	147	282	0.04
Tyrosine	274	1420	303	0.21
Tryptophan	279.8	5600	348	0.29

by derivativization with dansyl chloride and other fluorescent reagents at a defined site or as a ligand in equilibrium attachment to a binding site. The choice of an extrinsic fluorophore depends on its affinity for the site of interest, on its absorption and emission spectrum, its quantum yield and on its lifetime. The difference between the absorption and emission spectrum of a fluorophore was first observed in 1852 by Sir George Stokes (see section 1.1) with a solution of quinine. The so-called Stokes shift results in the emission occurring at longer wavelength (lower energy) than absorption. The excitation and emission spectra, as well as the magnitude of the Stokes shift, depend on the fluorophore and the environment. The latter is determined by the position of the fluorophore on the surface of the protein, when the influence of the solvent dominates, or within a cleft of the protein molecule under the influence of the local environment. This is one of the reasons why fluorescence emission can be used as a monitor for the environment (dielectric constant and refractive index). A change from a polar to a non-polar environment results in a shift and enhancement of fluorescence. An example of this is the well-known change in tryptophan fluorescence during the unfolding or folding of proteins, when indole residues move between water and a non-polar environment (see p. 123). The choice of compounds with particular absorption and emission spectra has to be a balance between quenching due to other chromophores present and the potential use of energy transfer (see below) to obtain additional information.

Table 8.1 is a guide to fluorescence characteristics of some amino acids; λ is the wavelength in nm, ε is the extinction coefficient and ϕ is the quantum yield. The precise values depend on solvent composition and derivativization.

The fact that the quantum yield in fluorescence is less than one is due to the return of some molecules to the ground state via radiationless energy transfer (see p. 289) and to thermal (collisional) quenching. In photolysis a number of additional factors depending on the reaction mechanism can reduce the quantum yield. More comments on quantum yields will be found in connection with discussions of photolysis of caged compounds in section 8.4.

Fluorescence quenching

The term 'steady state fluorescence', applied to the methods discussed in this section, implies observations of changes in intensity, wavelength or polarization of the emitted light under conditions of continuous excitation. This is in distinction to observations during the fluorescence lifetime after excitation by a single flash of modulated light. A summary of the principles involving the use of fluorescence signals to follow rate processes should include those in common use as well as those with potential. Even so there are, so far, few examples of their application in the literature. Most uses of steady state fluorescence methods in kinetic investigations involve the recording of changes in emission of unspecified origin during the time course of the reaction. However, some examples of more heuristic uses of time resolved fluorescence changes will illustrate the principles involved. The discussion will also provide the background to some of the methods applied to experiments described in previous chapters.

Mechanisms of fluorescence quenching are usually divided into two groups: collisional and complex formation; these are also referred to as dynamic and static mechanisms, respectively. The distinction between dynamic and static mechanisms can get blurred, as can the distinction between collision complexes and subsequent more specific interactions (see discussion of diffusion controlled reactions in section 7.4). In addition to these mechanisms there can be internal quenching if the optical density of the solution is high. This last phenomenon, which is the so-called *inner filter effect*, is only a nuisance and provides no useful information.

As an introduction to any experiment which involves the quantitative use of a quenching phenomenon, data for a Stern–Volmer plot should be obtained. These involve the measurements of fluorescence over a range of quencher concentration (c_Q) and plotting F_0/F against c_Q. The Stern–Volmer equation is

$$F_0/F = 1 + k_Q\tau_0 c_Q = 1 + K_D c_Q, \tag{8.2.1}$$

where F and F_0 are fluorescence in the presence and absence of quencher, k_Q is the bimolecular association rate constant and τ_0 is the fluorescence lifetime in the absence of quencher. K_D is the Stern–Volmer quenching constant and can be interpreted, in the case of complex formation, as a dissociation constant. A linear plot indicates that a single type of fluorophore (identical lifetime) is involved with all of them equally accessible to the quencher. The distinction between collisional quenching and complex formation relies on the viscosity dependence of the former. As a diffusion controlled reaction the temperature dependence of collisions is also largely due to its effect on the viscosity of the solvent. Information obtained from collisional quenching can contribute to an understanding of diffusion controlled reactions and some of the criteria developed in section 7.4 apply.

The comparison of the effectiveness of a number of reagents, which are widely used for the study of collisional quenching, can give information about the ease of access to fluorophores and about their charge environment. For instance the bimolecular rate constant k_Q of quenching by I^- is considerably enhanced by a positive charge environment and vice versa by a negative one. The size of the quencher can give information on any structural barriers shielding the fluorophore. Acrylamide is often used as a quencher to test ready access to the fluorophore. The use of oxygen as a quencher has its attractions since its size and neutrality allows it to penetrate into structures. The application of this phenomenon to the study of molecular dynamics is described in section 7.3.

The intensity of fluorescence emission, under steady state illumination, is proportional to the number of molecules in the excited state, which, in turn is dependent on the lifetime of this state. Quenching by collision of molecules in the excited state acts by reduction of this lifetime (see discussion of parallel reactions in section 3.2). Dissolved oxygen will only have an appreciable effect on fluorescence emission when high pressure is used to attain an appreciable oxygen concentration. This is required for a reasonable probability of collisions between quencher (oxygen) and excited molecules during their lifetime. At saturation with air the oxygen concentration of aqueous solutions is 0.258 mM. With the expected second order rate constants for fluorophore–oxygen collisions of 10^9 to $10^{10}\,M^{-1}\,s^{-1}$ this can potentially result in the quenching of luminescence with a time constant of about 2.5 to 25 μs. This is slower than the fluorescence decay of most fluorophores in use, but it is fast compared to phosphorescence and to avoid quenching oxygen free solutions have to be used for such studies.

Energy transfer

There appear to be a number of not well-explained mechanisms responsible for static quenching, which involves distinct complex formation between the quencher and the molecule containing the fluorophore. Of special interest is radiationless energy transfer. The time resolved interpretation of fluorescence energy transfer from donor to accepter is, in theory, a very sensitive way of following intra- and intermolecular movement. This transfer, for instance from an excited tryptophan on a protein molecule (emission ≈ 348 nm) to NADH (absorbence 340 nm) bound in its proximity, occurs without the appearance of a photon and is primarily due to dipole–dipole interaction between donor and acceptor. This phenomenon, called Förster energy transfer (Förster, 1959), is distinct from and more interesting than radiative transfer. The latter depends on the optical properties of the whole sample rather than on the molecules. On account of its sensitivity to the distance between donor and acceptor non-radiative energy transfer has been called a spectroscopic ruler by Stryer & Hougland (1967). The rate of energy transfer is characterized by the time constant (fluorescence lifetime)

$$\tau_T/\tau_D = (R_0/r)^6 \tag{8.2.2}$$

where τ_T and τ_D are respectively the time constants for transfer to a specific acceptor and of the lifetime of the donor in the absence of acceptor. R_0 and r are the characteristic Förster distance at which the probability of transfer is 0.5. Equation (8.2.2) tells one that a change in distance will affect energy transfer with a power of 6, but it is only possible to quantify the absolute amount of quenching if the transfer rate constant, $k_T = \tau_T^{-1}$, has been evaluated from a more complex equation:

$$k_T = \frac{9000\,(\ln 10)\kappa^2\phi_d}{128\pi^5 n^4 N r^6 \tau_d} \int \frac{F_d(v)\varepsilon_a(v)}{v^4}\,dv \tag{8.2.3}$$

containing terms for the quantum yield of the donor, ϕ_d, the refractive index of the medium, n, the extinction coefficient of the acceptor, ε_a, the corrected fluorescence intensity of the donor, F_d, the wave number range, $v + dv$, and a factor κ describing the relative orientation in space of the transition dipoles of donor and acceptor. The last of these terms introduces the greatest uncertainty into the absolute value of the transfer rate. The integral in the above equation is referred to as the overlap integral J_v:

$$J_v = \int F_d(\lambda)\varepsilon_a(\lambda)\lambda^4 d\lambda \tag{8.2.4}$$

Two aspects of energy transfer fluorescence have practical consequences for kinetic investigations. First, the sensitivity to distance has been used for monitoring changes in conformation, as a consequence of a change in state, for instance in actin–myosin interaction (Trayer & Trayer, 1988). Although estimates of the absolute distances between donors and acceptors in complex molecules depend on fudge factors (especially orientation), the method has proved valuable for determining changes. The time resolved records of such changes are now beginning to find application to interesting problems (for instance Beechem, 1994). The appearance, in the programme for the 1994 meeting of the Biophysical Society, of several other abstracts on a variety of time resolved fluorescence studies promises a rapid expansion of the use of these methods. This approach could also give valuable information about the relation between collisions, encounters and binding (see section 7.3). Refinements of the method should make it possible to make a distinction between proximities in the two states and to follow the transition.

Another interesting application of fluorescence energy transfer was described by Katchalski-Katzir, Hass & Steinberg (1981). They studied the intramolecular mobility of polypeptides in solution using donor and acceptor labels respectively on the two end groups of the chains. Mobility and flexibility could then be judged from energy transfer as the chain length was increased and the ends of the same molecule could approach each other. Steinberg (1971) provided a clear discussion of the theory involved in these studies and presented a useful compilation of donor–acceptor pairs for use in energy transfer experiments.

A word of caution is necessary about the interpretation of fluorescence quenching for the determination of the degree and rate of binding of, say, NADH to a protein molecule. When there are several binding sites on one protein molecule one can obtain information about the geometric relation of these sites, but there is an added complexity to the form of the titration curve. Using the example of the tetrameric lactate dehydrogenase, which binds four NADH molecules, Holbrook (1972) found that the relative tryptophan fluorescence of molecules with 0, 1, 2, 3, 4 sites occupied by NADH is 1, 0.648, 0.398, 0.229, 0.12, respectively. The occupation of the first site results not only in the quenching of the fluorescence of the tryptophans of that subunit, but also quenches those of others, albeit to a lesser degree. As a result of this non-linear relationship between binding and fluorescence quenching the records of the time course of NADH binding are not exponential (see figure 8.1). This has led to erroneous interpretations of the kinetic mechanism.

Fluorescence polarization and anisotropy

Optical spectroscopic methods for studying rotational motion depend on creating an anisotropic molecular orientation by the process of photoselection. This is achieved by illuminating a randomly oriented population of

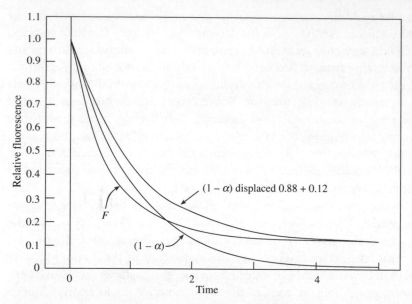

Figure 8.1   Simulation of protein fluorescence quenching (F) during NADH binding to lactate dehydrogenase. The curve for $(1-\alpha)$ represents the exponential time course of a pseudo first order reaction calculated from the fluorescence change as described in the text. The displaced trace is an exponential with the amplitude and time constant of the fluorescence change.

chromophores with linearly polarized light. This results in selective excitation of those chromophores whose transition dipole moment lies in or near the direction of polarization. Fluorescence, phosphorescence and other excitation to triplet states, can be used in a variety of ways to analyse the rate of randomization of the oriented dipoles. Chromophores with excited states with different lifetimes have to be used depending on the correlation time of the system under investigation.

In this section the application of steady state excitation of fluorescence to studies of anisotropy is described. Upon excitation of a fluorescent solution and observation at right angles, both to the direction of propagation of the exciting light and to the direction of the electric vector, the polarization, P, of the emitted light is defined as:

$$P = (I_{\parallel} - I_{\perp})/(I_{\parallel} + I_{\perp})$$

The subscripts $\parallel$ and $\perp$ stand, respectively, for intensities of emission observed through parallel and perpendicularly arranged polarizers. The anisotropy, r, of the emitted light is defined by

$$r = \frac{I_{\parallel} - I_{\perp}}{I_{\parallel} + 2I_{\perp}} \tag{8.2.5}$$

The anisotropy is related to the correlation time ϕ by equation (8.2.6).

$$r = \frac{r(0)}{1 + (\tau/\phi)} \qquad (8.2.6)$$

and the rotational correlation time is defined as

$$\phi = \eta V/RT \qquad (8.2.7)$$

where η is the viscosity, and V is the molecular volume, and τ is the lifetime of the fluorophore. Some more fundamental information about the relation between lifetimes and correlation times will be found in O'Connor & Phillips (1984). The rotational relaxation time is three times the rotational correlation time. Jameson & Sawyer (1994) present an analysis, due to Gregorio Weber, of the relation between these two rotational time constants. This paper presents extensive references and a historical review of the application of fluorescence anisotropy to the study of biomolecules.

Studies of anisotropy are applied to the investigation of hydrodynamic properties of large molecules containing fluorophores and the interaction of non-fluorescent macromolecules with small fluorescent ligands. The latter can be used to obtain equilibrium binding constants as well as kinetic information. While intrinsic or chemically attached fluorophores on large molecules are often sufficiently mobile to produce considerable independent depolarization, fluorescent ligands with reasonable binding constants, due to several points of contact, are likely to change from rapid rotation in free solution to rigid binding on a protein or nucleic acid. The fluorescence anisotropy of small fluorophores, such as NADH or flavin nucleotides, is therefore very sensitive to its state in solution – free or bound. In fact the pioneering experiments of Weber (1953) concerned the binding of flavin nucleotides to proteins.

The expression for ϕ is only really valid for a spherical molecule; ellipsoids and other shapes have multiple rotational relaxation times (Tanford, 1961, should be consulted for details). When considering the volume, for instance of a protein molecule, one also has to take into account the partial specific volume (about 0.75) and the hydration (about 30%). With these provisos dimensions of a molecule with a fixed chromophore can be estimated when the viscosity of the medium is known. Conversely, if the size of the molecule is known, the local viscosity, say in a membrane, can be estimated. However, in membrane systems rotations are on the microsecond time scale and long lifetime triplet state chromophores rather than fluorescence have to be used (see review by Cherry, 1992).

A representative list of lifetimes for different fluorophores is given in table 8.2. As to whether a particular molecular motion can be investigated

Table 8.2. *Lifetimes of some*
fluorescent compounds

Compound	$\tau \times 10^9$ s
NADH	0.4
N-acetyltryptophanamide	3.0
Fluorescein anion	5.1
Dansyl derivatives	9.0
Quinine bisulphate	18.9
Anthracein	250.0
Naphthalene	330.0
Riboflavin[a]	4.2
Pyridoxamine 5-phosphate[a]	4.3

[a] Visible spectrum.

by observation of fluorescence depolarization depends on the relation between the time constant of the movement and the lifetime of the fluorophore. We can consider two extreme cases for steady state anisotropy:

if $\phi \gg \tau$ then $r = r(0)$
if $\phi \ll \tau$ then $r = 0$

Anisotropy for the exploration of the mobility of domains of protein molecules is complemented with lifetime experiments briefly discussed in the next section. Another frequent use of static or time resolved steady state anisotropy measurements is the investigation of protein–protein and protein–nucleic acid interaction. An illustrative example is the association of dimeric to tetrameric lactate dehydrogenase from *Bacillus stearothermophilus* labelled with fluorescamine (Clarke *et al.*, 1985). In that case the rotational correlation time increases from 55 ns to 95 ns as the result of an increase in molecular weight from 80 000 to 160 000.

The following important point has to be taken into account during the quantitative interpretation of fluorescence anisotropy by the steady state method, as distinct from lifetime studies described in the next section. The anisotropies of the different molecular species present are additive, but it is the fractional intensities which are additive; differences in emission intensities between reactants and products have to be taken into account. If the fluorescence intensity as well as the anisotropy changes during a simple first order reaction, a plot of $r(t)$ against time will not be an exponential function. This is illustrated in the graph and legend of figure 8.2.

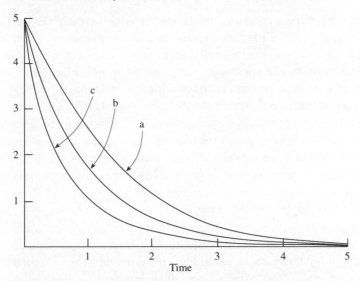

Figure 8.2   Simulated progress of anisotropy change from 0.5 to 0 during reactions accompanied by changes in fluorescence by factors of (a) 0.5, (b) 1 and (c) 2. In (b) an exponential record is obtained, while the other conditions clearly show deviation from true exponential decay (J. F. Eccleston, private communication).

8.3 The scope of fluorescence lifetime studies

The kinetics of fluorescence decay

The determination of the rates of decay (lifetime) of the excited states of fluorophores, initiated by a light flash, is itself an exercise in kinetics. Some of the methods for deconvoluting exponentials which were developed for this purpose, have influenced the approach to the analysis of other rate processes (see Brand & Johnson, 1992). In the present section those aspects of lifetime studies will be discussed which can contribute to our knowledge of the dynamics of proteins. While it was emphasized in section 2.1 that exponential decay should be characterized by the terms time constant or relaxation time, the time constant for the decay of fluorescence intensity after a flash of light is generally referred to as 'lifetime' in the literature on that subject. It represents the average amount of time a molecule remains in the excited state. The intensity of emission at any one time is proportional to the concentration of molecules in the excited state. This can be compared with the properties of exponentials discussed in section 2.1 and the law of mass action (section 3.1). The latter relates the rate of a reaction to the concentration of the reactant.

The subject of lifetime studies can be divided into two closely related problems. First, the decay of any fluorophore used for an investigation has to be characterized for the homogeneity and rate of decay(s). Second, lifetime measurements can be applied to studies of reactions during decay, such as molecular rotation and interactions with the environment which result in changes of emission. Some of the theory and analysis of changes, such as anisotropy or quenching, during decay of emission are also applicable when chromophores with long lifetime triplet states are used for the study of events which are appreciably slower than fluorescence decay (see table 8.2). This involved both changes in luminescence (phosphorescence) and changes in absorbence during the decay of dyes like eosin, which can be excited to the triplet state (Cherry, 1992).

Some of these applications are complementary to the steady state methods discussed in section 8.2. Investigations of fluorescence lifetimes and of anisotropy or fluorescence quenching phenomena in the lifetime mode, that is during the decay after a single flash, require more elaborate instrumentation and theory than steady state investigations. On the whole applications rather than detail of methods are discussed here. The use of the lifetime method for the study of molecular rotation, domain movement and more local dynamic events can often, some experts say always, provide additional information even for those problems which can be investigated with considerable success by steady state measurements.

Impulse response and harmonic response methods

The cognoscenti in the field of fluorescence are divided in their opinion about the preferred use of two different methods for the study of decay times and effects upon them. In this place it is only justified to make reference to the existence of two approaches to lifetime analysis and to point to the literature on the subject (see Gratton, Jameson & Hall, 1984; Jameson & Hazlett, 1991). The impulse response techniques yield direct records of the fluorescence intensity versus time after a brief exciting pulse. In the harmonic response technique fluorescence is excited by a light source with the intensity modulated sinusoidally at high frequency, typically in the megahertz range. In the latter case the fluorescence signal will also be modulated sinusoidally, but the finite persistence of the excited state will lead to a phase delay and demodulation of the fluorescence relative to the excitation. The true potential of the harmonic response analysis can only be achieved with a multifrequency phase fluorometer, which permits facile selection of modulation frequencies over a wide frequency range (from 10^5

to 10^8 hertz). Most kineticists are less familiar with the analysis of harmonic response than with exponential decay after a single step. The interpretation of oscillatory response has been referred to as a more specialized subject in section 6.2, when periodic pressure perturbation of equilibria was discussed.

The origins of multiple lifetimes of single fluorophores

In theory a single fluorophore in a homogeneous environment should decay with a single exponential time course. However, in practice this is rarely, if ever, the case. While many experiments are carried out with extrinsic fluorophores, covalently or non-covalently attached, it is obviously of considerable advantage if an intrinsic fluorophore can be used to study some property or reaction of a protein molecule. Both the favourable extinction coefficient and the quantum yield of tryptophan (see table 8.2) make this the most widely investigated of the fluorescent amino acid residues. In particular many proteins with single tryptophan residues per molecule, or per monomer of oligomeric molecules, have been selected for detailed study (Munro, Pecht & Stryer, 1979; Beechem & Brand, 1985). With the advent of genetic engineering it has become possible to replace all but one of the tryptophan residues of a protein molecule, or to introduce a single residue into proteins which contain none. Furthermore such single residues can be introduced in different positions along the polypeptide chain to probe the local environment and movement. In some cases it has been possible to do all these manipulations with tryptophan residues without affecting any of the functions of the protein (see for instance Matouschek *et al.*, 1990; Dunn *et al.*, 1991). Reference to such operations on enzyme molecules has already been made in several sections in connection with the use of fluorescence changes for monitoring conformation changes during substrate binding.

The above comments indicate why one of the principal interests in the study of fluorescence decay, as distinct from steady state observations, is the causes for multiple exponential decays of, for instance, tryptophan. Multiple time constants for the decay of fluorescence of a single tryptophan residue, located in one specified location of a protein molecule, are receiving much attention. They can be due to a number of different factors although, however, as recently discussed by Bajzer & Prendergast (1993) these are difficult to distinguish. Bajzer & Predergast write for the specialist, but they give a good review of the literature on the problem of multiple decay times from apparently homogeneous fluorophores. If the protein is in equilibrium between conformational states, which provide different micro-environments for the indole residue, multiple time constants for the fluorescence decay will be observed, provided the rates of interconversion of the two states is slow compared to the rates of decay. This can apply to a simple two state protein as well as to the concept of multiple conformational substates, discussed in section 7.3. In the latter case non-exponential decay would be observed, which is analogous to the non-exponential geminate recombination of CO with myoglobin after photodissociation at low temperatures (p. 255). When the interconversion between the conformational states is fast compared to fluorescence decay, an average decay time will be observed.

The investigation of the fluorescence decay of free tryptophan and of some of its derivatives demonstrated why even a single tryptophan in a fixed environment within a protein molecule will give complex results. Three component time constants have been observed from the analysis of the decay of the excited state of tryptophan. These time constants range from 0.5 to 10 ns. In contrast 3-methylindole and 3-methyltryptophana-mide decay mono-exponentially. These compounds are widely used as standards for testing instrumentation. The multiplicity of the decay times for free tryptophan is attributed to the existence of rotamers with different orientations of the indole ring with respect to the free amino group. The photophysics of tryptophan and of its derivatives is discussed in detail by Creed (1984). The time resolved spectroscopy of tryptophan conformers has been studied in the gas phase by Phillips *et al.* (1988).

Further uses of lifetime measurements

In spite of the complexities described above, studies of the time resolved decay of tryptophan fluorescence are contributing much useful information about the structure and dynamics of protein molecules. From the point of view of a kineticist two aspects of such studies are of interest. First, the methods for the measurement and analysis of fluorescence decay provide a number of useful ideas for other types of kinetic investigations. Second, the results of such measurements can be used for temporal resolution of changes in the environment of the fluorophore and thus for the determination of rates of conformation changes. Lifetime measurements are made by averaging a large number of fast decays. With improving technology the number of such traces required for a good record is decreasing. Similarly the harmonic technique is providing satisfactory records more quickly. It is thus becoming possible to obtain information about time resolved changes in fluorescence lifetimes at shorter intervals. This is providing a new tool for kinetic investigations (see Beechem, 1994).

8.4 Photolabile (caged) compounds for the initiation of reactions

Introduction and range of applications

The methods and examples discussed in section 8.1 illustrated several applications of flash photolysis to the initiation of reactions. There the concern was with systems in which the photochemical process is part of a natural phenomenon such as the responses of visual pigment or bacterial rhodopsin and the primary reactions of photosynthesis. The reactions of haem proteins, where photosensitivity was used for the extensive kinetic investigations into the mechanism of the binary reactions of these proteins with the free ligands, have been discussed in section 7.3. The flash photolysis of haem proteins can, however, be used as a means to study another reaction or sequence of reactions. In the simplest form we have seen how the dissociation of CO from haemoglobin can be used to study the rate of association of oxygen with this protein. In a wider sense the reversal, by flash photolysis, of the inhibition of the reactions of haem proteins by bound CO, say within a sequence of electron transfer reactions, can initiate a reaction sequence in complex organized systems like mitochondria or microorganisms. This is an example of the use of a photolabile 'cage' for a reactant.

The search for methods which enable one to initiate reactions in organized systems, at a well-defined moment in time and within a defined space, led Kaplan, Forbush & Hoffmann (1978) to the use of what have become known as caged compounds. It must be pointed out that the term caged compounds means something different to chemists, who use it to describe clathrates. Many substrates, inhibitors and messengers can be made biologically inert by attachment of a photolabile protecting (caging) group, although it must not be assumed that attachment of such a protecting group to a biological molecule will necessarily prevent binding to a receptor or abolish activity. The use of caged compounds for the initiation of reactions by photolysis has become so wide a subject that it deserves treatment separate from that of other kinetic techniques involving flash photolysis (see Corrie & Trentham, 1993). There is, of course, considerable overlap between the technical and theoretical aspects of this

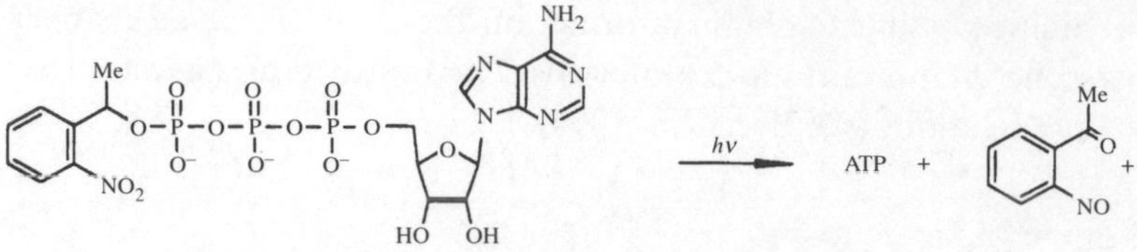

Figure 8.3   The overall reaction of the photolysis of 1-(2-nitrophenyl)ethyl P³-ester of ATP, which is abbreviated in the text to npe-ATP.

method and those described in the rest of this chapter. Photochemical cross linking reagents have also been used to start such reactions at a particular moment in time and at a specific location.

Apart from substrates with photolabile groups blocking the function, there are several other types of compounds and systems which permit the photochemical initiation of reactions with precise temporal and/or spatial definition. Montal, Lester and others (see for instance Gurney & Lester, 1987) have used photoisomerization of an inactive isomer to the active form of a substrate (messenger) to initiate reactions. The reviews by McCray & Trentham (1989), Corrie & Trentham (1993) and others quoted below, will complement the present discussion of the development of this subject. Trentham and his colleagues treat in particular detail the kinetics of the liberation of phosphate compounds from their inactive precursor.

The development of some caged compounds

The term caged compounds has been used more specifically for compounds which are inactive due to chemical coupling with a photolabile group rather than for photoisomerizable reagents. The paradigm for this is the first photolabile derivative of an important biological substrate: P³ 1-(2-nitrophenyl)ethyl ester of ATP, abbreviated to npe-ATP throughout this section (Kaplan *et al.*, 1978). The overall photolyic reaction is shown in figure 8.3. The ubiquitous role of ATP and other nucleoside phosphates in many reactions has resulted in the wide use of this compound for the initiation of physiological processes by flash photolysis. Kaplan *et al.* (1978) originally used npe-ATP for the study of a relatively slow reaction involved in the sodium pump of erythrocytes. McCray *et al.* (1980) determined the rate of liberation of ATP from npe-ATP following photolysis by a single flash with a frequency doubled ruby laser (347 nm). Their results opened up a new method with wide applications for the study of rapid reactions, which is expanding as an ever increasing range of photolabile caged substrates is

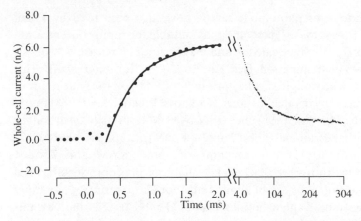

Figure 8.4   Whole cell recording after flash photolysis of 400 μM N-(7a-carboxy-2-nitrobenzyl)carbamoylcholine in the presence of clonal mammalian BC$_3$H1 cells, which contain nicotinic acetylcholine receptors. Photolysis was at 328 nm with a 600 ns pulse from a frequency doubled dye laser. The solid line represents a fit to the exponential rise with a time constant of 467 μs (from Matsubara *et al.*, 1992).

described and improved time resolution is achieved with novel caging groups.

The above blocking group for caging ATP is now used for many physiological substrates containing phosphate groups, for example ortho-phosphate, ADP, cAMP, GTP, cGMP and IP$_3$, to mention just a few. Walker, Reid & Trentham (1989) recently developed a (relatively) simple and rapid method for the synthesis of caged nucleotides and other phosphates. A much improved time resolution (see below) was achieved with the synthesis of the 3-phosphate-3',5'-dimethoxybenzoin ester of ATP (DMB-caged ATP) (Corrie & Trentham, 1992), but this compound has some shortcomings, which are discussed below. These developments together with the design of an inexpensive (compared to a laser) flash lamp by Rapp & Guth (1988) have widened considerably the possible uses of photolabile compounds.

The preparation and properties of caged choline derivatives and other pharmacologically active compounds have been described by Walker, McCray & Hess (1986) and Milburn *et al.* (1989). Gurney & Lester (1987) presented a comprehensive review of compounds available for neurobiological studies up to that time. More recently Matsubara, Billington & Hess (1992) used a nitrophenylacetate derivative of carbamoylcholine [N-(α-carboxy-2-nitrobenzyl)carbamoylcholine] for the photolytic production of carbamoyl choline in the microsecond time region (see figure 8.4).

The techniques for photolabile caging have also been used by Tsien, Kaplan and others, to design compounds suitable for rapid concentration jumps of ions (for instance calcium and magnesium) by release or sequestration following illumination (see Kaplan, 1990, for references). This application is clearly of great value for kinetic studies on the wide range of systems regulated by divalent cations. Kaplan & Ellis-Davies (1988) found that 1-(2-nitro-4,5-dimethoxyphenyl)-N,N,N',N'-tetrakis[(oxycarbonyl)-methyl]-1,2-ethanediamine (abbreviated as DM-nitrophen) binds Ca^{2+} and Mg^{2+} with dissociation constants of about 5 nM and 2.5 μM respectively. On exposure to a 347 nm laser flash the photoproducts have an overall dissociation constant of about 3 mM for Ca^{2+}, resulting in a rapid release of the ion. Kaplan & Ellis-Davies demonstrated that, by this technique, calcium induced contraction of skinned muscle fibres can be studied on a time scale of milliseconds. They also suggest that DM-nitrophen can be used for the rapid initiation of the many enzyme reactions which require magnesium as an essential cofactor for reactions involving nucleoside phosphate substrates. Another interesting application of caging, which has been developed recently, involves blocking of fluorescence by a photolabile group. This permits limited local initiation of fluorescence by photolysis and subsequent observation of the movement of the fluorescent compound, which is the reverse of FRAP, fluorescence recovery after photobleach by a fluorescent reporter.

Time resolution and other physical parameters

There are two distinct types of experiments with caged compounds to be considered here: those concerned with the development and assessment of new compounds and those connected with specific applications to different physiological processes. We shall discuss the two problems in succession. Detailed descriptions of methods for assessing the rates and quantities of the photolytic formation of active substrates from different caged compounds are presented by Walker *et al.* (1988) and Milburn *et al.* (1989). Some aspects of the decomposition of npe-ATP will serve as an example for such investigations. The broad absorption band of 1-(2-nitrophenyl)ethyl phosphate in the near UV, allows the use of a wavelength for photolysis which results in minimum interference with that used for monitoring the reaction. The rapid ($k > 10^5 \text{ s}^{-1}$) initial formation of the presumed acinitro intermediates (with extinction coefficients of $6.3 \times 10^3 \text{ M}^{-1} \text{ cm}^{-1}$ and $9.1 \times 10^3 \text{ M}^{-1} \text{ cm}^{-1}$ at 380 nm and 406 nm respectively) and the concomitant release of a proton, are followed by the liberation of ATP and 2-nitrosoacetophenone. The amplitude and rate of decay of the absorbence at 406 nm can be used as a measure of the amount and time course of ATP liberated. This final step is acid catalysed and is affected by Mg^{2+} and ionic strength. For accurate information about the rate of formation of free ATP it is necessary to carry out measurements under the particular conditions (composition of reaction mixture) of the system under investigation. The approximate rate of formation of ATP from npe-ATP is $125 \, s^{-1}$ at pH 7 and directly proportional to (c_{H^+}) in the pH range 6 to 8.

One of the final photolysis products, 2-nitrosoacetophenone, reacts with thiols and other groups and can cause damage to the biological systems under investigation. This complication can be avoided by the addition of thiol reagents, for instance dithiothreitol (DTT), which sequester nitroso-ketones. It is in fact possible to determine the rate of formation of photolysis products, in the presence of sufficiently high concentrations of DTT, by observation of the disappearance of the nitroso band at 740 nm (extinction coefficient $50 \, M^{-1} \, cm^{-1}$). The second order rate constant for the reaction of 2-nitrosoacetophenone with DTT at pH 7 and 21 °C and ionic strength 0.18 M is $3.5 \times 10^3 \, M^{-1} \, s^{-1}$.

The time resolution of the photochemical release of caged substrates depends on two factors: the energy/time profile of the light source used for photolysis and the effective rate of liberation of active substrate. It should be remembered that the time scale of the potential application of a kinetic technique depends critically on the accuracy of the determination of zero time. Some practical details of the investigation of different parameters are illustrated below. Suffice it to say for the moment that the range of compounds presently available for different systems have time constants from 10 ms to 10 μs for the liberation of phosphates and amino derivatives (neurotransmitters).

The principles of design and interpretation of kinetic investigations with photolabile precursors can be best illustrated by some of the many uses of npe-ATP. The important parameters to be determined with each compound are the absorption spectrum, quantum yield, rate of appearance of active substrate, the possible binding (competitive inhibition) of the caged compound and the effects of any photoproducts. Additional to these compound specific parameters, instrumentation parameters have to be considered. These are concerned with the light source for photolysis and with the detection equipment. The latter can either monitor the photolytic reaction of the caged compound or the physiological process stimulated by the released reagent. Although we are not concerned with the technical detail of instrumentation, some of its features have to be taken into account in the planning and interpretation of kinetic experiments.

The choice of the actinic source depends on the absorption spectrum of the caged compound, its rate of decomposition and the shortest time constant which is expected in the system under investigation. As will be seen below, in some experiments intensity of illumination can compensate for a slow flash. Walker *et al.* (1988) used a frequency doubled ruby or dye lasers for their development work on caged phosphate compounds; some additional lasers were used by Milburn *et al.* (1989). These lasers provide 1 ns pulses, while xenon flash lamps can give illumination with step functions of about 10 ms, when coupled with two shutters. The shape of the pulse is of importance both for the time course of photolysis and for subsequent observation of a process by light detection (fluorescence, absorbence or light scattering). Uniform photolysis throughout the sample has to be assured through illumination of the whole cell and through a balance between maximum light absorption and homogeneous photolysis through the depth of the cell. Photolysis of a cylindrical section of a solution can be satisfactory, provided the subsequent reaction is fast compared to diffusion and observation is restricted to the same segment. These problems, as well as the determination of the quantum yield are common to all photochemical experiments. Some special aspects of quantum yields of caged compounds are discussed by McCray & Trentham (1989). Similarly the interference caused to electronic monitoring equipment by the operation of firing a laser or a xenon lamp is a universal technical problem.

The time required to reach a threshold concentration for a substrate can be controlled by changing the concentration of its photolabile precursor. This obvious point is illustrated in figure 8.5. The simulated profile of substrate liberation is for a reaction with a time constant of 10 ms. A physiological reaction requiring five concentration units for initiation will be satisfied in 0.5 ms if the precursor concentration is equal to 100 concentration units. For practical reasons there are upper limits to the concentrations of precursors. In addition to availability and solubility, precursors are potential competitive inhibitors. The relation between their light absorption and the energy of the actinic source must also be taken into account. If the specific optical density of the solution due to the caged compound is too low, very little active substrate is liberated. On the other hand, if it is too high, light absorption results in inhomogeneous photolysis throughout the cell. The optimum condition is an optical density of 0.7 at the wavelength of excitation. Using high concentrations of a caged

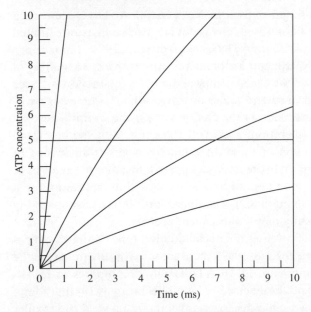

Figure 8.5   Simulation of rates of reaching a threshold concentration of ATP (5 mM) from caged ATP under conditions when the rate constant for ATP liberation is 100 s^{-1}. The traces (a) to (d) represent conditions of 5, 10, 20 and 100 mM initial caged-ATP concentrations.

compound, to achieve rapid production of threshold or saturation levels of substrate, is of course accompanied by a continuing increase in substrate concentration during the progress of the reaction under investigation.

Some general principles for the investigation of the time resolution obtainable by the use of a caged compound can be demonstrated with examples taken from experiments with npe-ATP. Variants of these experiments can be carried out to investigate other compounds. The principal problem is whether the active substrate is formed fast compared to the subsequent reaction. Slow appearance of substrate should show up as a lag phase (see McCray *et al.*, 1980). The noisy records of extinction, fluorescence, or light scattering changes often hide lag phases, which are then only expressed in the extraction of inaccurate rise times. The essential features of such tests are, therefore, seen more clearly from simulated experiments than from real ones. I am referring here to tests involving applications, like the interaction of ATP with actomyosin, rather than those involving observations of the photolytic and dark reactions of the decomposition of the caged compound. One useful test for caged ATP was employed in the original paper by McCray *et al.* (1980). The dissociation of the complex of actin with the myosin fragment S1, which can be followed by a decrease in light scattering or the fluorescence change of labelled actin, is rate limited by a second order reaction with ATP (see Walker *et al.*, 1988). Simulations and analysis of such experiments are of interest with respect to a number of aspects of the kinetic of the caged compound and its interaction with the kinetics of the biological system under investigation. The time course of three processes can interact: (1) the flash time; (2) the overall chemical reaction of substrate liberation; and (3) the reaction with the biological target. The deconvolution of these three kinetic events is analogous to several problems discussed in previous sections; for instance the analysis of separate fluorescence lifetimes (section 6.2) or of time constants for a sequence of reactions (section 3.2). The time course of the flash is readily determined independently of any subsequent process.

Although DMB-ATP releases the nucleotide 100 times faster than npe-ATP (Trentham, Corrie & Reid, 1992), there are some disadvantages in the use of the former compound. DMB-ATP slowly decomposes to DMB-phosphate and ADP and the efficiency of ATP liberation is 1/4 that found with npe-ATP. The low yield is due to a smaller quantum yield and smaller extinction coefficient at the wavelength of the laser.

Examples illustrating kinetic applications of caged compounds

Current efforts to enlarge the range of compounds and to shorten the time constants of photolysis are resulting in increasing potentialities of this approach to kinetic studies. The major advantage of the use of caged compounds, over that of flow and chemical relaxation techniques for the study of rapid reactions, is their wider application to organized systems. Examples of their use for investigations of rate processes in whole cells (intact or permeabilized), active transport vesicles, muscle fibres and crystals are given below. The first application of the ideas of Kaplan *et al.* (1978) to fast reactions was the study of the transients in solutions of muscle proteins (McCray *et al.*, 1980). Goldman *et al.* (1982) subsequently carried out the first of many investigations of muscle dynamics by the photolytic release of npe-ATP. An important aim of kinetic studies in biology is to ascertain whether reactions behave similarly in homogeneous solutions of purified components and in intact biological systems. If this is the case, then one can identify specific steps recognized in solution studies with isolated components, with functional steps in the organized system. Hibberd & Trentham (1986) have compared kinetic data from studies on solutions of muscle proteins with results obtained from investigations on muscle fibres with caged ATP. Some of these and similar studies are documented below.

In the study of structure–function relationships it is of importance to investigate reactions of proteins in crystals. Most structure information does come from X-ray crystallography; notwithstanding the success of NMR spectroscopy in the resolution of the three-dimensional structure of small molecules and of some domains of large ones. It is, therefore, important to investigate how enzymes function in crystals. When crystals are soaked in solutions of photolabile precursors of substrates, the reaction initiated by photolysis will be less affected by diffusion, compared with a reaction initiated by mixing crystals with a solution of substrate, especially in the first turnover. Laue diffraction has, in recent years, been used successfully in the study of time resolved structure changes of enzymes (see for instance Hajdu & Johnson, 1990). A recent exciting development was the use of caged-GTP for the photolytic initiation of the P-21 GTPase in crystals. This led to time resolved observations of protein structure changes during the single turnover hydrolysis of GTP (Schlichting *et al.*, 1990). These authors also use this technique to obtain evidence that the rate of hydrolysis in the crystal is similar to that in free solution. In this system the caged nucleotide binds to the protein, but not at the substrate binding site. GTP must move to the correct site after liberation from the cage. There can be advantages in bound caged compounds since their use can result in an initial first order rather than second order reaction between substrate and active site. In a recent review of time resolved crystallography Hajdu & Andersson (1993) also discuss a variety of possible experiments with caged enzymes.

In a chapter primarily intended to illustrate the use of caged compounds for kinetic studies it is more important to use some selected examples to discuss principles in some detail, rather than to survey all the available compounds and their applications, which are being constantly extended. The future emphasis is likely to be on compounds permitting more rapid release of a wider range of substrates for kinetic studies of muscle contraction, excitation contraction coupling and neuroreceptors (see for instance figure 8.4). The potential of the use of caged compounds for transient kinetic studies on enzyme reactions has not yet been tapped.

References

Ackerman, E. & Berger, R. L. (1963). Reaction of oxyhemoglobin with carbon monoxide. *Biophysical Journal*, **3**, 494–505.

Acton, F. S. (1970). *Numerical Methods that Work*. New York: Harper & Row.

Adam, G. & Delbruck, M. (1968). Reduction of dimensionality in biological diffusion processes. In *Structural Chemistry and Molecular Biology*, ed. A. Rich & N. Davidson, pp. 198–215. San Francisco: W. H. Freeman and Company.

Albery, A. & Knowles, J. R. (1976). Evolution of enzyme function and the development of catalytic efficiency. *Biochemistry*, **15**, 5631–40.

Anderson, B. F., Baker, H. M., Norris, G. E., Rumball, S. V. & Baker, E. N. (1990). Apolactoferrin structure demonstrates ligand-induced conformational changes in transferrins. *Nature (London)*, **344**, 784–7.

Anfinsen, C. B. (1972). The formation and stabilization of protein structure. *Biochemical Journal*, **128**, 737–49.

Anson, M. (1992). Temperature dependence and Arrhenius activation energy of F-actin velocity generated in vitro by skeletal myosin. *Journal of Molecular Biology*, **224**, 1029–38.

Arnis, S. & Hofmann, K. P. (1993). Two different forms of metarhodopsin II: Schiff base deprotonation precedes proton uptake and signaling state. *Proceedings of the National Academy of Sciences, USA*, **90**, 7849–53.

Arrhenius, S. (1889). Über die Reaktionsgeschwindigkeit bei der Inversion von Rohrzucker durch Sauren. *Zeitschrift für physikalische Chemie*, **4**, 226–48.

Arrhenius, S. (1915). *Quantitative Laws in Biological Chemistry*. London: Bell.

Atkins, P. W. (1982). *Physical Chemistry*, 2nd edn. Oxford: Oxford University Press.

Attwood, P. V. & Gutfreund, H. (1980). The application of pressure relaxation to the study of the equilibrium between metarhodopsin I and II from bovine retinas. *Federation of European Biochemical Societies Letters*, **119**, 323–6.

Austin, R. H., Beeson, K. W., Eisenstein, L., Frauenfelder, H. & Gunsalus, I. C. (1975). Dynamics of ligand binding to myoglobin. *Biochemistry*, **14**, 5355–73.

Austin, R. H., Beeson, K. W., Chan, S. S., Debrunner, P. G., Dowling, R., Eisenstein, L., Frauenfelder, H. & Nordlund, T. M. (1976). Logarithmic time base. *Reviews of Scientific Instruments*, **47**, 445–7.

Badcoe, I. G., Smith, C. J., Word, S., Halsall, D. J., Holbrook, J. J., Lund, P. & Clarke, A. R. (1991). Binding of chaperonins to the folding intermediates of

308

lactate dehydrogenase. *Biochemistry*, **30**, 9195–200.

Bagshaw, C. R. & Trentham, D. R. (1973). The reversibility of adenosine triphosphate cleavage by myosin. *Biochemical Journal*, **133**, 323–8.

Bajzer, Z. & Prendergast, F. G. (1993). A model for multiexponential tryptophan fluorescence intensity decay in proteins. *Biophysical Journal*, **65**, 2313–23.

Barman, T. E. & Gutfreund, H. (1964). A comparison of the resolution of chemical and optical sampling. In *Rapid Mixing and Sampling Techniques in Biochemistry*, ed. B. Chance, R. H. Eisenhardt, Q. H.Gibson & K. K. Lonberg-Holm, pp. 339–44. New York: Academic Press.

Barman, T. E. & Travers, F. (1985). The rapid-flow-quench method in the study of fast reactions in biochemistry: extension to subzero conditions. *Methods in Biochemical Analysis*, **31**, 1–59.

Barman, T. E., Hillaire, D. & Travers, F. (1983) Evidence for two step binding of ATP to myosin subfragment 1 by the flow-quench method. *Biochemical Journal*, **209**, 617–26.

Bateman, H. (1910). The solution of a system of differential equations occurring in the theory of radio-active transformations. *Proceedings of the Cambridge Philosophical Society*, **15**, 423–7.

Baylor, D. A., Hodgkin, A. L. & Lamb, T. D. (1974). Reconstruction of the electrical responses of turtle cones to flashes and steps of light. *Journal of Physiology*, **242**, 759–91.

Beechem, J. M. (1992). Global analysis of biochemical and biophysical data. In *Methods in Enzymology*, vol 210, ed. L. Brand & M. L. Johnson, pp. 37–53. New York: Academic Press.

Beechem, J. M. (1994). Real-time kinetics of gene transcription in multi-color time-resolved and stopped-flow fluorescence studies. *Biophysical Journal*, **66**, A233.

Beechem, J. M. & Brand, L. (1985). Time-resolved fluorescence of proteins. *Annual Review of Biochemistry*, **54**, 43–71.

Bender, M. L. (1971). *The Mechanism of Homogeneous Catalysis from Proton to Protein*. New York: Wiley.

Benkovic, S. J. (1979). Anomeric specificity of carbohydrate using enzymes. *Methods in Enzymology*, **63**, 370–9.

Berg, H. C. (1983). *Random Walks in Biology*. Princeton, NJ: Princeton University Press.

Berg, H. C. & Purcell, E. M. (1977). Physics of chemoreception. *Biophysical Journal*, **20**, 193–219.

Berg, O. G. & von Hippel, P. H. (1985). Diffusion-controlled macromolecular interactions. *Annual Review of Biophysics and Biophysical Chemistry*, **14**, 131–60.

Bernasconi, C. F. (1976). *Relaxation Kinetics*. New York: Academic Press.

Bevington, P. R. (1969). *Data Reduction and Error Analysis for the Physical Sciences*. New York: McGraw-Hill Book Company.

Biacheria, A. & Kegeles, G. (1957). Diffusion measurements in aqueous solutions of different viscosities. *Journal of the American Chemical Society*, **79**, 5908–12.

Biosca, J. A., Travers, F. & Barman, T. E. (1983). A jump in an Arrhenius plot can be the consequence of a phase transition. *Federation of European Biochemical Societies Letters*, **153**, 217–20.

Biosca, J. A., Travers, F., Hillaire, D. & Barman, T. E. (1984). Cryoenzymic studies on myosin subfragment 1: perturbation of an enzyme reaction by temperature and solvent. *Biochemistry*, **23**, 1947–55.

Birge, R. R. (1990). Photophysics and molecular electronic applications of the rhodopsins. *Annual Review of Physical Chemistry*, **41**, 683–733.

Blacklow, S. C., Raines, R. T., Lim, W. A., Zamore, P. D. & Knowles, J. R. (1988). Triosephosphate isomerase catalysis is diffusion controlled. *Biochemistry*, **27**, 1158–67.

Blake, C. C. F., Johnson, L. N., Mair, G. A., North, A. C. T., Phillips, D. C. & Sarma, V. R. (1967). Crystallographic studies of the activity of hen egg white lysozyme. *Proceedings of the Royal Society of London*, **B167**, 378–88.

Boas, M. L. (1966). *Mathematical Methods in the Physical Sciences*. New York: John Wiley & Sons Inc.

Brahm, H. (1977). Temperature dependent changes of chloride transport: kinetics in human red cells. *Journal of General Physiology*, **70**, 283–306.

Brand, L. & Johnson, M. L. (eds.) (1992). Numerical computer methods. *Methods in Enzymology*, **210**. New York: Academic Press.

Briggs, G. E. & Haldane, J. B. S. (1925). A note on the kinetics of enzyme action. *Biochemical Journal*, **19**, 338–9.

Britton, H. G. & Clarke, J. B. (1972). Mechanism of the 2,3-diphosphoglycerate dependent phosphoglycerate mutase from rabbit muscle. *Biochemical Journal*, **130**, 397–410.

Britton, H. G., Carreras, J. & Grisolia, S. (1972). Mechanisms of yeast phosphoglycerate mutase. *Biochemistry*, **11**, 3008–14.

Brooks, S. P. J. & Suelter, C. H. (1989). Practical aspects of coupling enzyme theory. *Analytical Biochemistry*, **176**, 1–14.

Brouwer, A. C. & Kirsch, J. F. (1982). Investigation of diffusion limited rates of chymotrypsin reactions by viscosity variations. *Biochemistry*, **21**, 1302–7.

Bruckert, F., Chabre, M. & Vuong, T. M. (1992). Kinetic analysis of the activation of transducin by photoexited rhodopsin. *Biophysical Journal*, **63**, 616–29.

Brune, M., Hunter, J. L., Corrie, J. E. T. & Webb, M. R. (1994). Direct, real-time measurements of rapid inorganic phosphate release using a novel fluorescent probe and its application to actomyosin subfragment 1 ATPase. *Biochemistry*, **33**, 8262–71.

Buchner, J., Schmidt, M., Fuchs, M., Jaenike, R., Rudolph, R., Schmid, F. X. & Kiefhaber, T. (1991). GroE facilitates refolding of citrate synthase by suppressing aggregation. *Biochemistry*, **30**, 1586–91.

Capellos, C. & Bielski, B. H. J. (1972). *Kinetic Systems*. NY: Wiley-Interscience.

Capellos, C. & Bielski, B. H. J. (1980). *Kinetic Systems*, reprint edition. Huntingdon, NY: Robert Krieger.

Cattell, M. (1934). Discussion on methods of measuring and factors determining the speed of chemical reactions. *Proceedings of the Royal Society of London*, **B116**, 206–7.

Cattell, M. (1936). The physiological effects of pressure. *Biological Reviews*, **11**, 441–76.

Chance, B. (1943). The kinetics of the enzyme-substrate compound of peroxidase. *Journal of Biological Chemistry*, **151**, 553–75.

Chance, B. (1952). Spectra and reaction kinetics of respiratory pigments of homogenized and intact cells. *Nature (London)*, **169**, 215–21.

Chance, E. M., Curtis, A. R., Jones, I. P. & Kirby, C. R. (1977). *FACSIMILE: A Computer Program for Flow and Chemistry Simulation, and General Initial Value Problems*. London: HMSO.

Chandrasekhar (1943). Stochastic problems in physics and astronomy. *Reviews of Modern Physics*, **15**, 1–89.

Chen, H. M., Martin, V. S. & Tsong, T. Y. (1992). pH-induced folding/unfolding of staphylococcal nuclease: determination of kinetic parameters by the sequential jump method. *Biochemistry*, **31**, 1483–91.

Cherry, R. J. (1992). Rotational diffusion of membrane proteins: studies of band 3 in the human erythrocyte membrane using triplet probes. In *Structural and Dynamic Properties of Lipids and Membranes*, ed. P. J. Quinn & R. J. Cherry, pp. 137–51. London and Chapel Hill, NC: Portland Press.

Chipman, D. M. & Schimmel, P. R. (1968). Dynamics of lysozyme–substrate interactions. *Journal of Biological Chemistry*, **243**, 3771–4.

Chock, P. B. & Gutfreund, H. (1988). Reexamination of the kinetics of the transfer of NADH between its complexes with glycerol-3-phosphate dehydrogenase and with lactate dehydrogenase. *Proceedings of the National Academy of Sciences, USA*, **85**, 8870–4.

Churchill, R. V. (1958). *Operational Mathematics*. 2nd edn. New York: McGraw-Hill.

Clarke, A. R., Smith, C. J., Hart, K. W., Birktoft, J. J., Banaszak, L. J., Wilks, H. M., Barstow, D. A., Atkinson, T., Lee, T. V., Chia, W. N. & Holbrook, J. J. (1987). Rational construction of a 2-hydroxyacid dehydrogenase with new substrate specificity. *Biochemical and Biophysical Research Communications*, **148**, 15–23.

Clarke, A. R., Waldman, A. D. B., Hart, K. & Holbrook, J. J. (1985). The rates of defined changes in protein structure during the catalytic cycle of lactate dehydrogenase. *Biochimica et Biophysica Acta*, **829**, 397 407.

Clegg, R. M. & Maxfield, B. W. (1976). Chemical kinetic studies by a new small pressure perturbation method. *Review of Scientific Instruments*, **47**, 1383–93.

Coates, J. H., Criddle, A. H. & Geeves, M. A. (1985). Pressure relaxations of actomyosin S1. *Biochemical Journal*, **232**, 351–6.

Collins, F. C. & Kimball, G. E. (1949). Diffusion controlled reaction rates. *Journal of Colloid Science*, **4**, 425–37.

Colquhoun, D. (1971). *Lectures on Biostatistics*. Oxford: Oxford University Press.

Colquhoun, D. & Hawkes, A. G. (1983). The principles of the stochastic interpretation of ion-channel mechanisms. In *Single Channel Recording*, ed. B. Sakmann & E. Neher. New York: Plenum Press.

Conte, S. D. & de Boor, C. (1981). *Elementary Numerical Analysis*, 3rd edn. New York: McGraw-Hill Book Company.

Cooper, A. (1983). Photoselection of conformational substates and the hypochromic photoproduct of rhodopsin. *Chemical Physics Letters*, **99**, 305–9.

Cornish-Bowden, A. & Luz Cardenas, M. (eds.) (1990). *Control of Metabolic Processes*. New York: Plenum Press.

Corrie, J. E. T. & Trentham, D. R. (1992). Synthetic, mechanistic and photochemical studies of phosphate esters of substituted benzoins. *Journal of the Chemical Society Perkin Transactions*, **I**, 2409–17.

Corrie, J. E. T. & Trentham, D. R. (1993). Caged nucleotides and neurotransmitters. In *Bioorganic Photochemistry Volume 2: Biological Applications of Photochemical Switches*, ed. H. Morrison, pp. 243–305.

Creed, D. (1984). Photochemistry and photophysics of near UV absorbing aminoacids. I. Tryptophan and its simple derivatives. *Photochemistry and Photobiology*, **39**, 537–83.

Crick, F. (1988). *What Mad Pursuit*. New York: Basic Books, Inc.

Daggett, V. & Levitt, M. (1993). Realistic simulations of native-protein dynamics in solution and beyond. *Annual Review of Biophysics and Biomolecular Structure*, **22**, 353–80.

Dalziel, K. (1975). Kinetics and mechanism of dehydrogenases. In *The Enzymes*, 3rd edn, vol. II, ed. P. D. Boyer, pp. 1–60. New York: Academic Press.

Danckwerts, P. V. (1969). Diffusion and mass transfer. In *Thomas Graham Memorial Symposium*. London: Gordon & Breach.

Darwin, F. (ed.) 1887. *The Life and Letters of Charles Darwin*. London: Murray.

Davis, J. S. (1988). Assembly processes in vertebrate skeletal thick filament formation. *Annual Review of Biophysics and Biophysical Chemistry*, **17**, 217–39.

Davis, J. S. (1993). Myosin thick filaments and subunit exchange: a stochastic simulation based on the kinetics of assembly. *Biochemistry*, **32**, 4035–42.

Davis, J. S. & Gutfreund, H. (1976). The application of pressure relaxation to the study of biochemical reactions. *Federation of European Biochemical Societies Letters*, **72**, 199–207.

Davis, J. S. & Harrington, W. F. (1993). A single order-disorder transition generates tension during the Huxley-Simmins phase 2 in muscle. *Biophysical Journal* , **65**, 1886–98.

Debrunner, P. G. & Frauenfelder, H. (1982). Dynamics of proteins. *Annual Review of Physical Chemistry*, **33**, 283–300.

Demchenko, A. P., Grycznski, I., Grycznski, Z., Wiczk, W., Malak, H. & Fishman, M. (1993). Intramolecular dynamics in the environment of the single tryptophan residue in staphylococcal nuclease. *Biophysical Chemistry*, **48**, 39–49.

Denbigh, K. (1951). *Thermodynamics of Steady States*. London: Methuen.

Denbigh, K. G., Hicks, M. & Page, F. M. (1948). The kinetics of open reaction systems. *Transations of the Faraday Society*, **44**, 479–94.

DeSa, R. J. & Gibson, Q. H. (1969). A practical automatic data aquisition system for stopped-flow spectrophotometry. *Computers and Biomedical Research*, **2**, 494–505.

Diebler, H., Eigen, M., Ilgenfritz, G., Maas, G. & Winkler, R. (1969). Kinetics and mechanism of reactions of main group metal ions with biological carriers. *Pure and Applied Chemistry*, **20**, 93–115.

Dixon, M. & Webb, E. C. (1964). *Enzymes*, 2nd edn. London: Longmans.

Douzou, P. (1977). *Cryobiochemistry*. London: Academic Press.

Dowd, J. E. & Riggs, D. S. (1965). A comparison of estimates of Michaelis–Menten kinetic constants from various linear transformations. *Journal of Biological Chemistry*, **240**, 863–9.

Dunford , B. & Hewson, W. D. (1977). Effect of mixed solvents on the formation of horseradish peroxidase compound I. The importance of diffusion controlled reactions. *Biochemistry*, **16**, 2949–57.

Dunn, M. F., Anguilar, V., Brzoviz, P., Drewe, W. F., Houben, K. F., Leja, C. A. & Roy, M. (1990). The tryptophan synthase bienzyme complex transfers indole between the α and β sites via a 25–30 Å long tunnel. *Biochemistry*, **29**, 8598–607.

Dunn, C. R., Wilks, H. M., Halsall, D. J., Atkinson, T., Clarke, A. R., Muirhead, H. & Holbrook, J. J. (1991). Design and synthesis of new enzymes based on the lactate dehydrogenase framework. *Philosophical Transactions of the Royal Society of London*, **B332**, 177–184.

Easton, D. M. (1978). Exponentiated exponential model (Gompertz kinetics) of Na^+ and K^+ conductance changes in squid giant axon. *Biophysical Journal*, **22**, 15–28.

Eaton, W. A., Henry, E. R. & Hofrichter, J. (1991). Application of linear free energy relations to protein conformational changes: The quaternary structural change of haemoglobin. *Proceedings of the National Academy of Sciences, USA*, **88**, 4472–5.

Eccleston, J. F., Gutfreund, H., Trentham, D. R. & Webb, M.R. (eds.) (1992). Nucleoside triphosphates in energy transduction, cellular regulation and information transfer in biological systems. *Philosophical Transactions of the Royal Society of London*, **Series B**, 3–112.

Edsall, J. T. & Gutfreund, H. (1983). *Bio-thermodynamics*. Chichester: John Wiley & Sons.

Eigen, M. (1954). Methods for investigation of ionic reactions in aqueous solutions with half times as short as 10^{-9} sec. Application to neutralization and hydrolysis reactions. *Discussions of the Faraday Society*, **17**, 194–205.

Eigen, M. (1963). Protonenübertragung, Saure-Base-Katalyse und enzymatische Hydrolyse.Teil I: Elementarvorgange. *Angewandte Chemie*, **75**, 489–508.

Eigen, M. (1966). Die Zeitmasstab der Natur. *Jahrbuch 1966 der Max-Planck-Gesellschaft*, 40–67.

Eigen, M. (1968). New looks and outlooks on physical enzymology. *Quarterly Reviews of Biophysics*, **1**, 3–33.

Eigen, M. (1971). Selforganization of matter and the evolution of biological macromolecules. *Naturwissenschaften*, **58**, 465–527.

Eigen, M. & de Maeyer, L. (1963). Relaxation methods. In *Techniques of Organic Chemistry*, vol. VIII/2, ed. S. L. Friess, E. S. Lewis & A. Weissberger, pp. 895–1054. New York: Wiley (Interscience).

Eigen, M. & de Maeyer, L. (1973). Theoretical basis of relaxation spectrometry. In *Techniques of Chemistry*, vol. VI/2, 3rd edn, ed. G. G. Hammes, pp. 90–143. New York: Wiley (Interscience).

Eigen, M. & Schuster, P. (1977). The hypercycle, a principle of natural self-organization Part A: emergence of the hypercycle. *Naturwissenschaften*, **64**, 541–65.

Eigen, M. & Schuster, P. (1978). The hypercycle, a principle of self-organization. Part B: the abstract hypercycle. *Naturwissenschaften*, **65**, 7–41.

Eigen, M., Kruse, W., Maas, G. & de Maeyer, L. (1964). Rate constants of protolytic reactions in aqueous solution. In *Progress in Reaction Kinetics*, vol. 2. Oxford: Pergamon Press.

Eigen, M, Kurtze, G. & Tamm, K. (1953). Zum Reactionsmechanismus der Ultraschallabsorption in wasserigen Electrolytlösungen. *Zeitschrift der Electrochemie*, **57**, 103–18.

Einstein, A. (1905). Eine Neue Bestimmung der Molekular Dimensionen. *Annalen der Physik*, **17**, 549–55.

Ellis, R. J. & van der Vies, S. M. (1991). Molecular chaperones. *Annual Review of Biochemistry*, **60**, 321–47.

Engelborghs, Y., Marsh, A. & Gutfreund, H. (1975). A quenched-flow study of the reaction catalysed by creatine kinase. *Biochemical Journal*, **151**, 47–50.

Englander, S. W. & Kallenbach, N. R. (1983). Hydrogen exchange and structural dynamics of proteins and nucleic acids. *Quarterly Reviews of Biophysics*, **16**, 521–655.

Epstein, H. F., Schechter, A. N., Chen, R. F. & Anfinsen, C. B. (1971). Folding of staphylococcal nuclease: kinetic studies of two processes in acid renaturation. *Journal of Molecular Biology*, **60**, 499–508.

Erickson, J., Goldstein, B., Holowka, D. & Baird, B. (1987). Effect of receptor density on the forward rate constant of ligand binding. *Biophysical Journal*, **52**, 657–62.

Fantania, H. R., Matthews, B. & Dalziel, K. (1982). On the mechanism of NADP$^+$-linked isocitrate dehydrogenase from heart mitrochondria. I. The kinetics of dissociation of NADPH from its enzyme complex. *Proceedings of the Royal Society of London*, B**214**, 369–87.

Faraday Society (1937). A general discussion on 'Reaction Kinetics'. *Transactions of the Faraday Society*, **33**.

Fersht, A. R. (1985). *Enzyme Structure and Mechanism*, 2nd edn. New York: Freeman.

Fierke, C.A., Johnson, K. A. & Benkovic, S. J. (1987). Construction and evaluation of the kinetic scheme associated with dihydrofolate reductase from *E. coli. Biochemistry*, **26**, 4085–97.

Finer, J. T., Simmons, R. M. & Spudlich, J. A. (1994). Single myosin molecule mechanics: piconewton forces and nanometer steps. *Nature (London)*, **368**, 113–19.

Finlayson, B. & Taylor, E. W. (1969). Hydrolysis of nucleoside triphosphates by myosin during the transient state. *Biochemistry*, **8**, 802–10.

Fischer, G. & Schmid, F. X. (1990). The mechanism of protein folding. Implications of *in vitro* refolding models for *de novo* protein folding and translocation in the cell. *Biochemistry*, **29**, 2205–13.

Forster, R. E., Edsall, J. T., Otis, A. B. & Roughton, F. J. W. (eds.) (1968). *CO_2: Chemical, Biochemical, and Physiological Aspects*. Washington, DC: National Aeronautics and Space Administration.

Förster, T. (1959). Transfer mechanisms of electronic excitation. *Discussions of the Faraday Society*, **27**, 7–17.

Forti, S., Menini, A., Rispoli, G. & Torre, V. (1989). Kinetics of phototransduction in retinal rods of the newt *Triturus cristatus. Journal of Physiology*, **419**, 265–95.

Frauenfelder, H. (1989). The Debye-Waller factor: from villain to hero in protein crystallography. *International Journal of Quantitative Chemistry*, **35**, 711–15.

Frauenfelder, H. & Wolynes, P. G. (1985). Rate theories and puzzles of hemeprotein kinetics. *Science*, **229**, 337–45.

Frauenfelder, H., Parak, F. & Young, R. D. (1988). Conformational substates in proteins. *Annual Review of Biophysics and Biochemistry*, **17**, 451–79.

Frauenfelder, H., Petsko, G. A. & Tsernoglou, D. (1979). Temperature-dependent X-ray diffraction as a probe of protein structure dynamics. *Nature (London)*, **280**, 558–63.

Fuortes, M. G. F. & Hodgkin, A. L. (1964). Changes in the time scale and sensitivity in the ommatidia of limulus. *Journal of Physiology*, **172**, 239–63.

Gavish, B (1986). Molecular dynamics and the transient strain model of enzyme catalysis. In *The Fluctuating Enzyme*, ed G. R. Welch. New York: John Wiley & Sons Inc.

Geeves, M. A. (1991). The dynamics of actin and myosin association and the crossbridge model of muscle contraction. *Biochemical Journal*, **274**, 1–14.

Geeves, M. A. & Gutfreund, H. (1982). The use of pressure perturbations to investigate rabbit muscle myosin subfragment 1 with actin in the presence of MgADP. *Federation of European Biochemical Societies Letters*, **140**, 11–15.

Geeves, M. A. & Halsall, D. J. (1987). Two-step ligand binding and cooperativity. *Biophysical Journal*, **52**, 215–20.

Geeves, M.A. & Ranatunga, K. W. (1987). Tension responses to increased hydrostatic pressure in glycerinated rabbit psoas muscle fibres. *Proceedings of the Royal Society of London*, B**232**, 217–26.

Geeves, M. A., Goody, R. S. & Gutfreund, H. (1984). Kinetics of acto-S1 interaction as a guide to a model for the crossbridge cycle. *Journal of Muscle Research and Cell Motility*, **5**, 351–61.

Geraci, G. & Gibson, Q. H. (1967). The reaction of liver dehydrogenase with reduced diphosphopyridine nucleotide. *Journal of Biological Chemistry*, **242**, 4275–8.

Gerwert, K. (1993). Molecular reaction mechanisms of proteins as monitored by time-resolved FTIR spectroscopy. *Current Opinion in Structural Biology*, **3**, 769–73.

Getzoff, E. D., Cabelli, D. E., Fisher, C. L., Parge, H. E., Viezolli, M. S., Banci, L. & Hallewell, R. A. (1992). Faster superoxide dismutase mutants designed by enhancing electrostatic guidance. *Nature (London)*, **358**, 347–51.

Getzoff, E. D., Tainer, J. A., Weiner, P. K., Kollman, P. A., Richardson, J. S. & Richardson, D. C. (1983). Electrostatic recognition between superoxide and copper, zinc superoxide dismutase. *Nature (London)*, **306**, 287–90.

Gibbons, B. H. & Edsall, J. T. (1963). Rate of hydration of carbon dioxide and dehydration of carbonic acid at 25°. *Journal of Biological Chemistry*, **238**, 3502–7.

Gibson, Q. H. (1956). An apparatus for flash photolysis and its application to the reactions of myglobin with gases. *Journal of Physiology*, **134**, 112–22.

Gibson, Q. H. (1990). Kinetics of ligand binding to haemoproteins. *Biochemical Society Transactions*, **18**, 1–6.

Gibson, Q. H. & Milnes, L. (1964). Apparatus for rapid and sensitive spectrophotometry. *Biochemical Journal*, **91**, 161–71.

Gibson, Q. H. & Roughton, F. J. W. (1955). The kinetics of dissociation of the first oxygen molecule from fully saturated oxyhaemoglobin in sheep blood solutions. *Proceedings of the Royal Society of London*, **B143**, 310–26.

Gibson, Q. H. & Roughton, F. J. W. (1957). The determination of dissociation of the velocity constants of the four successive reactions of carbon monoxide with sheep haemoglobin. *Proceedings of the Royal Society of London*, **B146**, 206–24.

Glaser, L. & Brown, D. H. (1955). Purification and properties of the D-glucose-6-phosphate dehydrogenase. *Journal of Biological Chemistry*, **216**, 67–79.

Glasstone, S., Laidler, K. J. & Eyring, H. (1941). *The Theory of Rate Processes*. New York: McGraw-Hill.

Goldberg, M. E., Rudolph, R. & Jaenike, R (1991). A kinetic study of the competition between renaturation and aggregation during the refolding of denatured-reduced eggwhite lysozyme. *Biochemistry*, **30**, 2790–7.

Goldman, Y. E., Hibberd, M. G., McCray, J. A. & Trentham, D. R. (1982). Relaxation of muscle fibres by photolysis of caged ATP. *Nature (London)*, **300**, 701–5.

Goldstein, B., Posner, R. G., Torney, D. C., Erikson, J., Holowka, D. & Baird, B. (1989). Competition between solution and cell surface receptors for ligand. *Biophysical Journal*, **56**, 955–66.

Gompertz, B. (1825). On the nature of the function expressive of the law of human mortality and on a new mode of determining the value of life contingencies. *Philosophical Transations of the Royal Society of London*, **2**, 513–85.

Graham, T. (1862). The rate of diffusion of small ions in gelatin. *Justus Liebigs Annalen der Chemie*, **21**, 1–70.

Gratton, E., Jameson, D. M. & Hall, R. D. (1984). Multifrequency phase and modulation fluorometry. *Annual Review of Biophysics and Bioengineering*, **13**, 105–24.

Guggenheim, E. A. (1926). On the determination of the velocity constant of a unimolecular reaction. *Philosophical Magazine*, **Series 7, 2**, 538–43.

Gurney, A. M. & Lester, H. A. (1987). Light-flash physiology with synthetic photosensitive compounds. *Physiological Reviews*, **67**, 583–617.

Gutfreund, H. (1955). Steps in the formation and decomposition of some enzyme-substrate complexes. *Faraday Society Discussions*, **20**, 1–8.

Gutfreund, H. (1959a). Structural and stereospecificity. In *The Enzymes*, 2nd edn, vol. 1, ed. P. D. Boyer, H. Lardy & K. Myrback, pp. 233–58. New York: Academic Press.

Gutfreund, H. (1959b). The reactions of reduced flavine nucleotides with oxygen. *Biochemical Journal*, **74**, 17p.

Gutfreund, H. (1965). *An Introduction to the Study of Enzymes*. Oxford: Blackwell Scientific Publications.

Gutfreund, H. (1972). *Enzymes: Physical Principles*. London: Wiley-Interscience.

Gutfreund, H. (1975). Kinetic analysis of the properties and reactions of enzymes. *Progress in Biophysics and Molecular Biology*, **29**, 161–95.

Gutfreund, H. (1987). Reflections on the kinetics of substrate binding. *Biophysical Chemistry*, **26**, 117–21.

Gutfreund, H. & Chock, P. B. (1991). Substrate channeling among glycolytic enzymes: fact or fiction. *Journal of Theoretical Biology*, **152**, 117–21.

Gutfreund, H. & Knowles, J. R. (1967). The foundations of enzyme action. *Essays in Biochemistry*, **3**, 25–72.

Gutfreund, H. & Sturtevant, J. M. (1956). The mechanism of chymotrypsin-catalysed reactions. *Proceedings of the National Academy of Sciences, USA*, **42**, 719–28.

Gutfreund, H & Sturtevant, J. M. (1959). Steps in the oxidation of xanthine to uric acid catalysed by milk xanthine oxidase. *Biochemical Journal*, **73**, 1–6.

Gutfreund, H. & Trentham, D. R. (1975). Energy changes during the formation and interconversion of enzyme-substrate complexes. In *Energy Transformation in Biological Systems*, Ciba Foundation Symposium 31 (new series). Amsterdam: Elsevier, Excerpta Medica, North-Holland.

Hajdu, J. & Andersson, I. (1993). Photoactivation of caged enzymes. *Annual Review of Biophysics and Biomolecular Structure*, **22**, 467–98.

Hajdu, J. & Johnson, L. N. (1990). Progress with Laue diffraction studies on protein and virus crystals. *Biochemistry*, **29**, 1669–78.

Haldane, J. B. S. (1930). *Enzymes*. Cambridge, MA: The MIT Press.

Halford, S. E. (1972). *Escherichia coli* alkaline phosphatase. *Biochemical Journal*, **126**, 311–22.

Halford, S. E. (1975). Stopped-flow fluorescence studies on saccharide binding to lysozyme. *Biochemical Journal*, **149**, 411–22.

Halford, S. E., Bennett, N. G., Trentham, D. R. & Gutfreund, H. (1969). Substrate induced conformation change in the reaction of alkaline phosphatase from *E. coli*. *Biochemical Journal*, **114**, 243–51.

Halford, S. E., Johnson, N. P. & Grinsted, J. (1979). Restriction enzyme two step hydrolysis of circular DNA. *Biochemical Journal*, **179**, 353–65.

Hall, R. L. (1982). Genetics of the nervous system in *Drosophila*. *Quarterly Reviews of Biophysics*, **15**, 223–480.

Hall, R. L., Vennesland, B. & Kczdy, F. J. (1969). Glyoxylate carboligase of *E. coli*: identification of carbon dioxide as the primary product. *Journal of Biological Chemistry*, **244**, 3991–8.

Hammes, G. G. (1974). Temperature-jump methods. In *Investigation of Rates and Mechanisms of Reactions; part II*, 3rd edn, ed. G. G. Hammes, pp. 147–85. New York: John Wiley & Sons.

Hammond, B. R. & Gutfreund, H. (1955). Two steps in the reaction of chymotrypsin with acetyl-L-phenylalanine ethyl ester. *Biochemical Journal*, **61**, 187–9.

Hartridge, H. (1923). Coincidence method for the wave length measurement of absorption bands. *Proceedings of the Royal Society of London*, A**102**, 575–87.

Hartridge, H. & Roughton, F. J. W. (1923a). A method for measuring the velocity of very rapid chemical reactions. *Proceedings of the Royal Society of London*, A**104**, 376–94.

Hartridge, H. & Roughton, F. J. W. (1923b). The kinetics of haemoglobin. II. The velocity with which oxygen dissociates from its combination with haemoglobin. *Proceedings of the Royal Society of London*, A**104**, 395–430.

Hastings, J. W., Mitchell, G., Mattingly, P. H., Blinks, J. R. & van Leeuwen, M. (1969). Response of aequorin bioluminescence to rapid changes in calcium concentration. *Nature (London)*, **222**, 1047–50.

Helmholtz, H. L. F. (1850). Measurements on the time of twitching of animal muscles and the velocity of propagation of nerve impulses. *Müllers Archiv für Anatomie und Wissenschaftliche Medizin*, 276–364.

Helmholtz, H. L. F. (1856). *Handbook of Physiological Optics*.

Helmholtz, H. L. F. (1954). *On the Sensations of Tone*, 2nd English edn. New York: Dover Publications, Inc.

Henderson, L. J. (1958). *The Fitness of the Environment*. Boston: Beacon Press.

Herbert, D., Elsworth, R. & Telling, R. C. (1956). The continuous culture of bacteria; a theoretical and experimental study. *Journal of General Microbiology*, **14**, 601–22.

Hess, G. P., Udgaonkar, J. B. & Olbricht, W. L. (1987). Chemical kinetic measurements of transmembrane processes using rapid reaction techniques. *Annual Review of Biophysics and Biophysical Chemistry*, **16**, 507–34.

Hessling, B., Souvignier, G. & Gerwert, K. (1993). A model independent approach to assigning bacteriorhodopsin's intramolecular reactions to photocycle intermediates. *Biophysical Journal*, **65**, 1929–41.

Hibberd, M. G. & Trentham, D. R. (1986). Relationships between chemical and mechanical events during muscular contractions. *Annual Review of Biophysics and Bioengineering*, **15**, 119–62.

Hill, A. V. (1909). The mode of action of nicotine and curari, determined by the form of the contraction curve and the method of temperature coefficients. *Journal of Physiology*, **39**, 361–73.

Hill, A. V. (1934). Discussion on methods of measuring and factors determining the speed of chemical reactions. *Proceedings of the Royal Society of London*, B**116**, 185–207.

Hill, A. V. (1948). On the time required for diffusion and its relation to processes in muscle. *Proceedings of the Royal Society of London*, B**135**, 446–53.

Hill, A. V. (1965). *Trails and Trials in Physiology*. Baltimore: Williams & Wilkins Co.

Hill, T. L. (1977). *Free Energy Transduction in Biology*. New York: Academic Press.

Hille, B. (1992). *Ionic Channels of Excitable Membranes*, 2nd edn. Sunderland, MA: Sinauer Associates Inc.

Hinshelwood, C. N. (1946). *The Chemical Kinetics of the Bacterial Cell*. Oxford: Clarendon Press.

Hirschfelder, J. O. (1983). My adventures in theoretical chemistry. *Annual Review of Physical Chemistry*, **34**, 1–29.

Ho, C. & Sturtevant, J. M. (1963). The kinetics of the hydration of carbon dioxide at 25°. *Journal of Biological Chemistry*, **238**, 3499–501.

Hodgkin, A. L. (1973). Address of the president at the anniversary meeting, 30.11.1972. *Proceedings of the Royal Society of London*, **B183**, 1–19.

Hodgkin, A. (1992). *Chance & Design*. Cambridge: Cambridge University Press.

Hofrichter, J., Sommer, J. H., Henry, E. R. & Eaton, W. A. (1983). Nanosecond absorption spectroscopy of haemoglobin: elementary processes in kinetic cooperativity. *Proceedings of the National Academy of Sciences, USA*, **80**, 2235–9.

Holbrook, J. J. (1972). Protein fluorescence of lactate dehydrogenase. *Biochemical Journal*, **128**, 921–31.

Holbrook, J. J. & Gutfreund, H. (1973). Approaches to the study of enzyme mechanisms: lactate dehydrogenase. *Federation of European Biochemical Societies Letters*, **31**, 157–69.

Holler, E., Rupley, J. A. & Hess, G. P. (1969). Kinetics of lysozyme-substrate interactions. *Biochemical and Biophysical Research Communications*, **37**, 423–9.

Hood, J. J. (1885). On the influence of heat on the rate of chemical change *Philosophical Magazine*, **20** (**5th series**), 323–8.

Horowitz, P. & Hill, W. (1980). *The Art of Electronics*. Cambridge: Cambridge University Press.

Huang, C. Y. (1979). Derivations of initial velocity and isotope exchange rate equations. *Methods in Enzymology*, **63**, 55–85.

Hull, W. E., Halford, S. E., Gutfreund, H. & Sykes, B. D. (1976). ^{31}P nuclear magnetic resonance study of alkaline phosphatase: the role of inorganic phosphate in limiting the enzyme turnover rate at alkaline pH. *Biochemistry*, **15**, 1547–61.

Hutchinson, G. E. (1978). *An Introduction to Population Ecology*. New Haven: Yale University Press.

Huxley, A. F. & Simmons, R. M. (1971). Proposed mechanism of force generation in striated muscle. *Nature (London)*, **233**, 533–8.

Huxley, H. E. (1969). The mechanism of muscular contraction. *Science*, **164**, 1356–66.

Huxley, H. E. (1990). Sliding filaments and molecular motile systems. *Journal of Biological Chemistry*, **265**, 8347–50.

Huxley, H. E. & Faruqi, A. R. (1983). Time-resolved X-ray diffraction studies on vertebrate striated muscle. *Annual Review of Biophysics and Bioengineering*, **12**, 381–417.

Huxley, H. E., Simmons, R. M., Faruqi, A. R., Krebs, M. & Bordas, J. (1983). Changes in X-ray reflections from contracting muscle during rapid mechanical transients and their structural implications. *Journal of Molecular Biology*, **169**, 469–506.

Hyde, C. C., Ahmed, S. A., Padlan, E. A., Miles, E. W. & Davies, D. R. (1988). Three-dimensional structure of the tryptophan synthase $\alpha_2\beta_2$ multienzyme complex from *Salmonella typhimurium*. *Journal of Biological Chemistry*, **263**, 17857–71.

Hynes, J. T. (1985). Chemical reaction dynamics in solution. *Annual Review of Physical Chemistry*, **36**, 573–97.

Irving, M., Lombardi, V. & Piazzesi, G. (1992). Myosin head movements are synchronous with the elementary force generating process in muscle. *Nature*

(London), **357**, 156–8.

Ishijima, A., Doi, T., Sakurada, K. & Yanagida, T. (1991). Sub-pico newton force fluctuations of actomyosin in vitro. *Nature (London)*, **352**, 301–6.

Jackson, A. P., Timmerman, M. P., Bagshaw, C. R. & Ashley, C. C. (1987). The kinetics of calcium binding to fura-2 and indo-1. *Federation of European Biochemical Societies Letters*, **216**, 35–9.

Jaenike, R. (1991). Protein folding: local structures, domains, subunits and assemblies. *Biochemistry*, **30**, 3147–61.

Jameson, D. M. & Hazlett, T. L. (1991). Time resolved fluorescence in biology and biochemistry. In *Biophysical and Biochemical Aspects of Fluorescence Spectroscopy*, ed. T. G. Dewey, pp. 105–33. New York: Plenum Publishing Corporation.

Jameson, D. M. & Sawyer, W. H. (1994). Fluorescence anisotropy applied to bimolecular interactions. *Methods in Enzymology*, **in press**.

Jencks, W. P. (1969). *Catalysis in Chemistry and Enzymology*. New York: McGraw-Hill.

Jencks, W. P. (1989). How does a calcium pump pump calcium? *Journal of Biological Chemistry*, **264**, 18855–8.

Johnson, F. H., Eyring, H. & Stover, B. J. (1974). *The Theory of Rate Processes in Biology and Medicine,*. New York: John Wiley & Sons.

Johnson, I. A. & Goldspink, G. (1975). Thermodynamic activation parameters of fish myofibrillar ATPase enzyme and evolutionary adaptations to temperature. *Nature (London)*, **257**, 620–2.

Jones, C. M., Ansari, A., Henry, E. R., Christoph, G. W., Hofrichter, J. & Eaton, W. A. (1992). Speed of intersubunit communication in proteins. *Biochemistry*, **31**, 6692–702.

Josephs, R. & Harrington, W. F. (1968). On the stability of myosin filaments. *Biochemistry*, **7**, 2834–47.

Kahlert, M. & Hofmann, K. P. (1991). Reaction rate and collisional efficiency of the rhodopsin system in intact retinal rods. *Biophysical Journal*, **59**, 375–86.

Kao, J. P. Y. & Tsien, R. Y. (1987). Ca^{2+} binding kinetics of fura-2 and azo-1 from temperature jump relaxation measurements. *Biophysical Journal*, **53**, 635–9.

Kaplan, J. H. (1990). Photochemical manipulation of divalent cation levels. *Annual Review of Physiology*, **52**, 897–914.

Kaplan, J. H. & Ellis-Davies, G. C. R. (1988). Photolabile chelators for rapid photorelease of divalent cations. *Proceedings of the National Academy of Sciences, USA*, **85**, 6571–5.

Kaplan, J. H., Forbush, B. & Hoffmann, J.F. (1978). Rapid photolytic release of adenosine 5′ triphosphate from a protected analogue: utilization by the Na:K pump of human red cells. *Biochemistry*, **17**, 1929–36.

Karplus, M. & McCammon, J. A. (1983). Dynamics of proteins: elements and function. *Annual Review of Biochemistry*, **52**, 263–300.

Karplus, M. & McCammon, J. A. (1986). The dynamics of proteins. *Scientific American*, **254 No 4**, 42–67.

Katchalski-Katzir, E., Hass, E. & Steinberg, I. Z. (1981). Intramolecular mobility of polypeptides in solution. *Annals of the New York Academy of Science*, **366**, 44–61.

Kauzmann, W. (1959). Factors in the interpretation of protein denaturation. *Advances in Protein Chemistry*, **14**, 1–63.

Keilin, D. (1966). *The History of Cell Respiration and Cytochrome*. Cambridge: Cambridge University Press.

Keilin, D. & Mann, T. (1937). On the haematin compound of peroxidase. *Proceedings of the Royal Society of London*, **B122**, 119–33.

Khundar, L. R. & Zewail, A. H. (1990). Ultrafast molecular reaction dynamics in real-time: progress over a decade. *Annual Review of Physical Chemistry*, **41**, 15–60.

Kihara, H., Takahashi-Ushijima, E., Amemiya, Y., Honda, Y., Vachette, P., Tauc, P., Barman, T. E., Jones, P. T. & Moody, M. F. (1987). Kinetics of structure and activity changes during the allosteric transition of aspartate transcarbamylase. *Journal of Molecular Biology*, **198**, 745–8.

Kim, P. S. & Baldwin, R. L. (1982). Specific intermediates in the folding reactions of small proteins and the mechanism of protein folding. *Annual Review of Biochemistry*, **51**, 459–89.

King, E. L. & Altman, C. (1956). A schematic method of deriving the rate laws for enzyme-catalysed reactions. *Journal of Physical Chemistry*, **60**, 1375–88.

King, P. J. & Gutfreund, H. (1984). Kinetic studies on the formation and decay of metarhodopsins from bovine retinas. *Vision Research*, **24**, 1471–5.

Kirschner, K., Eigen, M., Bittman, R. & Voigt, B. (1966). The binding of nicotinamide-adenine dinucleotide to yeast glyceraldehyde 3-phosphate dehydrogenase: temperature jump relaxation studies on the mechanism of an allosteric enzyme. *Proceedings of the National Academy of Sciences, USA*, **56**, 1661–7.

Kleutsch, B. & Frehland, E. (1991). Monte-Carlo-simulations of voltage fluctuations in biological membranes in the case of small numbers of transport units. *European Biophysics Journal*, **19**, 203–11.

Knoche, W. (1974). Pressure-jump methods. In *Investigation of Rates and Mechanisms of Reactions; part II*, 3rd edn, ed. G. G. Hammes, pp. 187–210. New York: John Wiley & Sons.

Knowles, J. R. (1976). Whither enzyme mechanisms?. *Federation of European Biochemical Society Letters*, **62** Supplement, E53–61.

Knowles, J. R. (1980). Enzyme-catalysed phosphoryl transfer reactions. *Annual Review of Biochemistry*, **49**, 877–919.

Knowles, J. R. (1989). The mechanism of biotin-dependent enzymes. *Annual Review of Biochemistry*, **58**, 195–221.

Knowles, J. R. (1991). To build an enzyme.... *Philosophical Transactions of the Royal Society of London*, **B332**, 115–21.

Knowles, J. R. & Albery, W. J. (1977). Perfection in enzyme catalysis: the energetics of triosephosphate isomerase. *Accounts of Chemical Research*, **10**, 105–11.

Kohl, B. & Hofmann, K. P. (1987). Temperature dependence of G-protein activation in photoreceptor membranes. *Biophysical Journal*, **52**, 271–87.

Kramer, H. A. (1940). Brownian motion in a field of force. *Physica (Utrecht)*, **7**, 284–304.

Krebs, H. A. & Roughton, F. J. W. (1948). Carbonic anhydrase as a tool in studying the mechanism of reactions involving H_2CO_3, CO_2, or HCO_3^-. *Biochemical Journal*, **43**, 550–5.

Kvassman, J. & Pettersson, G. (1979). Effect of pH on coenzyme binding to liver alcohol dehydrogenase. *European Journal of Biochemistry*, **100**, 115–23.

Kvassman, J. & Pettersson, G. (1989a). Evidence that 1,3-bisphosphoglycerate dissociation from phosphoglycerate kinase is an intrinsically rapid reaction step. *European Journal of Biochemistry*, **186**, 261–4.

Kvassman, J. & Pettersson, G. (1989b). Mechanism of 1,3-bisphosphoglycerate transfer from phosphoglycerate kinase to glyceraldehyde-3-phosphate dehydrogenase. *European Journal of Biochemistry*, **186**, 265–72.

Kvassman, J., Pettersson, G. & Ryde-Pettersson, U. (1988). Mechanism of glyceraldehyde-3-phosphate transfer from aldolase to glyceraldehyde-3-phosphate dehydrogenase. *European Journal of Biochemistry*, **172**, 427–31.

Lakowicz, J. R. (1983). *Principles of Fluorescence Spectroscopy*. New York: Plenum.

Lakowicz, J. R. & Weber, G. (1973a). Quenching of fluorescence by oxygen: a probe for structural fluctuations in macromolecules. *Biochemistry*, **12**, 4160–70.

Lakowicz, J. R. & Weber, G. (1973b). Quenching of protein fluorescence by oxygen: detection of structural fluctuations in proteins on the nanosecond timescale. *Biochemistry*, **12**, 4171–9.

Lamb, T. D. (1984). Electrical response of photoreceptors. In *Recent Advances in Physiology*, no. 10, ed. P. F. Baker. Edinburgh: Churchill Livingstone.

Lamb, T. D. & Pugh, E. N. Jr. (1992). A quantitative account of the activation steps involved in phototransduction in amphibian photoreceptors. *Journal of Physiology*, **449**, 719–58.

Langmuir, I. (1918). The adsorption of gases on plane surfaces of glass, mica and platinum. *Journal of the American Chemical Society*, **40**, 1361–403.

Leatherbarrow, R. J. (1992). *GraFit Version 3.0*. Staines, U. K.: Erithacus Software Ltd.

Levy, H. M., Sharon, N. S. & Koshland, D. E. (1959). The temperature dependence of the walking rate of ants. *Proceedings of the National Academy of Sciences, USA*, **45**, 758–9.

Liebman, P. A., Jagger, W. S., Kaplan, M. W. & Bargoot, F. G. (1974). Membrane structure changes in the rod outer segments associated with rhodopsin bleaching. *Nature (London)*, **251**, 31–6.

Lotka, A. J. (1956). *Elements of Mathematical Biology*. New York: Dover Publications, Inc.

Lowry, T. M. & John, W. T. (1910). Studies of dynamic isomerism. Part XII. The equations for two consecutive unimolecular changes. *Journal of the Chemical Society*, **97**, 2634–45.

Lymn, R. W. & Taylor, E. W. (1970). Transient state phosphate production in the hydrolysis of nucleotide triphosphates by myosin. *Biochemistry*, **9**, 2975–83.

Mannherz, H. G., Schenck, H. & Goody, R. S. (1974). Synthesis of ATP from ADP and inorganic phosphate at the myosin-subfragment I active site. *European Journal of Biochemistry*, **48**, 287–95.

Margenau, H. & Murphy, G. M. (1943). *The Mathematics of Physical Chemistry*. New York: D. van Nostrand Co. Inc.

Marquardt, D. W. (1963). An algorithm for least-squares estimation of nonlinear parameters. *Journal of the Society of Industrial and Applied Mathematics*, **11**, 431–41.

Martin, S. R., Schilstra, M. J. & Bayley, P. M. (1993). Dynamic instability of microtubules: Monte Carlo simulation and application to different types of microtubule lattice. *Biophysical Journal*, **65**, 578–96.

Matouschek, A., Kellis, J. T., Serrano, L., Bycroft, M. & Fersht, A. R. (1990). Folding intermediates characterized by protein engineering. *Nature (London)*, **346**, 440–5.

Matouschek, A., Kellis, J. T., Serrano, L. & Fersht, A. R. (1989). Mapping the transition state and pathway of protein folding by protein engineering. *Nature (London)*, **340**, 122–6.

Matsubara, N., Billington, A. P. & Hess, G. P. (1992). How fast does an acetylcholine receptor channel open? Laser-pulse photolysis of an inactive precursor of carbamoyl choline in the microsec time region with BC3H1cells. *Biochemistry*, **31**, 5507–14.

Maynard Smith, J. (1968). *Mathematical Problems in Biology*. Cambridge: Cambridge University Press.

Maynard Smith, J. (1974). *Models in Ecology*. Cambridge: Cambridge University Press.

McCalla T. R. (1967). *Introduction to Numerical Methods and FORTRAN Programming*. New York: John Wiley & Sons Inc.

McCammon, J. A. & Harvey, S. C. (1987). *Dynamics of Proteins and Nucleic Acids*. Cambridge: Cambridge University Press.

McCray, J. A. & Smith, P. D. (1977). Application of lasers to molecular biology. *Laser Applications*, **3**, 1–25. New York: Academic Press.

McCray, J. A. & Trentham, D. R. (1989). Properties and uses of photoreactive caged compounds. *Annual Review of Biophysics and Biophysical Chemistry*, **18**, 239–70.

McCray, J. A., Herbette, L., Kihara, T. & Trentham, D. R. (1980). A new approach to time-resolved studies of ATP-requiring biological systems: Laser flash photolysis of caged ATP. *Proceedings of the National Academy of Sciences, USA*, **77**, 7237–41.

McKillop, D. F. A. & Geeves, M. A. (1993). Regulation of the interaction between actin and myosin subfragment 1: evidence for three states of the filament. *Biophysical Journal*, **65**, 693–701.

Michaelis, L. & Menten, M. L. (1913). Die Kinetik der Invertinwirkung [The kinetics of invertin activity]. *Biochemische Zeitschrift*, **49**, 333–69.

Middleford, C. F., Gupta, R. K. & Rose, I. A. (1976). Fructose 1,6 bisphosphate: isomeric composition, kinetics and substrate specificity of aldolase. *Biochemistry*, **15**, 2178–87.

Milburn, T., Matsubara, N., Billington, A. P., Udgaonkar, J. B., Walker, J. W., Carpenter, B. K., Webb, W. W., Marque, J., Denk, W., McCray, J. A. & Hess, G. P. (1989). Synthesis, photochemistry, and biological activity of a caged photolabile acetylcholine receptor ligand. *Biochemistry*, **28**, 49–55.

Millar, N. C. & Geeves, M. A. (1988). Protein fluorescence changes associated with ATP and adenosine 5'[γ-thio] triphosphate. *Biochemical Journal*, **249**, 735–43.

Millhauser, G. L , Salpeter, E. E. & Oswald, R. E. (1988). Diffusion models of ion-channel gating and the origin of power-law distributions from single-channel recording. *Proceedings of the National Academy of Sciences, USA*, **85**, 1503–7.

Mills, D. R. , Peterson, R. L. & Spiegelman, S. (1967). An extracellular Darwinian experiment with a self duplicating nucleic acid molecule. *Proceedings of the National Academy of Sciences, USA*, **58**, 217–24.

Moffat, K. A. (1989). Time-resolved macromolecular crystallography. *Annual Review of Biophysics and Biophysical Chemistry*, **18**, 309–32.

Mollon, J. D., Bowmaker, J. K. & Jacobs, G. H. (1984). Variations of colour vision in a New World primate can be explained by polymorphism of retinal photopigments. *Proceedings of the Royal Society of London*, **B222**, 373–99.

Monod, J. (1950). La technique de culture continue; theorie et applications. *Annals Institut Pasteur*, **79**, 390–401.

Monod, J., Changeux, J.-P. & Jacob, F. (1963). Allosteric proteins and cellular control systems. *Journal of Molecular Biology*, **6**, 306–29.

Monod, J., Wyman, J. & Changeux, J.-P. (1965). On the nature of allosteric transitions: a plausible model. *Journal of Molecular Biology*, **12**, 88–103.

Moore, W. J. & Pearson, R. G. (1981). *Kinetics & Mechanisms*, 3rd edn. New York: John Wiley & Sons Inc.

Moser, C. C., Keske, J. M., Warnicke, K., Farid, R. S. & Dutton, P. L. (1992). Nature of biological electron transfer. *Nature (London)*, **355**, 796–802.

Mukamel, S. (1990). Femtosecond optical spectroscopy: a direct look at elementary chemical events. *Annual Review of Physical Chemistry*, **41**, 647–81.

Mulkey, R. M. & Zucker, R. S. (1991). Action potential must admit calcium to evoke transmitter release. *Nature (London)*, **350**, 153–5.

Munro, I., Pecht, I. & Stryer, L. (1979). Nanosecond motions of tryptophan residues in proteins. *Proceedings of the National Academy of Sciences, USA*, **76**, 56–60.

Nagy, K. (1991). Biophysical processes in invertebrate photoreceptors: recent progress and a critical overview based on Limulus photoreceptors. *Quarterly Review of Biophysics*, **24**, 165–226.

Nolte, H.-J., Rosenberry, T. L. & Neumann, E. (1980). Effective charge on acetylcholinesterase active sites determined from the ionic strength dependence of association rate constants with cationic ligands. *Biochemistry*, **19**, 3705–11.

Norrish, R. G. W. & Porter, G. (1954). The application of flash techniques to the study of fast reactions. *Discussions of the Faraday Society*, **17**, 40–6.

Northrup, S. H. & McCammon, J. A. (1984). Gated reactions. *Journal of the American Chemical Society*, **106**, 930–4.

Novick, A. (1955). Growth of bacteria. *Annual Review of Microbiology*, **9**, 99–110.

Novick, A. & Szilard, L. (1950). Experiments with the chemostat on spontaneous mutation of bacteria. *Proceedings of the National Academy of Sciences, USA*, **36**, 708–19.

O'Connor, D. V. & Phillips, D. (1984) *Time Correlated Single Photon Counting*. London: Academic Press.

Ogden, D. (1988). Answer in a flash. *Nature (London)*, **336**, 16–17.

Olson, J. S., Ballon, D. P., Palmer, G. & Massey, V. (1974). The mechanism of xanthine oxidase. *Journal of Biological Chemistry*, **249**, 4363–82.

Pais, A. (1982). '*Subtle is the Lord. . .*'. *The Science and Life of Albert Einstein*. Oxford: Clarendon Press.

Parkes, J. H. & Liebman, P. A. (1984). Temperature and pH dependence of the metarhodpsin I-metarhodopsin II kinetics and equilibria in bovine rod disk membrane suspensions. *Biochemistry*, **23**, 5054–61.

Parnas, I., Parnas, H. & Dudel, J. (1986). Neurotransmitter release and its facilitation in crayfish. *Pfluegers Archive European Journal of Physiology*, **406**, 121–30.

Perham, R. N., Duckworth, H. W. & Roberts, G. C. K. (1981). Mobility of polypeptide chain in the pyruvate dehydrogenase complex revealed by proton NMR. *Nature (London)*, **292**, 474–7.

Permyakov, E. A., Ostrovsky, A. V. & Kalinichenko, L. P. (1987). Stopped-flow kinetic studies of Ca(II) and Mg(II) dissociation in cod paralbumin and bovine α-lactalbumin. *Biophysical Chemistry*, **28**, 225–33.

Perutz, M. F. (1987). Molecular anatomy, physiology, and pathology of hemoglobin. In *Hemoglobin*, ed. G. Stamatoyannopoulos, A. W. Nienhuis, P. Leder & P. W. Majerus, pp. 127–78. Philadelphia: W. B. Saunders Co.

Perutz, M. F. (1989). Mechanisms of cooperativity and allosteric regulation in proteins. *Quarterly Review of Biophysics*, **22**, 139–236.

Perutz, M. (1992). *Protein Structure: New Approaches to Disease and Therapy.* New York: W. H. Freeman and Company.

Perutz, M. F. & Matthews, F. S. (1966). An X-ray study of azide methaemoglobin. *Journal of Molecular Biology*, **21**, 199–202.

Petrich, J. W., Lamberg, J. C., Balasubramania, S., Lambright, D. G., Boxer, S. G. & Martin, J. L. (1994). Ultrafast measurements of geminate recombination of NO with site specific mutant human myoglobin. *Journal of Molecular Biology*, **238**, 437–44.

Pettersson, G. (1976). The transient-state kinetics of two substate enzyme systems operating by an ordered ternary complex mechanism. *European Journal of Biochemistry*, **69**, 273–8.

Pettersson, G. (1989). Effect of evolution on the kinetic properties of enzymes. *European Journal of Biochemistry*, **184**, 561–6.

Pettersson, G. (1991). No convincing evidence is available for metabolite channelling between enzymes forming dynamic complexes. *Journal of Theoretical Biology*, **152**, 65–9.

Phillips, H. O., Marcinowsky, A. E., Sachs, S. B. & Kraus, K. A. (1977). Properties of organic-water mixtures. 11. Self diffusion coefficients of Na^+ in polyethylene glycol – water mixtures at 25 °C. *Journal of Physical Chemistry*, **81**, 679–83.

Phillips, L. A., Webb, S. P., Martinez, S. J., Phleming, G. R. & Levy, D. H. (1988). Time resolved spectroscopy of tryptophan conformers in a supersonic jet. *Journal of the American Chemical Society*, **110**, 1352–5.

Pielou, E. C. (1969). *An Introduction to Mathematical Ecology.* New York: Wiley Interscience.

Pompliano, D. L., Peyman, A. & Knowles, J. R. (1990). Stabilization of a reaction intermediate as a catalytic device: definition of the functional role of the flexible loop in triosephosphate isomerase. *Biochemistry*, **29**, 3186–94.

Porras, A. G., Olson, J. S. & Palmer, G. (1981). Reaction of reduced xanthine oxidase with oxygen. *Journal of Biological Chemistry*, **256**, 9096–103.

Porter, G. (1967). Quick as a flash. *Proceedings of the Royal Institution*, **42 no. 196**, 193–206.

Press, W. H., Flannery, B. P., Teukolsky, S. A. & Vetterling, W. T. (1986). *Numerical Recipes.* Cambridge: Cambridge University Press.

Prigogine, I. & Stengers, I (1984). *Order out of Chaos.* London: Heinemann.

Prinz, H. & Striessnig, J. (1993). Ligand-induced accelerated dissociation of (+)-cis-Diltiazem from L-type Ca^{2+} is simply explained by competition for individual attachment points. *Journal of Biological Chemistry*, **268**, 18580–5.

Privalov, P. (1982). Stability of proteins:proteins which do not present a single cooperative system. *Advances in Protein Chemistry*, **35**, 1–103.

Prod'hom, B., Pietrobon, D. & Hess, P. (1987). Direct measurement of proton transfer rates to a group controlling the dihydropyridine-sensitive Ca_2 channel. *Nature (London)*, **329**, 243–6.

Pryse, K. M., Bruckman, T. G., Maxfield, B. W. & Elson, E. L. (1992). Kinetics & mechanism of the folding of cytochrome C: using repetitive pressure perturbation. *Biochemistry*, **31**, 5127–36.

Radford, S. E., Dobson, C. M. & Evans, P. A. (1992). The folding of hen lysozyme involves partially structured intermediates and multiple pathways. *Nature (London)*, **358**, 302–7.

Randazzo, P. A., Kahn, R. A. & Northup, J. K. (1991). Activation of a small

GTP-binding protein by nucleoside diphosphate kinase. *Science*, **254**, 850–3.

Randazzo, P. A., Kahn, R. A. & Northup, J. K. (1992). Nucleoside diphosphate kinase: conclusion withdrawn. *Science*, **257**, 862.

Rapp, G. & Guth, K. (1988). A low cost high intensity flash device for photolysis experiments. *Pfluegers Archive European Journal of Physiology*, **411**, 200–3.

Rasmussen, B. F., Stock, A. M., Ringe, D. & Petsko, G. A. (1992). Crystalline ribonuclease loses function below the dynamical transition at 220K. *Nature (London)*, **357**, 423–4.

Rayment, I., Rypniewski, W.R., Schmidt-Base, K., Smith, R., Tomchick, D. R., Benning, M. M., Winkelmann, D. A., Wesenberg, G. & Holden, H. (1993). Three-dimensional structure of myosin subfragment-1: a molecular motor. *Science*, **261**, 50–64.

Reynolds, S. J., Yates, D. W. & Pogson, C. I. (1971). Dihydroxyacetone phosphate. *Biochemical Journal*, **122**, 285–97.

Riggs, A. D., Bourgeois, S. & Cohn, M. (1970). The lac repressor-operator interaction: III Kinetic studies. *Journal of Molecular Biology*, **53**, 401–17.

Ringe, D. & Petsko, G. A. (1984). Fluctuations in protein structure from X-ray diffraction. *Annual Review of Biophysics and Bioengineering*, **13**, 331–72.

Roberts, D. V. (1977). *Enzyme Kinetics*. Cambridge: Cambridge University Press.

Rose, I. A. (1975). Mechanism of the aldose-ketose isomerase reactions. *Advances in Enzymology*, **43**, 491–517.

Rose, I. A., Kuo, D. J. & Warns, J. V. B. (1991). A rate-determining proton relay in the pyruvate kinase reaction. *Biochemistry*, **30**, 722–6.

Rosenfeld, S. S. & Taylor, E. W. (1985). Kinetic studies of calcium and magnesium binding to troponin C. *Journal of Biological Chemistry*, **260**, 242 51.

Roughton, F. J. W. (1935). Recent work on carbon dioxide in the blood. *Physiological Reviews*, **15**, 241–95.

Roughton, F. J. W. & Chance, B. (1963). Rapid reactions. In *Techniques of Organic Chemistry*, 2nd edn, vol. 8 part II, ed. S. L. Freiss, E.S. Lewis & A. Weissberger, pp. 703–98. New York: Wiley-Interscience.

Roughton, F. J. W. & Rossi-Bernardi, L. (1968). Studies on the pK and rate of dissociation of the glycylglycine-carbamic acid molecule. In *CO₂: Chemical, Biological and Physiological Aspects*, ed. R. E. Forster, J. T. Edsall, A. B. Otis & F. J. W. Roughton, pp. 41–5. Washington, DC: National Aeronautical and Space Administration.

Rubinow, S. I. (1975). *Introduction to Mathematical Biology*. New York: Wiley-Interscience.

Rutherford, E. (1897). The velocity and rate of the recombination of the ions of gases exposed to Röntgen radiation. *Philosophical Magazine*, **44** (**5th series**), 422–40.

Rutherford, E., Chadwick, J. & Ellis, C. D. (1930). *Radiation from Radioactive Substances*, pp. 10–23. Cambridge: Cambridge University Press.

Sack, J. S., Saper, M. A. & Quiocho, F. A. (1989). Periplasmic binding protein structure and function: refined X-ray structures of the leucine/isoleucine/ valine-binding protein and its complex with leucine. *Journal of Molecular Biology*, **206**, 171–91.

Sakmann, B. & Neher, E. (1983). *Single Channel Recording*. New York: Plenum Press.

Sali, D., Bycroft, M. & Fersht, A. R. (1988). Stabilization of protein structure by interaction of α-helix dipole with a charged side chain. *Nature (London)*, **335**, 740–3.

Schleicher, A. & Hofmann, K. P. (1987). Kinetic study of the equilibrium between membrane bound and free photoreceptor G-protein. *Journal of Membrane Biology*, **95**, 269–79.

Schlichting, I., Almo, S. C., Rapp, G., Wilson, K., Petratos, K., Lentfer, A., Wittinghofer, A., Kabsch, W., Pai, E. F., Petsko, G. A. & Goody, R. S. (1990). Time resolved X-ray crystallographic study of the conformation change on Ha-Ras p21 protein on GTP hydrolysis. *Nature (London)*, **345**, 309–15.

Schmidt, D. E. & Westheimer, F. H. (1971). pK of the lysine amino group at the active site of lysine. *Biochemistry*, **10**, 1249–53.

Schrödinger, E. (1944). *What is Life?*. Cambridge: Cambridge University Press.

Schurr, J. M. (1976). Relaxation of rotational and internal modes of macromolecules determined by dynamic scattering. *Quarterly Review of Biophysics*, **9**, 109–34.

Segal, H. L. (1959). The development of enzyme kinetics. In *The Enzymes*, 2nd edn, vol. 1, ed. P. D. Boyer, H. Lardy & K. Myrback, pp. 1–48. New York: Academic Press.

Segel, I. H. (1975). *Enzyme Kinetics*. New York: Wiley-Interscience.

Segel, L. E. (1984). *Modeling Dynamic Phenomena in Molecular and Cellular Biology*. Cambridge: Cambridge University Press.

Sheetz, M. P. & Spudich, J. A. (1983). Movement of myosin-coated fluorescent beads on actin cables in vitro. *Nature (London)*, **303**, 31–5.

Shoham, M. & Steitz, T. A. (1982). The 6-hydroxymethyl group of a hexose is essential for the substrate induced closure of the cleft in hexokinase. *Biochimica et Biophysica Acta*, **705**, 380–4.

Shore, J. D. & Gutfreund, H. (1970). Transients in the reactions of liver alcohol dehydrogenase. *Biochemistry*, **9**, 4655–9.

Silvermann, D. N. & Tu, C. K. (1975). Buffer dependence of carbonic anhydrase catalysed oxygen-18 exchange at equilibrium. *Journal of the American Chemical Society*, **97**, 2263–9.

Sines, J. J., Allison, S. A. & McCammon, J. A. (1990). Point charge distributions and electrostatic steering on enzyme/substrate encounter: Brownian dynamics of modified copper/Zn superoxide dismutase. *Biochemistry*, **29**, 9403–12.

Smith, C. J., Clarke, A. R., Chia, W. N., Irons, L. I., Atkinson, T. & Holbrook, J. J. (1991). Detection and characterization of folding. *Biochemistry*, **30**, 1028–36.

Smith, J. C. (1991). Protein dynamics: comparison of simulations with inelastic neuron scattering experiments. *Quarterly Reviews of Biophysics*, **24**, 227–92.

Smoluchowski, M. von (1916). Versuch einer mathematischen Theorie der Koagulationskinetik kolloider Lösungen. *Zeitschrift für physikalische Chemie*, **92**, 129–68.

Somlyo, A. P., Walker, J. W., Goldman, Y. E., Trentham, D. R., Kobayashi, S., Kitazawa, T. & Somlyo, A. V. (1988) Inositol triphosphate, calcium and muscle contraction. *Philosophical Transactions of the Royal Society of London*, **B320**, 399–414.

Somogyi, B., Norman, J. A., Zempel, L. & Rosenberg, A. (1988). Viscosity and transient solvent accessibility of Trp-63 in the native conformation of lysozyme. *Biophysical Chemistry*, **32**, 1–13.

Sorensen, S. P. L. (1912). Über die Messung und Bedeutung der Wasserstoff Ionen Konzentration bei biologischen Prozessen. *Ergebnisse der Physiologie*, **12**, 393–532.

Spiegelman, S. (1971). An approach to the experimental analysis of precellular evolution. *Quarterly Reviews of Biophysics*, **4** , 213–54.

Srivastava, D. K. & Bernhard, S. A. (1986). Metabolite transfer via enzyme–enzyme complexes. *Science* , **234**, 1081–6.

Steinbach, P. J. (1991). Ligand binding to hemeproteins: connection between dynamics and function. *Biochemistry*, **30**, 3988–409.

Steinberg, I. Z. (1971). Long-range nonradioactive transfer of electronic excitation energy in proteins and polypeptides. *Annual Review of Biochemistry*, **40**, 81–114.

Stent, G. S. & Callender, R. (1978). *Molecular Genetics*. New York: Freeman.

Stryer, L. (1988). *Biochemistry*, 3rd edn. New York: W. H. Freeman and Company.

Stryer, L. (1991). Visual excitation and recovery. *Journal of Biological Chemistry*, **266**, 10711–14.

Stryer, L. & Hougland, R. P. (1967). Energy transfer: a spectroscopic ruler. *Proceedings of the National Academy of Sciences*, **58**, 719–26.

Stuehr, J. (1974). Ultrasonic methods. In *Investigation of Rates and Mechanisms of Reactions: part II*, 3rd edn, ed. G. G. Hammes, pp. 237–83. New York: John Wiley & Sons.

Tanford, C (1961). *Physical Chemistry of Macromolecules*. New York: Wiley.

Tanford, C. (1968). Protein denaturation. *Advances in Protein Chemistry*, **23**, 122–282.

Templeton, W., Kowlessur, D., Thomas, E. W., Topham, C. M. & Brocklehurst, K. (1990). A re-appraisal of the structural basis of stereochemical recognition in papain. *Biochemical Journal*, **266**, 645–51.

Tesi, C., Kitagishi, K., Travers, F. & Barman, T. E. (1991). Cryoenzymic studies on actomysin ATPase: kinetic evidence for communication between actin and ATP sites on myosin. *Biochemistry*, **30**, 4061–7.

Theorell, H. & Chance, B. (1951). Studies on liver alcohol dehydrogenase. *Acta Chemica Scandinavica*, **5**, 1127–44.

Thompson, D. W. (1948). *On Growth and Form*, American edn, 4th reprint. Cambridge: Cambridge University Press.

Thompson, S. P. (1914). *Calculus Made Easy*, 2nd edn. London: Macmillan & Co.

Tilton, R. F., Dewan, J. C. & Petsko, G. A. (1992). Effects of temperature on protein structure and dynamics: X-ray crystallographic studies on the protein ribonuclease-A at nine different temperatures from 98 to 320 K. *Biochemistry*, **31**, 2469–81.

Ting, T. D. & Ho, Y.-K. (1991). Molecular mechanism of GTP hydrolysis by bovine transducin: pre-steady-state kinetic analyses. *Biochemistry*, **30**, 8996–9007.

Török, K. & Trentham, D. R. (1994). Mechanism of 2-chloro(ε-amino-Lys75)-6-(4-N,N-diethylaminophenyl)-1,3,5-triazin-4-yl-calmodulin (TA calmodulin) interactions with smooth muscle myosin light chain kinase and derived peptides. *Biochemistry*, **33**, 12807–20.

Travers, F., Barman, T. E. & Bertrand, R. (1979). Transient-phase studies on the creatine reaction. *European Journal of Biochemistry*, **100**, 149–55.

Trayer, H. R. & Trayer, I. P. (1988). Fluorescence resonance energy transfer within the complex formed by actin and myosin subfragment 1. Comparison between weakly and strongly attached states. *Biochemistry*, **27**, 5718–27.

Trentham, D. R. & Gutfreund, H. (1968). The kinetics of the reaction of nitrophenyl phosphates with alkaline phosphatase from *Escherichia coli*.

Biochemical Journal, **106**, 455–60.

Trentham, D. R., Corrie, J. E. T. & Reid, G. P. (1992). A new caged ATP with rapid photolysis kinetics. *Biophysical Journal*, **61**, A295.

Trentham, D. R., Eccleston, J. F. & Bagshaw, C. R. (1976). Kinetic analysis of ATPase mechanisms . *Quarterly Reviews of Biophysics*, **9**, 217–81.

Trentham, D. R., McMurray, C. H. & Pogson, C. I. (1969). The active chemical state of D-glyceraldehyde 3-phosphate in its reactions with D-glyceraldehyde 3-phosphate dehydogenase, aldolase and triose phosphate isomerase. *Biochemical Journal*, **114**, 19–24.

Tsuruta, H. & Sano, T. (1990). A fluorescence temperature-jump study on Ca^{2+}-induced conformational changes in calmodulin. *Biophysical Chemistry*, **35**, 75–84.

Tsuruta, H., Sano, T., Vachette, P., Tauc, P., Moody, M. F., Wakabayashi, K., Amemiya, Y., Kimura, K. & Kihara, H (1990). Structural kinetics of the allosteric transition of aspartate transcarbamylase produced by physiological substrates. *Federation of European Biochemical Societies Letters*, **263**, 66–8.

Vanoni, M. A., Wong, K. K., Ballon, D. P. & Blanchard, J. S. (1990). Glutathione reductase: comparison of steady-state and rapid reaction primary kinetic isotope effects exhibited by the yeast, spinach and coli enzymes. *Biochemistry*, **29**, 5790–96.

von Bertalanffy, L. (1971). *General System Theory*. Harmondsworth, Middlesex: Penguin Books Ltd.

von Hippel, P. H., Bear, D. G., Morgan, W. D. & McSwiggen, J. A. (1984). Protein-nucleic acid interactions in transcription: a molecular analysis. *Annual Review of Biochemistry*, **53**, 389–446.

Walker, J. W., McCray, J. A. & Hess, G. P. (1986). Photolabile protecting groups for an acetylcholine receptor ligand. Synthesis and photochemistry of a new class of *o*-nitrobenzyl derivatives and their effects on receptor function. *Biochemistry*, **25**, 1799–805.

Walker, J. W., Reid, G. P. & Trentham, D. R (1989). Synthesis and properties of caged nucleotides. *Methods in Enzymology*, **172**, 288–312.

Walker, J. W., Reid, G. P., McCray, J. A. & Trentham, D. R. (1988). Photolabile 1-(2-nitrophenyl)ethyl phosphate esters of adenine nucleotide analogues. Synthesis and mechanism of photolysis. *Journal of the American Chemical Society*, **110**, 7170–7.

Walmsley, A. R. & Bagshaw, C. R. (1989). Logarithmic timebase for stopped-flow data aquisition and analysis. *Analytical Biochemistry*, **176**, 313–18.

Walsh, C. (1977). *Enzyme Mechanisms*. New York: Freeman.

Walsh, C. (1979). *Enzymatic Reaction Mechanisms*. New York: Freeman.

Webb, M. R. (1992). A continuous spectrophotometric assay for inorganic phosphate and for measuring phosphate release kinetics in biological systems. *Proceedings of the National Academy of Sciences, USA*, **89**, 4884–7.

Weber, G. (1953). Rotational Brownian motion and polarization of the fluorescence of solutions. *Advances in Protein Chemistry*, **8**, 415–59.

Westheimer, F. H. (1961). Kinetic isotope effects. *Chemical Reviews*, **61**, 265–90.

Whitaker, J. R., Yates, D. W., Bennett, N. G., Holbrook, J. J. & Gutfreund, H. (1974). The identification of intermediates in the reaction of pig heart lactate dchydrogenase with its substrates. *Biochemical Journal*, **139**, 677–97.

Wilkinson, G. N. (1961). Statistical estimations in enzyme kinetics. *Biochemical Journal*, **80**, 324–32.

Winkler-Oswatitsch, R. & Eigen, M. (1979). The art of titration. *Angewandte Chemie (international edition)*, **18**, 20–35.

Winsor, C. P. (1932). The Gompertz curve as a growth curve. *Proceedings of the National Academy of Sciences, USA*, **18**, 1–8.

Witt, H. H. (1991). Functional mechanism of water splitting photosynthesis. *Photosynthesis Research*, **29**, 55–77.

Wittenberg, J. B. & Wittenberg, B. A. (1990). Mechanisms of cytoplasmic hemoglobin and myoglobin function. *Annual Review of Biophysics and Biophysical Chemistry*, **19**, 217–41.

Woledge, R. C., Curtin, N. A. & Homsher, E. (1985). Energetic aspects of muscle contraction. *Monographs of the Physiological Society No. 41*. New York: Academic Press.

Wong, J. T.-F. (1975). *Kinetics of Enzyme Mechanisms*. London: Academic Press.

Wu, X., Gutfreund, H. & Chock, P. B. (1992). Kinetic method for differentiating mechanisms for ligand exchange reactions: application to test for substrate channeling in glycolysis. *Biochemistry*, **31**, 2123–8.

Wu, X., Gutfreund, H., Lakatos, S. & Chock, P.B. (1991). Substrate channeling in glycolysis: a phantom phenomenon. *Proceedings of the National Academy of Sciences, USA*, **88**, 497–501.

Wyman, J. & Gill, S. J. (1990). *Binding and Linkage*. Mill Valley, CA: University Science Books.

Yamada, W. M. & Zucker, R. S. (1992). The time course of transmitter release calculated from simulations of a calcium diffusion model. *Biophysical Journal*, **61**, 671–82.

Zewail, A. H. (1990). The birth of molecules. *Scientific American*, **263, no. 5**, 40–6.

Zhang, X.-Z., Strand, A. & White, H. D. (1989). A general pre-steady-state solution to complex kinetic mechanisms. *Analytical Biochemistry*, **176**, 427–31.

Zierler, K. (1981). A critique of compartmental analysis. *Annual Review of Biophysics and Bioengineering*, **10**, 531–62.

Zimm, B. H. & Bragg, J. K. (1959). Theory of the phase transition between helix and random coil in polypeptide chains. *Journal of Chemical Physics*, **31**, 526–35.

Index

Page numbers in italic refer to illustrations. Page numbers in bold refer to a main section on a particular subject.